AF782412

ULTRAFAST PHOTONICS

ULTRAFAST PHOTONICS

Proceedings of the Fifty-Sixth
Scottish Universities Summer School in Physics
St Andrews, September 2002

A NATO Advanced Study Institute

Edited by

A Miller — University of St Andrews
D T Reid — Heriot-Watt University
D M Finlayson — University of St Andrews

Series Editor

P Osborne — University of Edinburgh

Co-published by
Scottish Universities Summer School in Physics &
Institute of Physics Publishing, Bristol and Philadelphia

Copyright © 2004
The Scottish Universities Summer School in Physics

No part of this book may be reproduced in any form
by photostat, microfilm or any other means without
written permission from the publishers.

British Library cataloguing-in-Publication Data:

*A catalogue record for this book is available
from the British Library*

ISBN 0-7503-0904-0

Library of Congress Cataloging-in-Publication Data are available.

Co-published by

SUSSP Publications
School of Physics, Edinburgh University,
The King's Buildings, Mayfield Road, Edinburgh EH9 3JZ, Scotland.

and

Institute of Physics Publishing, wholly owned by
The Institute of Physics, London.
Institute of Physics Publishing, Dirac House, Temple Back, Bristol BS1 6BE, UK.
US Office: Institute of Physics Publishing, The Public Ledger Building,
Suite 929, 150 Independence Mall West, Philadelphia, PA 19106, USA.

Printed in the UK by MPG Books Ltd, Bodmin, Cornwall.

SUSSP Proceedings

1	1960	Dispersion Relations
2	1961	Fluctuation, Relaxation and Resonance in Magnetic Systems
3	1962	Polarons and Excitons
4	1963	Strong Interactions and High Energy Physics
5	1964	Nuclear Structure and Electromagnetic Interactions
6	1965	Phonons in Perfect and Imperfect Lattices
7	1966	Particle Interactions at High Energy
8	1967	Methods in Solid State and Superfluid Theory
9	1968	Physics of Hot Plasmas
10	1969	Quantum Optics
11	1970	Hadronic Interactions of Photons and Electrons
12	1971	Atoms and Molecules in Astrophysics
13	1972	Properties of Amorphous Semiconductors
14	1973	Phenomenology of Particles at High Energy
15	1974	The Helium Liquids
16	1975	Non-linear Optics
17	1976	Fundamentals of Quark Models
18	1977	Nuclear Structure Physics
19	1978	Metal Non-metal Transitions in Disordered Solids
20	1979	Laser-Plasma Interactions: 1
21	1980	Gauge Theories and Experiments at High Energy
22	1981	Magnetism in Solids
23	1982	Laser-Plasma Interactions: 2
24	1982	Lasers: Physics, Systems and Techniques
25	1983	Quantitative Electron Microscopy
26	1983	Statistical and Particle Physics
27	1984	Fundamental Forces
28	1985	Superstrings and Supergravity
29	1985	Laser-Plasma Interactions: 3
30	1985	Synchrotron Radiation
31	1986	Localisation and Interaction
32	1987	Computational Physics
33	1987	Astrophysical Plasma Spectroscopy
34	1988	Optical Computing
35	1988	Laser-Plasma Interactions: 4
36	1989	Physics of the Early Universe

/continued

SUSSP Proceedings (continued)

37	1990	Pattern Recognition and Image Processing
38	1991	Physics of Nanostructures
39	1991	High Temperature Superconductivity
40	1992	Quantitative Microbeam Analysis
41	1992	Spatial Complexity in Optical Systems
42	1993	High Energy Phenomenology
43	1994	Determination of Geophysical Parameters from Space
44	1994	Simple Quantum Systems
45	1994	Laser-Plasma Interactions 5: Inertial Confinement Fusion
46	1995	General Relativity
47	1995	Laser Sources and Applications
48	1996	Generation and Application of High Power Microwaves
49	1997	Physical Processes in the Coastal Zone
50	1998	Semiconductor Quantum Opto-electronics
51	1998	Muon Science
52	1998	Advances in Lasers and Applications
53	1999	Soft and Fragile Matter
54	2000	The Restless Universe
55	2001	Heavy Flavour Physics
56	2002	Ultrafast photonics
57	2003	Large Hadron Collider Phenomenology
58	2004	Hadron Physics
59	2004	Soft Condensed Matter in Molecular and Cell Biology

Lecturers

Donal Bradley	Imperial College of Science Technology and Medicine
Takis Hadjifotiou	Nortel Networks
Jorn Hvam	Technical University of Denmark
Jaroslav Kovac	Slovak University of Technology
Thomas F Krauss	University of St Andrews
Alan Miller	University of St Andrews
David A B Miller	Stanford University
Jean-Louis Oudar	CNRS Laboratoire Photonique et Nanostructure
Richard V Penty	University of Cambridge
Derryck T Reid	Heriot-Watt University
Peeter Saari	University of Tartu
Solomon M Saltiel	Sofia University
Ifor D W Samuel	University of St Andrews
Wilson Sibbett	University of St Andrews
George Stegeman	University of Central Florida
Osamu Wada	Kobe University
Aleksei M Zheltikov	M V Lomonosov Moscow State University

Executive Committee

Prof Alan Miller	University of St Andrews	*Director and Co-Editor*
Dr Tom Brown	University of St Andrews	*Secretary*
Dr Derryck Reid	Heriot-Watt University	*Treasurer and Co-Editor*
Dr David Finlayson	University of St Andrews	*Co-Editor*

International Advisory Committee

Prof David A B Miller	Stanford University, USA
Prof George Stegeman	University of Central Florida, USA
Prof Aleksei M Zheltikov	M V Lomonosov Moscow State University, Russia

Preface

This book is the Proceedings of the 56th Scottish Universities Summer School in Physics (SUSSP56), on 'Ultrafast Photonics' held at the University of St Andrews, Scotland between 1st and 14th September 2002.

Ultrafast photonics has become an interdisciplinary topic of high international research interest because of the spectacular development of compact and efficient lasers producing optical pulses with durations in the femtosecond time domain. The scope of the School ranged from the relevant fundamental physics of optical interactions and material properties through to new system concepts for ultrahigh capacity data transfer/processing using femtosecond pulses. Areas of scientific breakthrough and subsequent technological advance were emphasised. The School included researchers from major collaborative research initiatives in ultrafast photonics in Europe, USA and Japan.

'Photonics' encompasses optics-based devices, components and systems for information technology. Photonics has been the foundation upon which the major advances in communications have been made. Present day long-haul telecommunication systems are almost entirely based on the transmission of short bursts or pulses of laser light along silica optical fibres with Erbium-doped fibre amplifiers inserted to regenerate the signals periodically. Optical fibres have the advantage of low loss, high reliability, huge information carrying capacity and, most important, cost. The phenomenal success of optical fibre communications has created a substantial opto-electronics industry that provides efficient laser sources, high-speed modulators, optical amplifiers and detectors.

Although optical data transmission is now well established, the full potential has not been reached to meet the ever-increasing demands of data-communication. Indeed, photonics-related component and system developments are still relatively immature in terms of data switching, routing and component integration. Current system approaches will eventually cease to meet the ever increasing demand of information transfer. Novel approaches and ideas are needed; these require a rethink of the fundamental issues of light generation, new optoelectronic materials, laser-semiconductor interactions, ultrashort pulse propagation, nanometric structures, device miniaturisation and integration. The explosion of the internet will demand that existing wavelength division multiplexing (WDM) schemes will need to be complemented with time-domain multiplexing (TDM) techniques involving digital optical signals using ultrashort laser pulses. Ultimately, photonics involving ultrashort pulses of femtosecond duration offers exciting opportunities at multi-Tb/s data rates with the goal of complete optical-electronic systems combined on a single chip. SUSSP56 was timely in bringing world experts in this field together with young people who are excited by this very active area of research.

SUSSP56 was held in the School of Physics and Astronomy and John Burnet Hall on the North Haugh of the University of St Andrews and close to the ancient town's shops, pubs, beaches and historical monuments and adjacent to the famous 'Old Course', the very first golf course. The School was supported as a NATO Advanced Study Institute and a European Community Summer School. The 116 participants from 20 countries attended 44 lectures presented by 17 lecturers over 11 working days. A number of discussion sessions created stimulating debate on the key issues and students presented their own work in 67 poster papers in two very lively sessions. The School was further enriched by the visit of a group of 6 Japanese scientists who gave overviews of research activities in ultrafast photonics at FESTA, AIST and Tohoku University.

A busy social programme kept everybody well occupied outside of the formal sessions. Participants sampled the Scottish scenery with a climb up Ben Vrackie, trips to Stirling Castle, the Wallace (of Braveheart fame) Monument, and a tour of St Andrews. The university gave a reception and a formal dinner provided a taste of haggis. 'Pub crawls' of the numerous St Andrews establishments were organised by the local post-graduate students when many were introduced to the effects (good and not so good) of malt whisky. An 'International Evening' allowed students to sample food and drinks from the many nations represented. Some participants braved a swim in the North Sea from the adjacent West Sands Beach while others attempted golf, putting or just walking the links. A highlight for many was the final night ceilidh including traditional Scottish dancing. It is our hope that the friendships that formed during the School will lead to life-long bonds and links with fruitful research collaborations spanning many international borders and cultures.

The organisers are grateful for the help of many organisations and individuals in the success of the school. We would like to thank our sponsors; NATO and the European Community, the Engineering and Physical Sciences Research Council, the National Science Foundation (USA), Scottish Enterprise, the Scottish Branch of LEOS, Institute of Physics Publishing, and SUSSP. The Director and all participants greatly appreciated the time and effort put in by the local organising team led by Tom Brown as School Secretary and Derryck Reid as Treasurer. David Finlayson deserves special thanks for providing his expert guidance from over 40 years of experience with SUSSP Schools, presenting his specialist knowledge of malt whiskies, and contributing many hours of diligent converting and editing the chapters of this book.

The staff of John Burnet Hall, headed by Jackie Mathews, again provided a very high quality and friendly residential service. Thanks are also due to the staff and students of the School of Physics and Astronomy for providing willing assistance in many tasks that contributed to the overall success of the School.

Alan Miller
St Andrews, September 2003

Editors' Note

The chapters, written by internationally renowned experts in the field of ultrafast photonics, are based on the lectures presented at the school. They provide tutorial-style introductions to the various topics, suitable for post-graduate research students and others with a basic knowledge of photonics. The book progresses from identifying some of the systems issues through fundamental optical physics to future devices and applications. The editors would like to thank the authors for their excellent and timely contributions to this volume. Michael Mazilu is acknowledged for his expert help in handling the many different submitted file types and converting them into the LaTex format of the final version.

Contents

Ultrafast digital processing

David A B Miller

Stanford University, USA

1 Introduction

This chapter will deal with some possible applications of ultrafast photonic technology. We will discuss some of the key features of ultrafast photonics that might be useful in digital information processing, and go on to discuss some specific applications currently being researched, including optically assisted analog-to-digital conversion, ultrafast optical gates, optical clock distribution, and optical interconnects. In each case we will give some of the background motivation for the application, as well as giving a specific research example.

It is worth pointing out at the outset that making the match between the exciting opportunities of a new technology, such as ultrafast photonics, and practical uses that will displace existing technology in real applications is a challenging task. The practical world is no respecter of how erudite or beautiful is the physics with which we are working, or, for that matter, how smart we think we are! The apparently most minor of practical details can prevent a new technology from actually being used. Existing technologies can be very difficult to displace because existing systems are often optimized round the strengths and limitations of that technology, because the investments have often been made that make that existing technology inexpensive, and because engineers and technicians are already trained to design, install and maintain them.

On the positive side, we should note that new technologies certainly do displace existing ones, and also find applications the existing technologies could not support. In so doing, it is often surprising what aspects of a new technology are the key to its successful use. The reason for the dominance of silicon electronics over germanium or gallium arsenide is the properties of the silicon, silicon-oxide interface, with its very low density of surface states, which allows very large numbers of transistors to be made reliably. The important parameter for applications was not, as scientists have often imagined, the performance of the individual transistor, but rather the material properties that enable us to make large numbers. It is also often surprising that the final dominant applications of a revolutionary technology have little to do with the applications imagined by the original scientists or business people. (For example, Thomas Watson, then chairman of IBM, said in 1943 that he thought there was a world market for maybe five computers, and

Ken Olson, president, chairman, and founder of Digital Equipment Corporation, stated in 1977 that there was no reason anyone would want a computer in their home.) In the early days of the laser, who would have imagined that the dominant volume use of lasers would be in audio products (compact disk players) and information storage (CD-ROM)?

Finding the applications that can exploit a novel and potentially revolutionary technology like ultrafast photonics is an entrepreneurial activity that calls for creativity and for the open-mindedness to discard cherished ideas, no matter how smart, for ideas that work, no matter how simple.

Prospective technologies that offer capabilities that go vastly beyond current technology in at least one major capability are obvious fertile grounds for revolutionary applications. Ultrafast photonics is such an area. It has orders of magnitude of potential improvement over existing technology in many ways (e.g., speed, bandwidth), it has a well developed scientific base, and we can truly envisage that the core technologies of ultrafast photonics could be manufactured in forms that are ultimately inexpensive and convenient if we can find the applications that would drive the necessary investment.

The application of ultrafast photonics in digital processing is still speculative, and is a subject of research. In what follows, we will first summarize the possible features of ultrafast photonics that might be exploited for applications in digital processing. Then we will examine in turn three possible example areas of application. In each case, we will attempt to summarize the underlying problem to be solved, and then describe examples of research on approaches to the problem using ultrafast photonics. The examples we discuss here will be, first, optically-assisted analog-to-digital conversion, second, ultrafast functional optoelectronic devices, by which we mean devices that perform some processing on high speed data, and, finally, ultrafast photonics in digital electronic systems, including both optical clock injection and optical interconnects.

2 Ultrafast photonics for digital processing

Before discussing particular example applications, we will look at some of the attributes of ultrafast photonics that might be useful in digital processing. There are perhaps more different attributes than one might imagine.

2.1 Very high speed digital processing

Likely the first attribute of ultrafast optics that would occur to someone thinking about their use in digital processing would be that the optical pulses can be very short, in the picosecond or femtosecond range, and hence we might think of doing something very fast with them, such as performing very fast processing with very high speed devices.

This remains an interesting area for future possibilities, though if one's goal was to process the maximum total amount of information, it is unlikely this is the route one would choose. Electronic logic devices are extremely efficient in total energy used per logic operation, and it is typically this dissipation that will limit the overall number of logic operations that can be performed. The scale of such dissipations is a few femtojoules of total energy (not just the energy required at the input of the device) for small electronic gates. The core of the energy dissipation in an electronic gate is that one has to charge or discharge the capacitance of the gate, by a charge of the order of fC, over the logic voltage swing of $\sim$ 1V, hence the femtojoule energies. Electronic gates can also be made very reliably in very large numbers, at very low cost per gate: a transistor in an integrated circuit costs less a single character of text in a newspaper! Silicon CMOS (complementary metal-oxide-silicon) integrated circuit technology continues to improve (Semiconductor Industry Association 2001), and is likely to do so for some time to come, making it a formidable competitor for any proposed logic technology. It is also the case that transistor logic gates themselves, even in silicon CMOS (which is not the fastest transistor technology), have internal operating speed limits as fast as tens of picoseconds, with scaling into the picosecond range likely in future technologies. Electronic logic technologies are, however running into increasing problems with interconnection, because the electrical interconnects do not scale to keep pace with the logic, and the area of optical interconnections is one of growing research interest (Miller 1997, 2001). (We will discuss some uses of short pulse in optical interconnections below.)

Optical logic technologies tend to have the opposite characteristics from electrical ones. Most methods of making optical logic gates tend to have high overall energies (e.g., 100fJ–100pJ) per logic operation, and are not readily made in large numbers. Many forms of optical logic gate require the operating power to be supplied optically, which creates formidable optical power requirements if the gates are to be run at any substantial duty cycle (one gate operating with 1pJ every picosecond would require 1W of optical power). On the other hand, optical logic gates do not have a basic problem with having the interconnects keep pace with the logic, and some of them undoubtedly can operate extremely fast, at least for individual logic operations. (See Miller 1999 for a general discussion of use of optics in digital systems.)

The strengths and weaknesses of optical gates compared to electrical ones suggest that we might consider optical gates for simple operations where the information is already in the form of light, and/or where extremely high speed is required. The growing speed of electronic gates also suggests that we might consider hybrid gates that may internally be electronic, but that have optical inputs and outputs. Such gates might offer the best of both worlds, at least for operation into the picosecond regime.

There are no current major applications of ultrafast optical gates, though some may emerge for use with optical networks at high data rates. Whether we will see optical networks with very high data rates in the time domain is open to question, because wavelength-division multiplexing has opened up the entire bandwidth of optical fibres without the need for extremely high speeds. The whole field of flexible networks with high total data rates is an exciting one, and ultrafast optical gates may well find a role, such as the wavelength conversion discussed below. Many format conversions can also be imagined based on ultrafast photonics, such as converting from time to space (i.e., time-division de-multiplexing), or time to wavelength, and the inverses of these conversions.

2.2 Precision of timing

Mode-locked lasers intrinsically generate picosecond or femtosecond pulses at a well-defined repetition rate. Such repetition rates can vary from 10's of MHz to 100's of GHz. Variation in delay between pulses in a pulse train, or in the phase of oscillators generally, is called jitter. The variation in the time delay between successive pulses (pulse-to-pulse jitter) is typically much less than the pulse width of the mode-locked laser, basically because, in a simple way of thinking about mode-locked lasers, the pulses circulate on the average many times in the cavity before being emitted, or equivalently, only a small fraction of the circulating energy in the cavity is emitted with each pulse, and hence there is an averaging effect that tends to define the timing quite accurately. The jitter of mode-locked lasers is not fundamentally better than the best electronic oscillators, but mode-locked lasers may turn out to be quite good practical oscillators for low jitter in the range of GHz frequencies and above, partly because they use optical resonators to define the frequency, and optical resonators (in contrast to simple electrical ones) can have high quality factors (ratio of stored energy to energy loss per cycle) in this frequency range.

Perhaps more important than the jitter of modelocked lasers is the ability of optics to deliver very sharp timing pulses. It is difficult electrically to propagate short pulses on wires over any substantial distance and have those pulses remain sharp. Loss in wires increases substantially as the frequency increases, and short pulses have very high frequency components. As a result, short electrical pulses broaden substantially as they propagate down wires. The effective delay of the rising edge of an electrical signal is also quite strongly temperature dependent because the resistivity of normal electrical conductors is also quite temperature dependent. Optical pulses down to picosecond lengths broaden little over metres or more of optical fibre, and even shorter pulses can be propagated with care. The delay in propagation in optical fibres also varies relatively little with temperature (a fractional change $\sim 10^{-5}$/degree Celsius). Hence short pulse optics is an attractive way to deliver precise timing over substantial distances.

The ability to deliver precise timing is potentially important in analogue-to-digital conversion, in optical clock injection in digital electronics, and possibly also in re-timing in optical interconnects, all of which will be discussed below.

2.3 Broad spectral bandwidth

One other characteristic of short optical pulses is that they naturally must have substantial spectral bandwidth (see Figure 1). For a pulse of length τ, the bandwidth in Hertz is inversely proportional to τ (with the precise proportionality constant depending on definitions). In practice this means that pulses $\sim$100fs in length have spectral bandwidth

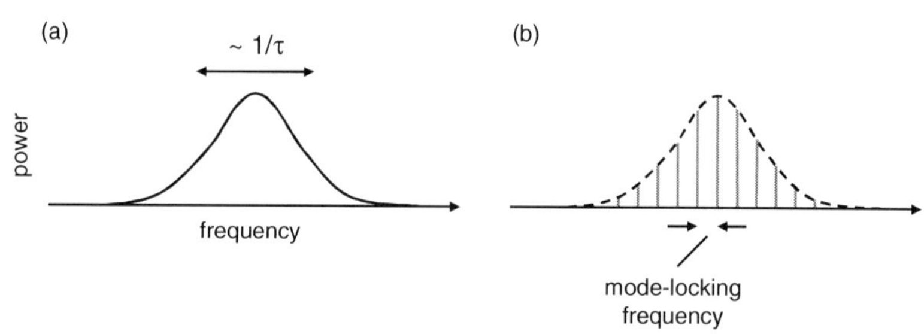

Figure 1. *Spectrum of (a) a single short pulse of length τ and (b) a modelocked train of short pulses*

~10nm (depending on wavelength). Unlike most broad band sources of light (such as an ordinary tungsten light bulb), short optical pulses can be completely coherent sources, that is, as coherent as a single frequency laser, and can have all of the focusing properties of conventional laser beams. They therefore can be particularly interesting as broad band sources for applications since all of their power can be efficiently coupled, for example, into optical fibres.

A mode-locked laser pulse train also has the very odd, and potentially useful, property that it is actually a source of multiple continuous wave light beams at frequencies spaced evenly by the mode-locking repetition frequency. That is, if we were to put the mode-locked laser through a spectrometer with sufficient resolution, we would find a "comb" of frequencies (see Figure 1), and if we looked at only one of those frequencies, we would find that it was actually a continuous beam! This means that a mode-locked laser could actually replace multiple continuous wave lasers as, for example, the source for a wavelength-division multiplexed optical communications system. To be useful for that application, it is likely that we would need a laser to be mode-locked at, for example, 25GHz, 50GHz, or even 100GHz, frequencies that are high for mode-locked lasers, but by no means unheard of (see, e.g., Chen and Wu 1992, and, for a recent example, Krainer 2002). It is also possible to devise ways of mode-locking lasers so that they produce groups of modes, spaced by larger, uniform spacings (Mielke *et al.* 2002). In the time domain, this means that we can generate multiple beams at different centre wavelengths, each of which is also a relatively short pulse laser beam. Such sources could be interesting for wavelength-division multiplexed optical interconnects, in which one would get both some of the advantages of the wavelength-division multiplexing for combining multiple signals in one spatial beam and the timing advantages of short optical pulses.

We will discuss some of the possible uses of broadband short pulse light sources for optical interconnects and networks below.

3 Optically-assisted analogue-to-digital conversion

Analogue-to-digital conversion (ADC) is a key function for interfacing sensing systems to digital processors, and is used in everything from audio products to cellular telephone systems and radar. Digital processing can be much more sophisticated than analogue processing in terms of the complexity of functions that can be performed, and, if the ADC is good enough in the initial conversion, digital processors can retain the quality of the signal no matter how much processing is performed.

ADC's vary greatly in their performance, depending on the application. Audio systems require large numbers of bits of precision (for example, 16–20), but need not operate very fast, say 50 kS/s (kilo samples per second)). Radio frequency systems such as communications and radar require high speed converters, for example, into the GS/s (giga samples per second) range, usually with as many bits as possible of resolution (though less than the 16–20 bits of the audio systems). Laboratory data acquisition systems might have intermediate requirements. ADC's and the causes of their limitations have been comprehensively reviewed by Walden (1999).

Though ADC's vary, the product of the sampling rate, f_{samp}, and the number of levels that the ADC can resolve, that is, 2^N, where N is the number of bits of resolution of the

ADC, is practically bounded for converters faster than $\sim 10^6$ S/s, with

$$2^N f_{\mathrm{samp}} < \quad 1 \times 10^{11}\mathrm{s}^{-1} \quad \text{to} \quad 4 \times 10^{11}\mathrm{s}^{-1}, \tag{1}$$

and this practical performance limit has changed only relatively slowly in recent years. A limit like this, incidentally, means that, if we want a converter to operate twice as fast, we have to give up one bit of resolution.

As we might guess from the dimensions of this product, the apparent source of this limit is the precision of the sampling time, or, in the language of the ADC field, the aperture uncertainty, (which is chosen to be defined as the root mean square uncertainty in aperture timing). In fact, as has been discussed by Walden (1999), the limit on the number of bits of resolution from aperture uncertainty is

$$N_{\mathrm{aperture}} = \log_2 \left(\frac{2}{\sqrt{3}\pi f_{\mathrm{samp}} \tau_a} \right) - 1, \tag{2}$$

or equivalently

$$2^{N_{\mathrm{aperture}}} f_{\mathrm{samp}} \simeq \frac{0.184}{\tau_a}, \tag{3}$$

which means that the practical limits for this product of between 1×10^{11} and 4×10^{11} correspond to aperture uncertainties τ_a of ~2ps and ~0.5ps. It might seem surprising that sampling at, for example, 1 MS/s could require picosecond precision in the sampling time, but, for resolutions of 16–17 bits, this is exactly what would be required. Uncertainty of 2ps in the sampling time only has to make a change in the sampled signal of about one part in 10^5 to correspond to a resolution limit of 17 bits.

There is some debate about why electronic ADC's appear to be limited by timing uncertainty in the sampling. Electrical sampling clocks can be extremely precise in their phase, at least locally. One known phenomenon is that electrical sampling gates may effectively open and close at slightly different times depending on the level of signal they are sampling, which will give an effect equivalent to sampling jitter. This issue does create a possible opportunity for ultrafast photonics if the very precise timing of optically delivered sampling pulses can be exploited, and this might hold the key for radical improvements in ADC performance in the future. This area is the subject of considerable current research (see, e.g., Bhushan *et al.* 2002, Juodawlkis *et al.* 2001, Urata *et al.* 2001a).

One approach is to use ultrafast photoconductive gates to sample the electrical signal (Urata *et al.* 2001a). In this approach, illustrated in Figure 2, a short pulse laser hits a photoconductive gate. This gate, which may be made of GaAs purposely grown at a low temperature (LT GaAs) such as 250°C, will conduct for a short period of time. The LT GaAs has a short carrier lifetime (e.g., picoseconds) because arsenic inclusions thought to form during the low temperature growth act as recombination centres for the photogenerated carriers. This time is long enough to charge up a small capacitor to the voltage present on the electrical input line. The use of the short optical pulse gives very precise sampling of the electrical signal. The voltage on the capacitor can then be converted to a digital signal with a conventional electronic ADC.

To make a system that has a high overall sampling rate, the idea in this approach is to replicate the switches, capacitors and electronic ADC's, to perhaps as many as 100 such

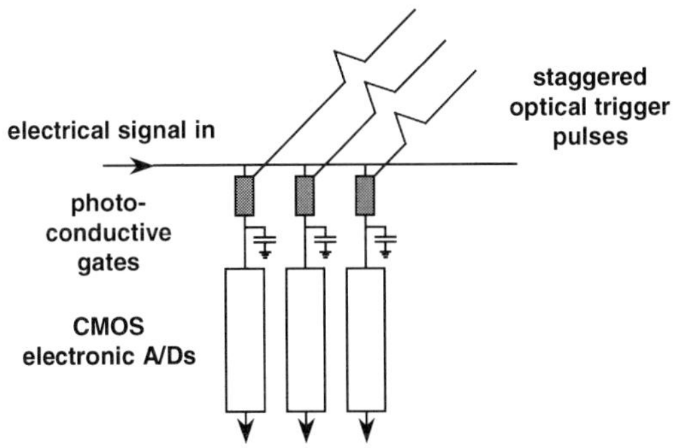

Figure 2. *Schematic of the ADC system with multiple photoconductive gates and CMOS ADC converters.*

samplers and ADC's. Then the different sampling gates are triggered one after the other in rapid succession. With 100 samplers and ADC's, an aggregate sampling rate of 100GS/s could be achieved while allowing each conventional ADC 1ns to perform its conversion. CMOS electronics is likely to be capable of making many low power ADC's on a single chip, making such an approach viable as long as the sampling gates and hold capacitors are well integrated with the CMOS electronics. The size of the hold capacitors required to operate at the 100 GS/s rate is of the order of a few 10's of fF. Larger capacitors would take too long to charge up when driven by the typical 50ohm impedance of an electrical line. The sampling with these gates has been demonstrated to have a bandwidth greater than 50GHz (Urata *et al.* 2001b). Such capacitances are feasible if the sampling gates and hold capacitors are integrated with the CMOS. Figure 3 shows a picture of sampling gates solder bonded to a CMOS input circuit (Urata *et al.* 2002).

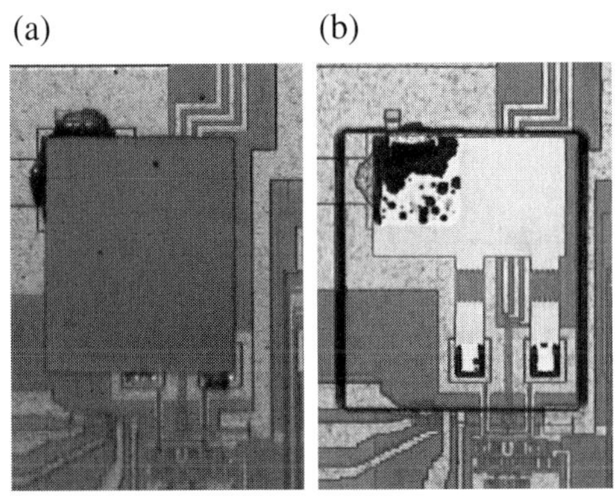

Figure 3. *(a) Picture of a pair of LT GaAs photoconductive sampling gates solder-bonded to an input buffer amplifier. The central rectangular area is the LT GaAs layer after removal of the GaAs substrate used for growth. (b) A similar structure in which the LT GaAs layer has been intentionally removed to show the pattern of the contacts that form the photoconductive switches on the other side of the LT GaAs layer*

4 Ultrafast functional optoelectronic devices

It is a common view to believe that it is costly and slow to convert from optics to electronics and back to optics, and that such systems will necessarily consume a lot of power. That is the case in practice if the optical, optoelectronic, and electronic technologies are not well integrated. If, however, the integration is very close, for example, with little extra or stray capacitance, with small devices, and with the right kinds of devices that operate efficiently, then we can contemplate many quite efficient and fast systems that potentially could have low cost. This gives us the option to mix electrical and optical functions. We can make devices that, for the information flow, have optical inputs and outputs, while also allowing electronic control and readout, and the potential to build in moderately complex functions if we choose to incorporate more electronics internally.

The quantum-confined Stark effect (Miller *et al.* 1985), an electro-absorption mechanism seen in quantum wells, is an efficient mechanism for giving optical outputs, and has the important feature here that it is also apparently very fast. The likely speed limit is in the range of ~100fs, related to the width of the exciton absorption peak in the material (Knox *et al.* 1986, Knox *et al.* 1989, Schmitt-Rink *et al.* 1990). This mechanism gives a strong shift of the optical absorption edge in quantum well structures as an electric field is applied perpendicular to the quantum well layers. The mechanism is usable for optical modulators both in waveguide structures and also in structures of micron thickness for light propagation perpendicular to the surface. The devices are made by incorporating the quantum wells in the intrinsic region of a p-i-n diode, and reverse biasing the diode to apply the field to the quantum wells. See Miller (1999) for a summary of the physics of these devices and their use in digital systems in general.

One way of making a fast optically-controlled optical gate with quantum well electro-absorption is to combine the quantum well modulator with a photodetector, with the photodetector output controlling the modulator. This general class of devices, known as SEEDs (for) has been considered for digital optical switching (Miller 1990, Lentine and Miller 1993), and is also the genesis for optical interconnects to silicon integrated circuits with quantum well modulators (Goossen *et al.* 1995), which will be discussed in the next section.

To make particularly fast devices, for example, into the regime of a few picoseconds, some novel approaches have to be taken. One need with any device in which the voltage must change, as it must do in a modulator, is to find a way of recharging the capacitance, for example, of modulator and photo-detector, after switching, so that the device can recover rapidly and be ready for its next use. A capacitor can be recharged through a resistor, but how can one make the resistive-capacitive time constant sufficiently short, for example, on a picosecond timescale? Any external connection of a resistor will likely involve a large amount of stray capacitance. One way of effectively achieving a rapid recovery of the capacitor voltage is through diffusive conduction (Livescu *et al.* 1989). This mechanism can readily be engineered to operate on picosecond or even faster timescales. The concept of diffusive conduction is illustrated in Figure 4. The p-i-n diode can be viewed as a capacitance between two partially conducting planes, and can be modelled as a mesh of resistors and capacitors as shown. A laser pulse incident on the diode (which typically will have a Gaussian spatial cross section) will generate electrons and holes inside the intrinsic (i) region, which will then separate in the presence of the reverse bias

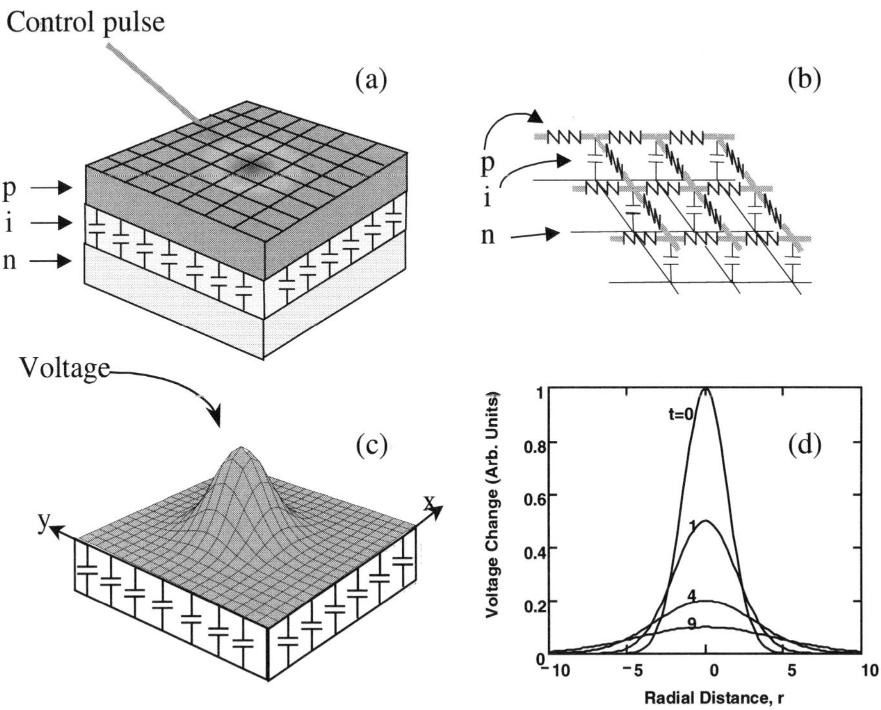

Figure 4. *(a) Schematic of p-i-n diode viewed as a mesh of resistors and capacitors. (b) equivalent circuit diagram of resistor-capacitor mesh. (c) voltage change on the resistor-capacitor mesh after exciting with a Gaussian laser pulse that has created carriers that have moved to screen the field (shown as an increase for convenience, though the actual value of the reverse bias voltage will locally decrease). (d) evolution of the voltage distribution in time, showing relaxation of the local voltage change*

field that will have been applied to the diode.

Because the carriers separate, they will locally partially screen the electric field in the region of the exciting laser pulse, leading to a voltage distribution across the device that is approximately Gaussian as shown in Figure 4(c). This non-uniform voltage across the capacitance cannot, of course, be stable, and it will relax in time as shown in Figure 4(d). By considering the mesh of resistors and capacitors, this voltage relaxation can be shown to obey an equation that is like a diffusion equation, that is,

$$\frac{\partial V}{\partial t} = D\nabla_{xy}^2 V, \tag{4}$$

where

$$D = \frac{1}{R_{\mathrm{sq}} C_A}, \tag{5}$$

with C_A being the capacitance per unit area of the device and R_{sq} being the sum of the resistances per square of the n and p layers. (Resistance per square is the resistance between two opposite sides of a square of the layer. Such a resistance is readily seen not to depend on the size of the square.) For an initially Gaussian spatial shape of laser pulse, we would obtain an initial voltage distribution of the form

$$V(r,t=0) = V_M \exp\left(\frac{-r^2}{w_0^2/2}\right), \tag{6}$$

and the subsequent voltage would evolve as

$$V(r,t) = V_m\left(\frac{\tau}{t+\tau}\right)\exp\left(\frac{-r^2}{4D(t+\tau)}\right), \tag{7}$$

with a characteristic time τ of

$$\tau \equiv \left(\frac{w_0}{\sqrt{2}}\right)^2 \frac{1}{4D}. \tag{8}$$

Note that this time gets shorter for smaller laser spots and for smaller R_{sq} and C_A. It can be engineered to be picoseconds or shorter.

Note too that, though this process behaves like a diffusion process, it is not diffusion of the carriers themselves, but rather of the voltage distribution, and, perhaps surprisingly, this voltage diffusion can be much faster than the carrier diffusion. Though it may seem counter-intuitive for the voltage to move faster than the electrons, this is a routine phenomenon in, for example, electrical transmission lines, where the signals move near to the velocity of light, much faster than the movement of the electrons themselves. It may be rationalized by noting that there is quite a large carrier concentration in the p and n layers, and only small local movements of this large carrier concentration are sufficient to cause large movements in the voltage pulse. A similarly rapid mechanism of diffusion in layered device structures has been more recently described as giant ambipolar diffusion; it has, however, been verified recently that the two mechanisms are essentially the same despite their different descriptions (Yairi and Miller 2002).

Devices can be made using a single quantum well diode functioning as both a photodetector and a modulator (Yairi *et al.* 1999), as in the simple example described above for illustrating diffusive conduction. In this case, the quantum well diode is initially reverse biased, and is absorbing at the photon energy of interest. A control pulse hits the device, and is absorbed. The photogenerated electrons and holes separate, giving the local change in voltage as described above. This local voltage change leads to a reduced absorption, and another "signal" pulse arriving just after the control pulse will be passed by the device. The voltage change and hence the absorption of the device relax rapidly back to their initial value by diffusive conduction. Such a device therefore acts as an optically controlled optical gate; if a control pulse arrives at the right time, then a signal pulse is passed by the device, otherwise the signal pulse is absorbed. Such a mechanism can be used, for example, for wavelength conversion, to transfer modulation from one light beam (the control beam) to another (the signal beam). The wavelengths of the control and signal beams need not be identical as long as they both lie within the spectral range where the device can modulate light (a range that may be from a few nanometers to a few tens of nanometers depending on device design).

As a high speed switching device, this particular configuration has the disadvantage that the time taken to screen the field, and hence turn the device transparent, depends on

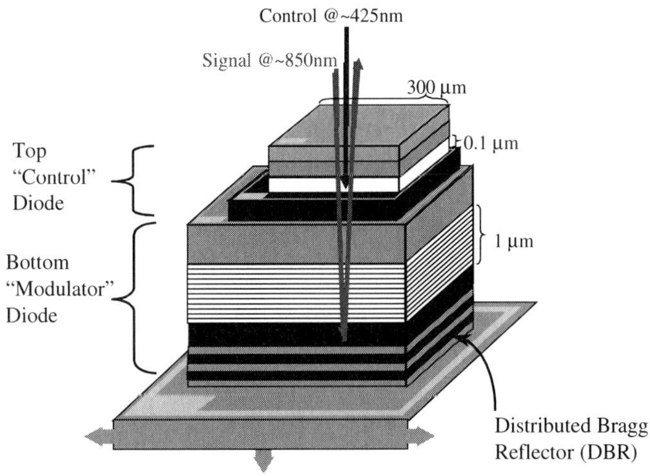

Figure 5. *Structure of a two diode optically controlled gate.*

the carriers escaping from the quantum wells, which can take many tens of picoseconds in some cases, depending on the quantum well design and the applied electric field (Cavailles *et al.* 1992). Another configuration, shown in Figure 5, uses two diodes, one of which is a quantum well diode, which will act as the modulator, and the other is a photodiode with a conventional bulk semiconductor as the absorbing region. Though there are two diodes in this structure, the diffusive conduction mechanism still operates in essentially the same fashion as before. If we presume for simplicity that the top and bottom layers of the whole structure are highly conducting, then we can presume as a first approximation, that the overall voltage is held constant everywhere over the whole device, and so any changes in the voltage locally across the top photodiode result in equal and opposite voltage changes across the modulator diode, giving an optically controlled optical gate again, but without relying on the extraction of carriers from the quantum well region for the turn-on of the gate.

This device was demonstrated to be able to transfer the modulation from a 50GHz burst of control pulses onto a signal beam, as shown in Figure 6 (Yairi *et al.* 2001). This burst operation demonstrates that the devices are rapidly ready for re-use. Running at continued high repetition rates would result in some voltage drop across the overall resistance of the device, including any external resistance, but it is expected that this would necessitate only a moderate change in the operating bias voltage of the device. Another reason for interest in these devices is the relatively low optical energy densities required to run the devices, of the order of a few 10's of $fJ/\mu m^2$.

The two-diode structure with both the control and signal beams coming from the top has the disadvantage that the control beam has to be at a significantly different wavelength from the signal beam, otherwise the signal beam would also be absorbed in the top diode. The contrast ratio that can practically be achieved in the modulation of the signal beam is also not particularly large (e.g., a factor of 2). Additionally, surface-normal modulators (those operated with light propagating perpendicular to the layers) are more

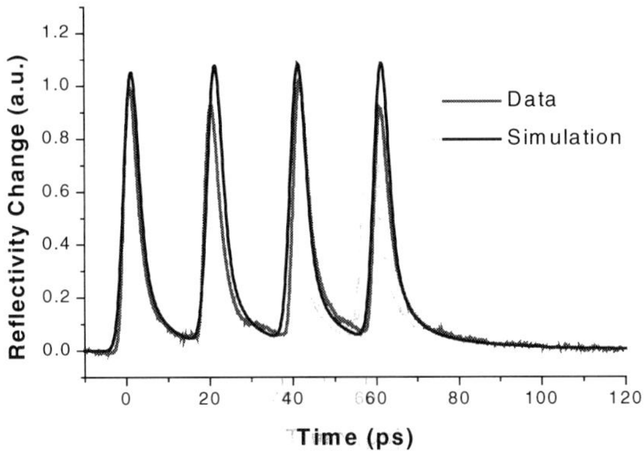

Figure 6. *Demonstration of the transfer of a 50GHz burst of pulses from a control to a signal beam. The signal beam reflectivity is plotted, together with simulations of the operation of the device.*

difficult to make at the wavelengths of interest for telecommunications (e.g. $\sim 1.5\mu$m) because the absorption and the resulting contrast ratio would be less in those materials. An attractive solution is to change to waveguide devices. Waveguide modulators can have long interaction lengths and hence high contrast ratios. It is also possible to propagate the signal light through the waveguide, and have the control light be incident from the top of the device, and so the control and signal light can be at the same or similar wavelengths, even in a two-diode device.

A first generation waveguide device is shown in Figure 7. This is a single diode device, and has successfully shown signal transfer from the control to the signal beam at 2.5GHz in a proof-of-principle demonstration (Sabnis *et al.* 2001). A second generation device would use a separate photodiode on top of the modulator diode, and would move the wavelength of operation to $\sim 1.5\mu$m by changing to, for example, InGaAs and InP materials. This structure may also be suitable for making a two dimensional matrix of switches, allowing a full cross-bar, wavelength-converting switch architecture.

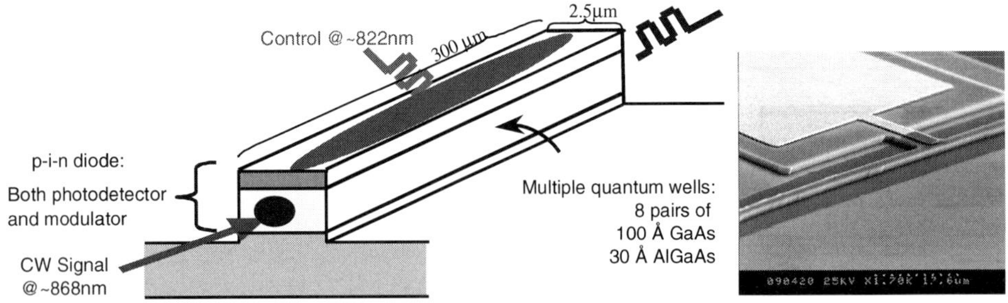

Figure 7. *An optically controlled optical gate for wavelength conversion, using a waveguide diode structure.*

5 Ultrafast photonics in digital electronic systems

There are several potential ways in which ultrafast photonics can help the performance of information processing systems that are otherwise conventionally electronic. Electronics, and silicon CMOS in particular, are likely to remain the dominant method of information processing of any substantial complexity for some time to come. We perhaps do not even have the beginnings of any technology that would supplant them for complex processing. There are, however, problems in electronic chip technology, quite beyond the continued scaling of the transistors themselves. Significant problems exist in interconnections, both within and between electronic processors; electronic wires do not scale to keep up with electronic transistors (Miller 1997, 2000), and it is now being appreciated that there may also be significant problems in achieving precise enough distribution of clock signals within electronic systems, especially once the digital clock rates exceed 10GHz. We will discuss below both the clocking issue with potential short-pulse optical solutions, and how short pulse optics could help in interconnects.

5.1 Optical clocking for electronics

Electronic digital systems rely not only on all signals corresponding to well-defined logic voltage levels, but also that all the signals arrive at well-defined times. The retiming of signals through the system to force them into the well-defined time slots is accomplished by clocking the system. Clocking often is used with latch circuits; the outputs from some set of logic stages are fed into latches where they are stored, and the latches are all read out at the same time by clocking the latches. Digital designers typically like to be able to presume that the timing of all of the different signals out of the different latches are precise to within 5–10% of the total bit time. This allows the construction of relatively complicated logical circuits that will operate reliably without generating false outputs because of timing errors. (See, e.g., Dally and Poulton 1998 for a discussion of design of digital logic systems.) Consequently the timing of the delivery of the clock signal needs to be similarly precise. For a 10GHz clock, therefore, we need timing precision of about 5ps, delivered to multiple points on the chip.

The delivery of electrical signals with this kind of precision is difficult. There are systematic or slowly varying changes in timing (known as "skew") between different parts of the chip or between different chips made to the same design, though often such skew can be controlled by good design and by the use of timing feedback loops such as phase-locked loops. More fundamentally troubling are the unpredictable and often rapid changes in timing, known as "jitter". These can arise from noise in the system, and can also arise from such effects as changes in supply voltage changing the delay of electronic circuits. Inductance in electrical lines, such as the power supply lines, causes changes in supply voltage as the current changes. Faster clocking leads to more rapid current changes, which in turn leads to larger inductive voltage drops. Such voltage drops are essentially not predictable because they depend on the switching of gates and output lines, which in turn depends on the specific data being processed.

We could imagine distributing the clock in the system by taking a short pulse laser and dividing it into multiple beams (e.g., by diffractive optical spot array generators) to deliver the clock to at least a moderately large number (e.g., 1000's) of points on a chip.

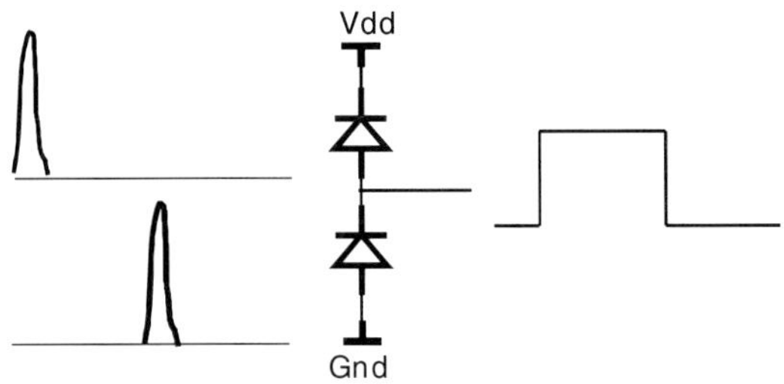

Figure 8. *Illustration of receiverless clock injection.*

Such an approach would eliminate at least some of the levels of the clock distribution tree in an electronic chip, or would allow us to deliver the clock to points in the chip where timing is particularly critical (such as serializer/deserializer circuits for high speed inputs and outputs). It might also enable us to reduce some of the power dissipation on the chip from the clock distribution. CMOS chips may have from 10–50% of their power consumption associated with the distribution of the clock. It is important in any such scheme that the amplifiers used in the optical receivers do not themselves have so many gain stages that they introduce more delay variation or power dissipation than they remove.

One approach is to use "receiver-less" optical clock injection (Debaes *et al.* 2002, Bhatnagar *et al.* 2002). This is illustrated in Figure 8. Two photodetectors are connected in series (sometimes called a "totem-pole" configuration). If the top photodetector is hit first with a short pulse, the voltage at the centre point rises rapidly. If the optical pulse is strong enough and the capacitance of the photodetectors is low enough, then the voltage can rise essentially to the upper supply voltage. If a subsequent pulse now hits the lower photo-diode, the output voltage will similarly drop essentially to the lower (0 V) supply voltage. Hence a clock signal with full logic voltage swing can be generated at the output without any voltage amplification. This approach has been shown to generate low jitter clock signals (e.g., less than 5–10ps, limited by measurement resolution) and improvement compared to electrical clock injection. Such schemes are therefore promising for the elimination of jitter in clocks for electronic digital processing systems.

5.2 Use of short pulses in optical interconnects

Interconnects in electronic systems have several fundamental difficulties. It has been understood increasingly in recent years that electrical wiring does not scale to keep up with transistors as they are scaled down in size (Miller 2000). One possible approach to this and other problems of electrical interconnects is to replace some electrical interconnects with optical interconnects. The physics of optical interconnects is substantially different (Miller 1997), and completely avoids physics such as the loss in electrical wires that is responsible for the limits on scaling of electrical wire information capacity.

Such optical interconnects have been the subject of growing research in recent years. A difficulty with optical interconnects, especially if they are to reach all the way to silicon chips, is that there is no good optical output device, either laser or modulator, that currently can be made entirely within silicon technology. Solder bonding of III-V optoelectronic devices has, however, been shown to be quite a viable possibility (Krishnamoorthy and Goossen 1998). Both vertical-cavity surface-emitting lasers (VCSELs) and quantum-well optical modulators have been demonstrated as possible output devices. The modulators have the advantages that they can likely operate to faster speeds (e.g., > 10GHz) than the VCSELs. Because they avoid the laser gain dynamics, they can operate with a range of input wavelengths, hence allowing wavelength-division multiplexing and they can operate with short optical pulses (e.g., picoseconds or even 100fs duration). The possibility of operating with short optical pulses opens up several additional features for optical interconnects.

Removal of signal skew and jitter

As signals pass through various levels of logic, different logical signal outputs may have slightly different timings. This variation in signal timing is known as signal skew. This is illustrated in Figure 9. Suppose that three outputs from an electrical circuit all have

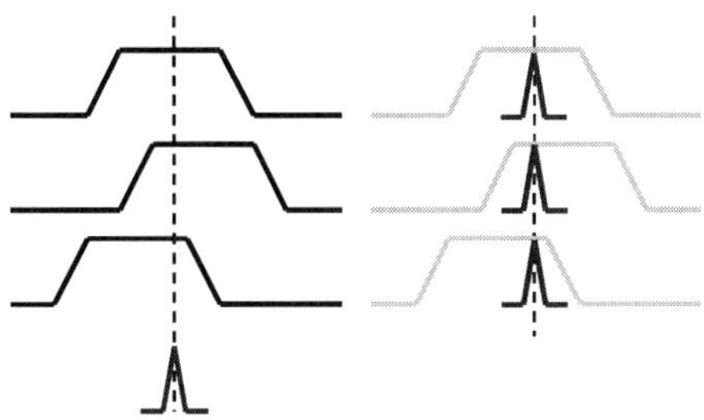

Figure 9. *Illustration of the removal of signal skew by reading out three modulators, each driven by a skewed signal, with three synchronized short optical pulses. The resulting three pulses now carry the information, but the skew has been removed.*

slightly different timings, and we use those to drive three optical modulators. We then read out the three modulators using three synchronized short pulses, one for each modulator. (It is relatively easy to ensure that three such optical pulses are synchronized by making them by dividing one short-pulse beam.) As long as the short optical pulses arrive during a time that all of the modulator voltages are valid simultaneously, the signal carried forward by the short pulses will correctly represent the logical signals at the output, but the skew will have been removed (Keeler *et al.* 2000).

It is similarly possible to remove jitter from a signal by reading out a modulator with a short pulse (Keeler *et al.* 2000). This is illustrated in Figure 10. A signal, with about 3/8th of a bit of jitter introduced, drives a modulator. The modulator is read out with a short optical pulse. The bottom trace is the optical output of a receiver circuit after receiving the short pulse readout and retransmitting the result with another modulator, so this data shows a complete system operation with the jitter successfully removed.

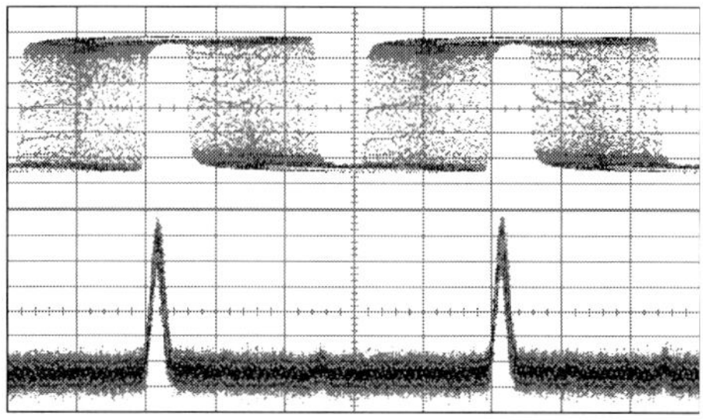

Figure 10. *Illustration of removal of jitter from a jittered signal (top trace) by reading out a modulator with a short pulse.*

Improvement of receiver sensitivity and interconnect latency

The use of short optical pulses to carry information gives rise to a situation never encountered in normal electrical systems, namely the input is driven with a pulse with much faster rising and falling edges than can be generated by the electrical circuits themselves. One therefore gets the impulse response of the receiver circuit, which can be faster than the response under conventional electrical drive. This leads to faster response from the receiver, and to greater sensitivity. Sensitivity improvements of 3dB have been demonstrated (Keeler *et al.* 2002).

For related reasons, the total delay (known as latency) in receivers can be substantially less with short pulse interconnects (Agarwal and Miller 2001). Latency is particularly important for shorter distance interconnects, and these latency reductions could make optical interconnects viable for some of the longer interconnects inside chips themselves.

Short pulse wavelength-division multiplexing

The use of ~100fs optical pulses opens up another possibility with optical interconnects, which is to use the bandwidth of the short pulse to give a wavelength-division multiplexed interconnect between chips. The concept for this is illustrated in Figure 11. A short optical pulse shines on a grating, which splits it into its spectral components. This spectrum lands on a linear array of optical modulators on a silicon chip. Each modulator modulates a specific piece of the spectrum, and reflects the modulated light back to the grating. The grating then recombines all of the spectral components into a single beam. This beam is then focused into a single fibre and the information is transferred to similar optics at the receiving end. There the different spectral components are distributed over a linear array of receivers, achieving a multichannel interconnect over a single fibre without electronic multiplexing. This system has been demonstrated (Nelson *et al.* 2002) in a proof of principle connection between two silicon CMOS chips.

This general approach of wavelength-division multiplexing directly to CMOS also suggests a possible future convergence of network interfaces and interconnects. It will not be very long before silicon CMOS is running at clock rates larger than 10GHz, and we could imagine wavelength-division multiplexed connections directly from optical networks to the silicon chip and back out again. Optical networks will be able to fill the entire fibre

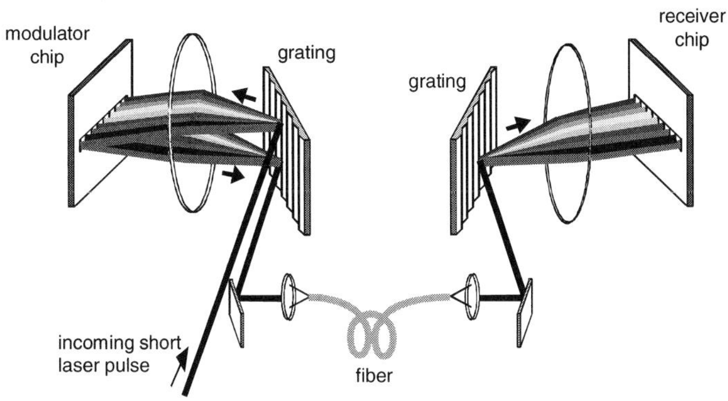

Figure 11. *Short pulse wavelength-division multiplexed interconnect.*

bandwidth solely with channels running at 10's of GHz rates. Indeed, it may be such network interfaces that drive the technology first, rather than interconnections within systems. There is the potential for solving major problems in large switching systems with technology of this kind also, possibly collapsing entire switching cabinets onto single silicon chips in the future.

The major challenges in such a wavelength-division multiplexed interconnect are in the optics. In particular it would be desirable to miniaturize the optical system, and improve losses. One possible way of obtaining miniaturized wavelength splitting devices is to exploit superprism effects. These are phenomena, first discovered in periodic optical structures (often known as photonic crystals) (Zengerle 1987, Kosaka *et al.* 1999, Nelson *et al.* 2000), and recently implemented in dielectric interference filters (Nelson *et al.* 2000), which can be considered to be one-dimensional photonic crystals for the case of simple Bragg reflector structures, for example. In the superprism effect, for certain specific angles of incidence and wavelengths, the angle of propagation of light inside the structure is strongly dependent on the incident wavelength, hence the term superprism. Recently, we have discovered that even larger and more controllable shifts in such structures can be obtained by using particular non-periodic structures (Gerken *et al.* 2002, Gerken and Miller 2002). These non-periodic structures show promise for giving shifts that are, for example, linear with wavelength, and are therefore very interesting for wavelength-division multiplexing and demultiplexing in general in compact, manufacturable structures.

6 Conclusions

We can see that there is a broad range of potential applications of ultrafast photonics in digital systems. Some of these, such as wavelength-division multiplexed interconnects, are surprising in that they have little to do with the very high speed character of ultrafast technology, yet ultrafast technology is still the key to implementing them. Others, such as optical clock injection, may seem rather simple uses of ultrafast technology but yet they might turn out to be the kind of mainstream application that might drive the practical development of the field. Still others, such as ultrafast optoelectronics, may be ways of marrying true ultrafast phenomena with practical applications in telecommunications.

Given the advanced state of our scientific understanding of ultrafast photonics, and the growing list of practical possibilities, we are at a key point for seeking out the practical applications, applications that will drive the next generation of scientific and technological investment. The joining of the applications and the opportunities will likely be limited mostly by our imagination, and is a field that calls for all our creativity and ingenuity to drive it forward.

Acknowledgments

I am pleased to acknowledge the contributions at Stanford of Diwakar Agarwal, Hatice Altug, Petar Atanackovic, Aparna Bhatnagar, Ray Chen, Henry Chin, Christof Debaes, Tim Drabik, Volkan Demir, Onur Fidaner, Martina Gerken, Noah Helman, Gordon Keeler, Helen Kung, Bianca Nelson, Rafael Piestun, Vijit Sabnis, Ryo Takahashi, Ryohei Urata, Micah Yairi, and Jun Zheng to the work discussed here.

I am pleased also to acknowledge the funding support of DARPA and of the Interconnect Focused Research Center (prime contractor: Georgia Institute of Technology), without which support this work would not have been possible.

References

Agarwal D and Miller D A B, 2001, 'Latency in Short Pulse-based Optical Interconnects', in *IEEE Lasers and Electro-Optics Society 2001 Annual Meeting, San Diego, California* Paper ThR3

Bhatnagar A, Debaes C, Chen R, Helman N C, Keeler G A, Agarwal D and Miller D A B, 2002, 'Receiver-less Clocking of a CMOS Digital Circuit Using Optical Short Pulses', *IEEE Lasers and Electro-Optics Society Annual Meeting, 2002, Glasgow, UK,* Paper MO3

Bhushan A S, Kelkar P V, Jalali B, Bohraz O and Islam M, 2002, '130-GSa/s photonic analog-to-digital converter with time stretch preprocessor', *IEEE Photonics Technol. Lett.* **14** 684

Cavailles J A, Miller D A B, Cunningham J E, Li Kam Wa P, Miller A, 1992, 'Simultaneous measurements of electron and hole sweep-out from quantum wells and modeling of photoinduced field screening dynamics', *IEEE J. Quantum Electron.* **28**, 2486

Chen Y K and Yu M C, 1992, 'Monolithic colliding-pulse mode-locked quantum-well lasers', *IEEE J. Quantum Electron.* **28**, 2176

Dally W J and Poulton J W, 1998, *Digital Systems Engineering*, (Cambridge University Press)

Debaes C, Agarwal D, Bhatnagar A, Thienpont H and Miller D A B, 2002, 'High-Impedance High-Frequency Silicon Detector Response for Precise Receiverless Optical Clock Injection', in SPIE Photonics West 2002 Meeting, San Jose, California, *Proc. SPIE* **4654** 78

Gerken M, Nelson B E and Miller D A B, 2002, 'Thin-film wavelength demultiplexer based on photonic crystal and group velocity effects', *OSA Conference on Integrated Photonics Research, Vancouver BC*, Paper IFH2

Gerken M and Miller D A B, 2002, 'Thin film (de)mux based on group velocity effects', *European Conference on Optical Communications, Copenhagen, Denmark*, Paper 11.3.3.

Goossen K W, Walker J A, D'Asaro L A, Tseng B, Leibenguth R, Kossives D, Bacon D D, Dahringer D, Chirovsky L M F, Lentine A L and Miller D A B, 1995, 'GaAs MQW Modulators Integrated with Silicon CMOS', *IEEE Photonics Technology Lett.* **7**, 360

Juodawlkis P W, Twichell J C, Betts G E, Hargreaves J J, Younger R D, Wasserman J L, O'Donnell F J, Ray K G and Williamson R C, 2001, 'Optically sampled analog-to-digital converters', *IEEE Trans. Microwave Theory and Techniques* **49** 1840

Keeler G A, Nelson B E, Agarwal D and Miller D A B, 2000, 'Skew and Jitter Removal Using Short Optical Pulses for Optical Interconnection', *IEEE Photonics Technol. Lett* **12** 714

Keeler G A, Agarwal D, Nelson B E, Helman N C and Miller D A B, 2002, 'Performance Enhancement of an Optical Interconnect Using Short Pulses from a Modelocked Diode Laser', in *CLEO 2002, Long Beach, California.* Paper CTuF5

Knox W H, Miller D A B, Damen T C, Chemla D S, Shank C V and Gossard A C, 1986, 'Subpicosecond Excitonic Electroabsorption in Room-Temperature Quantum Wells', *Appl. Phys. Lett* **48**, 864

Knox W H, Henry J E, Goossen K W, Li K D, Tell B, Miller D A B, Chemla D S, Gossard A C, English J, Schmitt-Rink S, 1989, 'Femtosecond Excitonic Optoelectronics', *IEEE J. of Quantum Electronics* **24**, 2586

Kosaka H, Kawashima T, Tomita A, Notomi M, Tamamura T, Sato T and Kawakami S, 1999, 'Superprism phenomena in photonic crystals: Toward microscale lightwave circuits', *J. Lightwave Techn.* **17** 2032

Krainer L, Paschotta R, Spuhler G G, Klimov I, Teisset C Y, Weingarten K J and Keller U, 2002, 'Tunable picosecond pulse-generating laser with repetition rate exceeding 10 GHz', *Electronics Lett.* **38**, 225

Krishnamoorthy A V and Goossen K W, 1998, 'Optoelectronic-VLSI: Photonics Integrated with VLSI Circuits', *IEEE J. Sel. Top. Quantum Electron.* **4**, 899

Lentine A L and Miller D A B, 1993, 'Evolution of the SEED technology: bistable logic gates to optoelectronic smart pixels', *IEEE J. Quantum Electron.* **29**, 655

Livescu G, Miller D A B, Sizer T, Burrows D J, Cunningham J, Gossard A C and English J H, 1989, 'High-speed absorption recovery in quantum well diodes by diffusive electrical conduction', *Appl. Phys. Lett.* **54**, 748

Mielke M, Alphonse G A and Delfyett P J, 2002, '60 channel WDM transmitter using multi-wavelength modelocked semiconductor laser' *Electronics Lett.* **38**, 368

Miller D A B, Chemla D S, Damen T C, Gossard A C, Wiegmann W, Wood T H and Burrus C A, 1985, 'Electric Field Dependence of Optical Absorption near the Bandgap of Quantum Well Structures', *Phys. Rev.* **B 32**, 1043

Miller D A B, 1990, 'Quantum-well self-electro-optic effect devices', *Optical and Quantum Electronics* **22**, S61

Miller D A B, 1997, 'Physical Reasons for Optical Interconnection', *Int. J. Optoelectronics* **11** 155

Miller D A B, 1999, 'Optics for Digital Information Processing' in *Semiconductor Quantum Optoelectronics*, pp 433-461, Eds. Miller A, Ebrahimzadeh M, and Finlayson D M, Proceedings of the Fiftieth Scottish Universities Summer School in Physics, St. Andrews (June 1998). (Publishers: The Scottish Universities Summer School in Physics, SUSSP Publications, and Institute of Physics Publishing)

Miller D A B, 2000, 'Rationale and Challenges for Optical Interconnects to Electronic Chips', *Proc. IEEE* **88** 728

Nelson B E, Gerken M, Miller D A B, Piestun R, Lin C C and Harris J S, Jr., 2000, 'Use of a Dielectric Stack as a One-Dimensional Photonic Crystal for Wavelength Demultiplexing by Beam Shifting', *Opt. Lett.* **25**, 1502

Nelson B E, Keeler G A, Agarwal D, Helman N C and Miller D A B, 2002, 'Demonstration of a Wavelength-division multiplexed Chip-to-Chip Optical Interconnect', in *CLEO 2002, Long Beach, California* Paper CTuF6

Sabnis V, Demir H V, Yairi M B, Miller D A B and Harris J H, Jr., 2001, 'Observation of Wavelength-Converting Optical Switching at 2.5 GHz in a Surface-Normal Illuminated Waveguide', in *IEEE Lasers and Electro-Optics Society 2001 Annual Meeting, San Diego, California* Paper TuCC2

Schmitt-Rink S, Chemla D S, Knox W H, Miller D A B, 1990, 'How fast is excitonic electroabsorption?' *Optics Lett.* **15**, 60

Semiconductor Industry Association 2001, International Technology Roadmap for Semiconductors, http://public.itrs.net/

Urata R., Takahashi R, Sabnis V A , Miller D A B, and Harris J S, 2001a, 'Ultrafast Differential Sample and Hold Using Low-Temperature-Grown GaAs MSM for Photonic A/D Conversion', IEEE Photonics Tech. Lett. 13, 717-719

Urata R, Takahashi R, Sabnis V A, Miller D A B, and Harris J S, 2001b, 'High-Speed Sample and Hold Using Low Temperature Grown GaAs MSM for Photonic A/D Conversion', *Conference on Lasers and Electro-Optics 2001, Baltimore, MD* Paper CMN4.

Urata R, Nathawad L Y, Ma K, Takahashi R, Miller D A B, Wooley B A and Harris J S, Jr., 2002, 'Ultrafast Sampling using Low Temperature Grown GaAs MSM Switches Integrated with CMOS Amplifier for Photonic A/D Conversion', *IEEE Lasers and Electro-Optics Society Annual Meeting, Glasgow* Paper ThT4

Walden R H, 1999, 'Analog-to-digital converter survey and analysis', *IEEE J. Selected Areas in Communications,***17** 539

Yairi M B, Coldren C W, Miller D A B and Harris J S, Jr., 1999, 'High-Speed, Optically-Controlled Surface-Normal Modulator Based on Diffusive Conduction', *Appl. Phys. Lett.* **75**, 597

Yairi M B, Demir H V and Miller D A B, 2001, 'Optically Controlled Optical Gate with an Optoelectronic Dual Diode Structure – Theory and Experiment', *Optical and Quantum Electronics* **33**, 1035

Yairi M B and Miller D A B, 2002, 'Equivalence of diffusive conduction and giant ambipolar diffusion', *J. Appl. Phys.* **91**, 4374

Zengerle R, 1987, 'Light propagation in singly and doubly periodic planar waveguides', *J. Mod. Opt.* **34** 1589

Femtosecond sources based on vibronic crystals

Tom Brown and Wilson Sibbett

University of St Andrews, Scotland

1 Introduction

At the heart of any ultrafast photonics-based communications system must lie a practical and versatile source of optical pulses. Most of the current technology is based on the exploitation of picosecond-pulses that are obtained from familiar semiconductor laser diodes. The ultrashort pulses from these lasers can be obtained using a variety of techniques including gain-switching, Q-switching, active modelocking and more sophisticated configurations may involve nonlinear mirrors based on optical fibre loops. The generation of femtosecond pulses for future datacommunication systems presents a somewhat different and perhaps greater challenge because it places more stringent conditions on the specifications of the laser source. This is due mainly to the requirement for a broad bandwidth gain material where a considerable portion of the gain can be coherently engaged to support these exceptionally short pulses. For example, a 100fs pulse centred around 1500nm requires a lasing bandwidth of 20nm.

Whilst the generation of sub-100fs pulses has been demonstrated using semiconductor-based sources (see Penty in this volume) and at moderately high pulse repetition rates (Khrushchev *et al.* 1998, Tsuchiya *et al.* 2001) the requirement to conduct post-generation temporal compression of the pulses adds greatly to system complexity. Fibre laser configurations also represent alternative practical sources of femtosecond pulses. Indeed, in certain cases, these lasers are capable of operation in the sub-100fs regime at MHz repetition rates(Tamura *et al.* 1994a, Tamura *et al.* 1994b, Cautaerts *et al.* 1997, Lefort *et al.* 2002). However, for the generation of pulses having durations in the few-femtosecond domain, vibronic crystals represent the preferred gain media. Such so-called vibronic lasers tend to be viewed as intrinsically complicated and for use mainly in research laboratories due to the extremely high cost of the CW argon-ion or Nd-based frequency-doubled solid-state pump lasers and the low pulse repetition rates available from their extended-cavity designs. This does not have to be the case and recent research has shown that femtosecond vibronic lasers can be made to be highly compact and relatively simple (Ramaswamypaye and Fujimoto 1994, Tomaru 2001). Such compact lasers can operate at pulse repetition rates comparable with their picosecond semiconductor counterparts, can give wide wavelength coverage and offer good electrical-to-optical efficiencies whilst generating pulses in the sub-100fs regime.

Most of the key advances in vibronic lasers have been driven by the development of novel gain materials, the continued evolution of laser diode pump sources, the application of tailored dispersion compensation and dedicated design strategies for the resonators. We believe that vibronic lasers will have a part to play in the development of the next generation of femtosecond networks and may, in hybrid configurations, have a component-related role in future telecommunication/datacommunication systems. This could be either in their own right as sources of optical pulses or in hybrid combination with semiconductor laser elements as a means for pulse compression and all-optical clock recovery.

An ultimate objective in research into practical femtosecond-source components must be the development of simple, electrically pumped semiconductor sources that offer suitably broad gain bandwidths. It is to be expected that many techniques demonstrated in vibronic laser systems will contribute directly to the adaptation of semiconductor-based laser systems for future network applications.

The remainder of this chapter consists of Section 2 on the generation of femtosecond pulses, Section 3 on the design of compact resonators for low pump-threshold lasers and Section 4 on a look towards future designs or implementations.

2 Producing femtosecond pulses

The shortest pulses are generated using the principle of mode locking where a phase relationship is established amongst the longitudinal modes of a laser resonator. In general, standing-wave (and travelling-wave) resonators support many modes that satisfy the condition that an integer number of half-wavelengths (or wavelengths) are contained within the cavity. This standing-wave condition in particular implies that a node must be formed on each of the resonator terminal mirrors allowing us to write an expression for the wavelength of the Nth mode, λ_N:

$$\lambda_N = 2nL/N, \tag{1}$$

where nL is the optical length of the cavity, n being the average value of the refractive index, and N is an integer. For lasing to occur, a mode must also fall within the above-threshold gain bandwidth of the laser. This allows us to compute the maximum number of modes that can be coherently coupled in a particular gain medium. For example, a Cr^{4+}:YAG laser having a 0.5m long standing-wave resonator operating around 1500nm will have a spacing between adjacent modes of approximately 0.0025nm allowing 4000 modes to be supported in an above threshold laser gain bandwidth of only 10nm.

To generate ultrashort pulses, a fixed phase relationship must be established between the longitudinal modes that reach threshold. Figure 1 illustrates the build-up of a pulse from longitudinal modes for an increasing number of longitudinal modes. The dark lines indicate the sum of the modes under the condition of a fixed phase relationship, whilst the light line represents the sum of the same number of modes that are randomly phased. It can be seen that as the number of modes involved in the process increases, two effects take place. The pulse becomes shorter and the amplitude/intensity of the pulse increases with

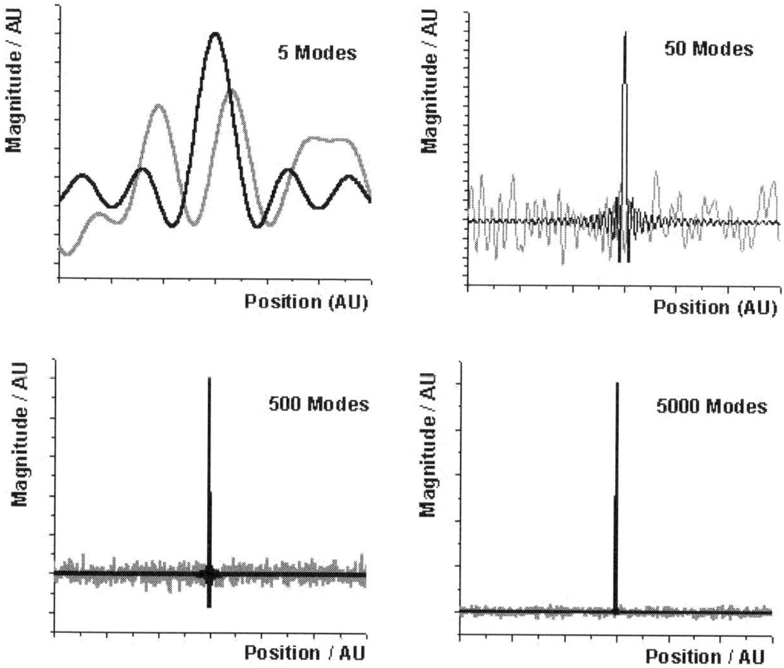

Figure 1. *The build up of an optical pulse through the coherent addition of longitudinal modes. The dark line represents a pulse where a fixed phase relationship exists between the modes. The light line is the sum of the same number of modes where a random phase relationship exists between the modes.*

respect to the incoherent case. The final steady-state condition for a modelocked laser is therefore one where a discrete pulse circulates in the laser cavity with a very high-peak power and a short duration. To obtain the shortest pulses, the largest possible number of modes must be locked together and so an extended gain bandwidth of the laser is of key importance.

A complementary approach to understanding the build up of modelocking is to consider the dynamic losses in a laser cavity. A laser will operate in a regime where loss is minimised. It is possible to design the laser system so that the losses are minimised when the laser is operating in a regime where ultrashort pulses are generated. As will be seen later in the chapter, this approach has been very successful in the development of some types of compact laser.

2.1 Active modelocking

A widely used technique for modelocking lasers is active modelocking. In this method, an acousto-optic modulator (AOM) is placed within the laser cavity and a modulation produced at the cavity repetition rate. We can consider that this device modulates the amplitude of a particular mode at the peak of the gain, giving rise to sidebands that correspond to the adjacent laser modes. In such a way, a transfer of power takes place between adjacent modes, locking their relative phases. Whilst active modelocking has proved successful, it is not suitable for the generation of the femtosecond pulses due to a combination of effects including the finite bandwidth over which the modulator can operate and phase drifts between the AOM and the laser modes. The AOM also increases cavity complexity and cavity losses. Therefore, to obtain femtosecond-pulses we rely on the influence and control of the intracavity radiation to phase-lock the cavity modes. This technique is known as passive modelocking.

2.2 Passive modelocking

The shortest pulse durations are obtained through passive modelocking. Whilst used originally in dye lasers during the 1970s and 1980s to obtain pulses with durations of $<$ 20fs (Finch *et al.* 1988), the technique has been transferred to solid state lasers with great success. There are several forms of passive modelocking including harmonic modelocking and additive pulse modelocking, but in this chapter we will concentrate on two methods that offer modelocked laser pulses in a straightforward manner.

2.2.1 Kerr-lens modelocking (KLM)

Kerr-lens modelocking was first demonstrated in a Ti:Sapphire laser by a research group in the University of St Andrews in 1989 (Spence *et al.* 1991). It has since proved itself to be a very versatile technique covering a wide variety of wavelength regions and producing pulses of sub-5fs duration(Sutter *et al.* 1999), the shortest pulses obtained directly from any laser oscillator. KLM has also been incorporated into a wide variety of commercial products and it is now possible to purchase lasers that produce ultrashort pulses having durations of 10fs.

The optical Kerr effect is well known and relates to an induced intensity-dependent refractive index in the medium through which the optical beam propagates. In a laser resonator, the pulse undergoes self-focussing as it passes through a non-linear element that, in the case of many different lasers, is the gain medium itself. A further important component in a KLM system is self-phase modulation (SPM). The Kerr effect produces a modulation of refractive index and induces a phase modulation that acts to produce a shift in the instantaneous phase of an evolving pulse. The intensity range in the profile of a pulse gives rise to phase shifts that results in a spectral broadening. As noted earlier, the shortest pulse durations require a wide gain bandwidth above threshold so that SPM can be fully exploited.

Pulse evolution in a KLM laser builds up from intracavity stochastic noise present when running CW. A sufficiently large noise spike is subject to the nonlinear optical processes described above to initiate the Kerr-lens mechanism. Once started, the pulse begins to form until a stable equilibrium is reached and femtosecond-pulses are produced ultimately from the laser oscillator.

To obtain modelocking in practice, an aperture can be placed in the laser cavity, such that light that is strongly self-focused passes through the aperture, whereas light which is not self-focussed suffers some loss on transmission through the aperture. This is known as 'hard aperture' KLM and is illustrated in Figure 2. It is also possible to obtain KLM

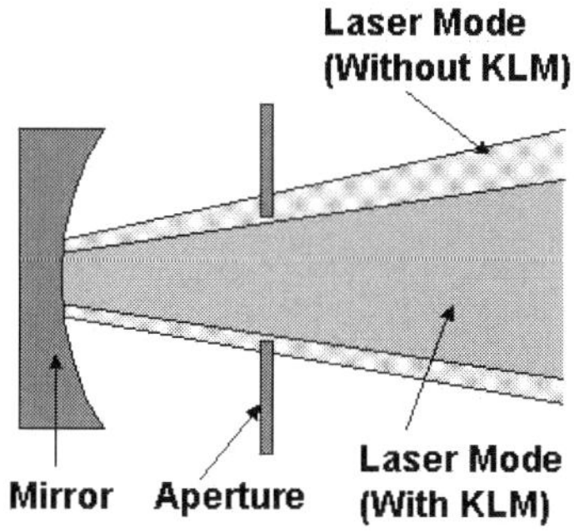

Figure 2. *Hard aperture modelocking*

without an external aperture through a subtle process known as 'soft-aperture' KLM. In this case, the self-focusing of the laser beam improves the overlap between pump and signal in the gain medium. This gives the self-focused light a higher gain (lower overall loss) than non-self-focused light, thus the laser will operate in a pulsed format; this is illustrated in Figure 3. The process of KLM, both in the hard and soft aperture configurations can be thought of as an amplitude modulation driven by the dynamics of the optical radiation itself.

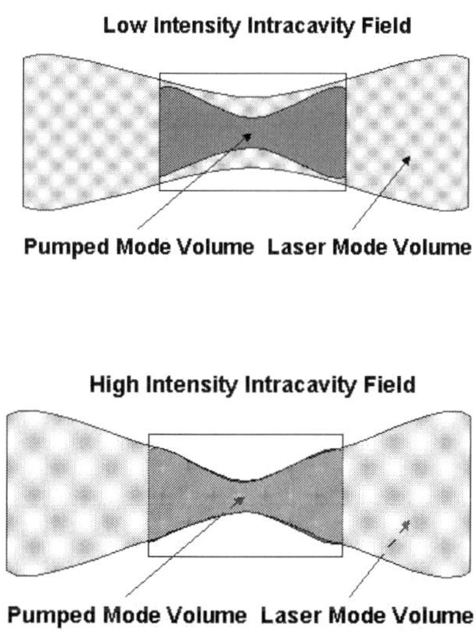

Figure 3. *Soft Aperture KLM. The pump mode (dark shading) and laser mode (light shading) illustrated in the upper diagram with low intensity intracavity field and hence no Kerr lens present; the lower diagram has a high intracavity field and hence there is a Kerr lens present. Note that in the lower diagram the overlap between pump and laser is enhanced giving preference for modelocked operation.*

In most cases, the KLM process described above is not self starting and noise spikes have to be produced in the laser cavity by, for example, moving one of the cavity mirrors, or tapping the bench upon which the laser is built. This is one of the major drawbacks of KLM and also tends to make the laser system rather sensitive to external perturbations.

Saturable absorption

A further method to initiate self-modelocking is to use a saturable absorber in the laser cavity. At its simplest, a saturable absorber is a cavity element that has a higher transmission for an intense pulse than for less intense continuous wave radiation. We thus can have a case where the loss is reduced for pulsed operation, so the laser operates preferentially in a mode-locked regime. One of the great advantages of modelocking based on saturable absorption is that pulsed operation is self-starting for well-designed laser cavities.

The use of saturable absorbers to modelock lasers was first pioneered in the 1960s (Mocker and Collins 1965; DeMaria *et al.* 1966; Smith 1970) and, by the mid 1970s the use of organic dye jets both as the gain medium and as saturable absorber allowed the generation of 500fs pulses (Shank and Ippen 1974). Further developments in mode-locking techniques and dispersion compensation (Fork *et al.* 1984) allowed pulses as short as 20fs to be generated by the late 1980s (Finch *et al.* 1988).

In the case of the organic-dye based lasers, the short steady-state net gain window necessary for femtosecond operation, was caused by a balance between the saturated loss and gain present in the gain medium (Valdmanis and Fork 1986). The gain window formed could be as short as the duration of the pulse itself. The long upper state lifetime and choice of saturable absorbing elements was thought to be a limitation for using saturable absorbers to modelock vibronic crystal media due to the Q-switching instabilities that were present. However the arrival of sophisticated semiconductor growth technologies has allowed the development of highly controlled saturable absorbers that can be designed to resist such instabilities. Such semiconductor-based devices, which combine the absorption effects of quantum-wells with the high reflectivity from semiconductor Bragg mirrors, were initially developed by Tsuda *et al.* (1996) with further major developments being made by Keller *et al.* 1996). These modelocking elements are known as saturable Bragg reflectors (SBRs) or semiconductor saturable absorbing mirrors (SESAMs) and are closely related devices. A typical design for an SBR is illustrated in Figure 4.

The SBR illustrated comprises a highly reflecting mirror formed by a Bragg stack of alternating refractive index semiconductor materials. In the top layer of the mirror, a single quantum well is placed where the electric field is calculated to be a maximum. The absorption dynamics of the quantum well can be used to effectively control the reflectivity of this composite mirror, that is, when the quantum well absorbs strongly, the mirror exhibits a higher loss than when the quantum well absorption is fully saturated. Thus a pulse that saturates the quantum well absorption experiences a lower loss and the laser will run preferentially in this mode of operation. The operation of the saturable absorber for the production of ultrashort pulses in vibronic media can be described in two cases that differ from the saturable loss / dynamic gain saturation process described above. Using the nomenclature outlined by Keller *et al.* (1996), these are called (a) fast saturable absorption and (b) soliton modelocking. A fuller description of the absorption processes involved is provided in Keller *et al.* (1996) and Paschotta and Keller (2001)

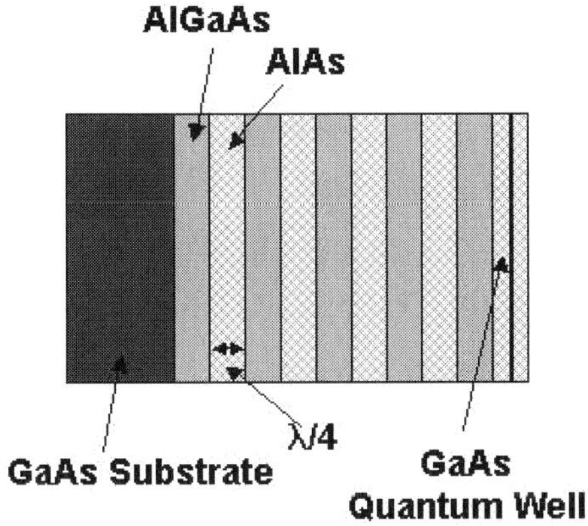

Figure 4. *Saturable absorber for operation near 850nm (Tsuda et al. 1996) . The lower part of the mirror is a highly-reflecting Bragg-stack comprising 30 AlAs / AlGaAs pairs. In the top layer, a GaAs quantum well has been placed where the electric field is maximised to provide a saturable absorption of between 0 –2% depending on position.*

though we will consider these techniques in turn below.

Case (a): fast saturable absorption

In this case, the fast dynamics of the saturable absorber lead to an effective net-gain window. These effects are independent of any gain saturation in the laser material and depend solely on the saturable absorber to produce short pulses. A fast saturable absorber must therefore possess very fast recovery times. The fast saturable absorption regime is often used to produce pulses of picosecond duration and has been used by Keller (1994) in a variety of familiar gain crystals for example, Nd:YAG, Nd:YLF and Nd:YVO_4.

A KLM laser may also be thought of as a laser modelocked by a 'synthetic' fast saturable absorption. In this case the optimised overlap between pump and laser is created by the Kerr-lens effect caused by the pulse. This creates an optimised gain window for the pulse that is only opened for the duration of the pulse itself, analogous to the behaviour of the fast saturable absorber described, but with no wavelength dependence due to the saturable absorption mechanism.

Case (b): soliton modelocking

In some cases, pulses produced by a saturable absorber are much shorter than would be implied by the recovery dynamics of a fast saturable absorber, indeed pulses of tens of femtoseconds duration can be produced relatively easily using SBR and SESAM structures. A hypothesis to explain this effect has been provided by Keller (1996). This model is based on the formation of solitons in the laser cavity and called 'soliton modelocking'. In this case, the saturable absorption is used to initiate and stabilise a soliton dominated by the balance between group velocity dispersion (GVD) and self-phase modulation within the optical cavity. Using this model, Keller has shown that the pulse durations produced can be as much as 10 times shorter than the net-gain window produced by the 'slow' saturable absorber. It is also clear that other non-linear optical processes take place within the formation and stabilisation of ultrashort pulses and the overall sustained modelocking behaviour is the result of a combination of these effects.

Whilst modelocking based on SBRs/SESAMs has proved to be very successful for obtaining self-starting generation of femtosecond-pulses, there are some limitations with this technique. To generate the shortest pulses, a very broadband high reflectivity characteristic must be provided by the Bragg mirror (Jung *et al.* 1997). The quantum well must also provide absorption at the operating wavelengths. This means that the SBR/SESAM must be designed for a specific operating wavelength limiting the tunability of the laser output. To design the low threshold cavities described later in this chapter it is also important to minimise the propagation loss of the structure.

2.3 Dispersion compensation.

Walmsley and Dorrer (2001) have shown that the dispersive properties of an optical system are given by the coefficients of an expansion of the spectral phase function, $\phi(\omega)$, for that system:

$$\phi(\omega) = \sum_n \frac{1}{n!}\phi^n(\omega_0)(\omega - \omega_0)^n, \tag{2}$$

where $\phi^n(\omega_0)$ is the n-th derivative of the phase function with respect to the frequency evaluated at ω_0. This quantity is specified in units of rad^{-1}fs^n, mostly simplified to fs^n.

The first four terms resulting from the expansion of Equation 2 can be interpreted in the following way: $\phi^0(\omega_0)$ represents the position of the reference frequency with respect to the pulse envelope. Whilst this parameter is unimportant in the pulse durations for the lasers described in this chapter, it becomes vastly more important in the study of the shortest optical pulses containing only one or two optical cycles and is thus the focus of much interesting and exciting research. $\phi^1(\omega_0)$ is related to the group velocity of the pulse and represents the speed with which the pulse envelope propagates through the medium. $\phi^2(\omega_0)$ is the quadratic spectral phase term and is often known as the group delay dispersion (GDD) parameter. This parameter, with units of fs^2, specifies the rate at which a pulse centred at the reference frequency increases in duration as it passes through the optical system. The terms 'positive dispersion' or 'negative dispersion' arise from the sign of this quantity. The higher-order terms resulting from this expansion, for example, third order dispersion (TOD) with units of fs^3 must also be considered in the design of resonators to produce the shortest optical pulses. This is because residual higher-order-dispersion, even in a cavity with well-compensated GDD, can cause variation in the spectral phase function across the pulse spectrum. This limits the available dispersion-compensated gain bandwidth and hence provides a lower limit for the time-bandwidth limited pulse duration.

Most of the vibronic laser crystals in use possess positive dispersion in the visible and near infrared spectral regions. It is therefore important to compensate for this dispersion to obtain the shortest pulses. Indeed, a Ti:sapphire laser operating without suitable dispersion compensation is generally limited to pulse durations of several picoseconds.

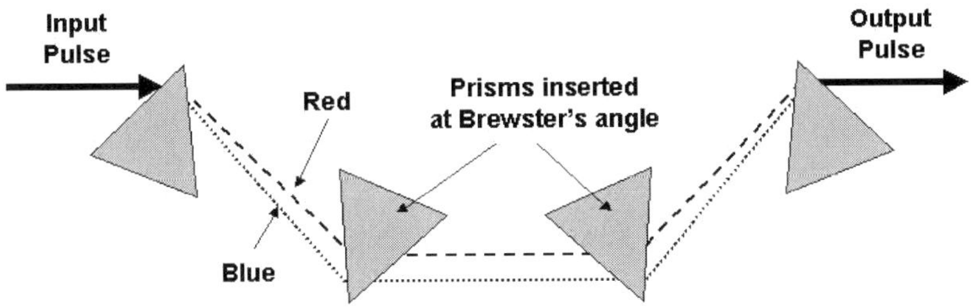

Figure 5. *Dispersion compensation using prisms. (Fork et al. 1984) Red-shifted components of the pulse (dashed line) travel through more of the prism material than blue-shifted components (dotted line) before being recombined. This in turn gives a longer optical path length for the red-shifted components relative to the blue-shifted components and thus acts as a source of negative dispersion in the cavity to counteract the positive dispersion in the gain medium. Careful adjusting of the prism insertion can therefore provide appropriate compression for a chirped input pulse.*

2.3.1 Prism-based dispersion compensation schemes

The traditional method for compensating dispersion in ultrashort-pulse lasers was first described by Fork *et al.* (1984) and relies on the use of pairs of prisms incorporated within the cavity at Brewster's angle. Whilst the angular dispersion in such a system may be relatively small, it is generally sufficient to compensate for the positive dispersion induced in the gain medium. The process can be understood with reference to the four-prism arrangement illustrated in Figure 5. The temporally dispersed, chirped pulse enters the first prism and its frequency components are dispersed in the normal way. The different wavelength components then travel different path lengths through the prism scheme thus allowing the chirped pulse to be compressed temporally. This system allows for careful tuning of the dispersion compensation through varying the amount of prism glass inserted into the cavity. The prism separation can also be varied and controlling this parameter carefully can give some control over the higher order dispersion within the laser. It is also found that the low-loss associated with the Brewster-angled prisms allows this methodology to be used for very low threshold cavities.

In efforts to reduce the component count and complexity of modelocked lasers having prism-based dispersion schemes, other designs have been developed that use only a single pair of prisms, or in some cases a single prism as will be discussed later in the chapter.

2.3.2 Mirror-based dispersion compensation schemes

Negative dispersion may also be introduced into the cavity through the use of mirror-based dispersion schemes. These fall into the two main categories of chirped mirror dispersion compensation and interferometer-based dispersion compensation. These two approaches will be discussed in more detail below, but the common advantage of mirror-based schemes is the exact control and tailoring of the dispersion compensation for a specific laser system and the ability with special design considerations to use chirped mirrors to compensate for the higher-order dispersion terms. Indeed, it is now possible to combine the dispersion compensation provided from a mirror-based scheme with the performance necessary for a cavity fold mirror to produce lasers that require no extra components for dispersion compensation. Thus, the very compact, low threshold lasers based on this technology described later in the chapter can become practicable.

2.3.3 Chirped-mirror based dispersion compensation.

A chirped mirror is a distributed Bragg-reflector (DBR) structure where the grating period has been varied throughout the mirror and a basic schematic of such a mirror is shown in Figure 6. The chirped Bragg grating results in different wavelengths being reflected from different points within the mirror stack. As applies to multiprism-based schemes discussed above, the path differences associated with the spectral extent of a chirped optical pulse can lead to temporal compression of the input pulse.

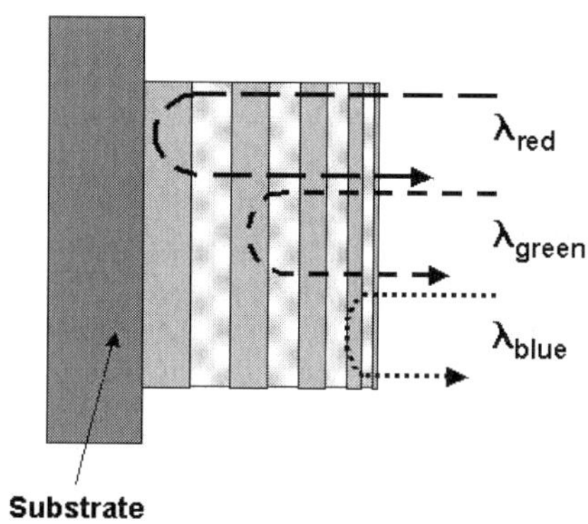

Figure 6. *Schematic of a chirped mirror. Due to the varying period throughout the Bragg stack that makes up the mirror, longer wavelengths travel further into the mirror structure than short wavelengths. This acts as a time delay for the red-shifted components of the pulse relative to the red-shifted components giving the desired producing pulse compression. Note that in most cases, several bounces are required to achieve the desired negative dispersion.*

Correctly designed chirped mirrors can also be used to compensate for the higher-order dispersion effects discussed above. Szipocs and Kohazikis (1997) have demonstrated mirrors that exhibit performance suitable to counteract both third- and fourth-order dispersion effects.

In practice, it is generally necessary to have several bounces from such a mirror to compensate fully for the dispersion. This places stringent requirements on the loss for such mirrors: a typical vibronic laser designed for low threshold, high repetition rate operation may well be operating with output couplings as low as 0.2%, and so any excess loss introduced by the dispersion compensation element and other cavity mirrors can have a very serious effect on overall laser performance.

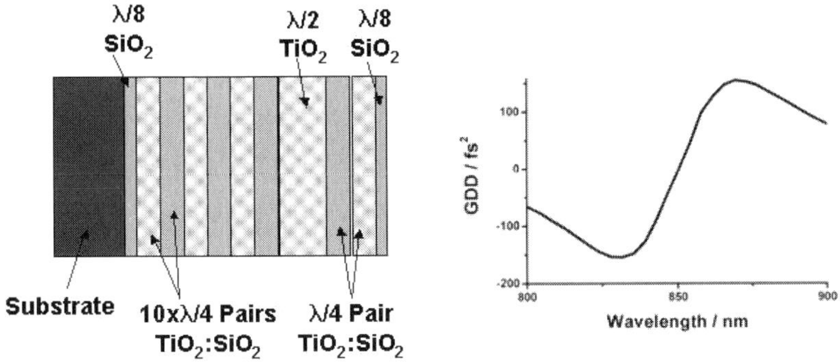

Figure 7. *(a.) The design of a dielectric-stack Gires-Tournois interferometer based mirror that is designed to be highly reflecting from 700-900nm. The interferometer is formed by the $\lambda/2$ layer within the mirror structure. Light near the resonance frequency of the interferometer is 'trapped' in the etalon producing the required dispersion performance. (b.) The calculated dispersion profile for the structure illustrated in Figure 7. When the laser is operated near 850nm, widely tunable dispersion performance can be obtained from this mirror. Similar to the chirped mirror in Figure 6, several bounces are normally required to achieve optimum modelocked laser performance.*

2.3.4 Interferometer-based mirror dispersion compensation.

An alternative approach is to use an interferometer combined monolithically with a DBR mirror to produce negative GDD within the laser cavity. These mirrors exhibit very low loss and are suitable for use in low-threshold cavities. Figure 7a shows a typical design for a mirror based on a single so-called Gires-Tournois interferometer (GTI) design (Szipocs *et al.* 2000). Figure 7b shows the calculated quadratic dispersion behaviour for this structure where it can be seen that operating at a suitable wavelength provides the negative dispersion necessary to counteract the positive dispersion introduced by the laser crystal.

In order to use the GTI design that is compatible with the broad wavelength range that is necessary for the generation of ultrashort pulses or to offer tunability, several such interferometers with slightly varying thickness can be combined to give a structure with a very high reflectivity and good dispersion performance over a wide range of wavelengths greater than 100nm. These mirrors are known as multiple-cavity GTI's (MCGTI)(Szipocs *et al.* 2000) and are now being used in a range of ultrashort pulse laser systems.

Thus far, we have considered the design of lasers that can produce femtosecond pulses. This requires careful consideration of the longitudinal mode structure of the laser, the method by which a common phase relationship can be established amongst the cavity modes and the management of the dispersion within such laser cavities. A variety of approaches have been discussed for managing these issues. In the next section, we will look at the design considerations for practical implementations of laser resonators that are suitable for the generation of femtosecond pulses. The emphasis will be on resonators that are compact, offer low pump power thresholds and give high pulse repetition rates, each of which brings different, but often related, challenges to the laser designer.

3 Compact, low threshold femtosecond lasers

The first criterion in the design of laser resonators for low pump threshold, pulsed operation is to minimise threshold of the laser in CW operation. This normally provides strong insights into the reduction of the modelocking threshold of the lasers. Valentine and Sibbett (1999) have shown that a relationship for the CW threshold of a laser, P_{th}, can be obtained following the expressions derived by Alfrey (1989):

$$P_{\text{th}} = \frac{(T+\Lambda)h\nu_p\pi}{4\sigma\tau\alpha_p \int_0^L \frac{\exp(-\alpha_p z)}{w_p^2(z)+w_c^2(z)}dz}. \tag{3}$$

T and Λ are the output coupled and parasitic losses respectively, $h\nu_p$ is the pump photon energy, σ and τ are the gain cross section and upper-state lifetime respectively, α_p is the pump absorption coefficient, L is the crystal length, z is the axial distance from the crystal face and $w_p(z)$ and $w_c(z)$ represent the diverging pump and cavity mode beams respectively.

From Equation 3 it can be seen that the threshold of the laser is linearly dependent on the cavity loss and pump energy, inversely proportional to the gain cross section/lifetime product and is strongly related to the overlap between pump and cavity modes in the laser crystal. It can be seen then that overall laser performance can thus be improved by optimising as many of these parameters as is possible without compromising the requirements for minimising the threshold for mode-locked operation.

3.1 Modelocked operation at low pump thresholds

Valentine and Sibbett (1999) have shown that the threshold for KLM is dependent on the position of the cavity focus in the laser gain medium. To operate at low threshold, it is important that intensity-dependent effects in the laser are accentuated; that is to say that both the strength of any nonlinear optical effect and the subsequent amplitude modulation effect arising must be enhanced by the choice of laser cavity design. In practical terms this requires that the cavity be designed such that a tight cavity focus is maintained at one end of the laser gain medium.

It is not normally possible to satisfy the condition described above in the 'classic' symmetric Z-cavity laser that has a Brewster-angled gain-medium. In this case, the cavity focus becomes distributed through the gain medium when attempts are made to position it on the end face of the rod. One suggestion for improving the threshold of such a resonator is to replace the Brewster-angled gain medium with an anti-reflection coated plane/plane crystal. Whilst this may be successful, the introduction of the AR coatings can lead to significant excess loss within the cavity. This increases the CW threshold and hence the modelocked threshold in the manner described above.

To address this problem, initial investigations concentrated on a three-mirror cavity similar to the one illustrated in Figure 8 (Burns *et al.* 1996). Within this cavity, both the tangential and sagittal foci can be placed coincident on the end of the rod. The success of this geometry was demonstrated by Valentine *et al.* (1997), where a Cr^{3+}:LiSAF laser generated 84fs pulses via KLM using only 104mW of pump power at 670nm from a broad

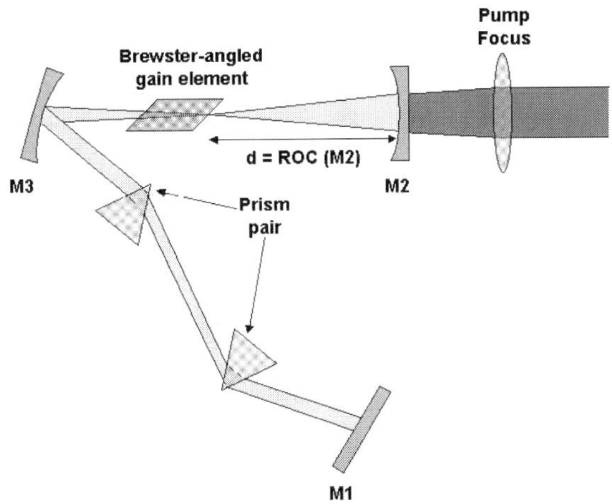

Figure 8. *Design for a 3-mirror low threshold femtosecond laser cavity. In this case, a pair of prisms is being used to provide dispersion compensation (Burns et al. 1996).*

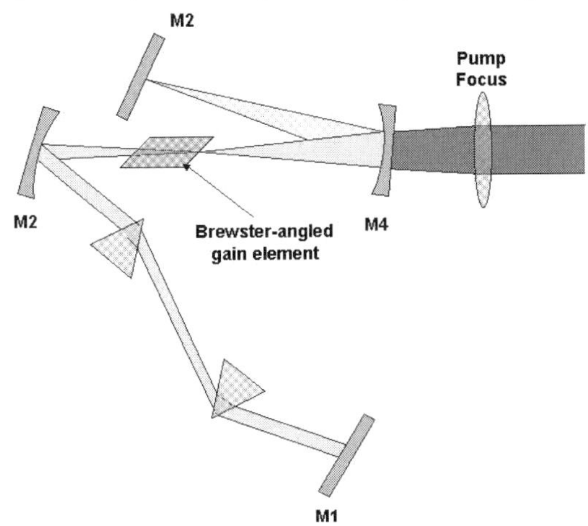

Figure 9. *A highly asymmetric four mirror laser cavity. This design allows low mode-locked pump thresholds whilst also allowing variation of the fluence on an SBR or SESAM placed in the short cavity arm to optimise performance (Valentine and Sibbett 1999).*

stripe laser diode. This result proved that laser operation in the femtosecond regime could be obtained successfully using relatively low diode pump powers.

One major disadvantage of the three-mirror resonator described above is that the laser output from the output coupler is divergent. Whilst this could be controlled with extra-cavity lenses etc, this adds to the component count. Further, this condition also makes hard-aperture KLM difficult as only a slit may be used to aperture the saggital plane. To counteract these problems a highly asymmetric four-mirror laser cavity has been developed as illustrated in Figure 9. This laser cavity has two stability regions; the low misalignment sensitivity region (LMS) where a tight focus is formed on the cavity mirror in the short arm and the high misalignment sensitivity region (HMS) where the beam is collimated in the shorter arm. The LMS region offers a further advantage when combined with a SESAM or SBR. This is that the spot size on the mirror may be varied in this arm allowing the fluence on the semiconductor mirror to be optimised for particular pump powers/output coupling.

For the more compact resonator geometries, careful thought must be given to dispersion compensation. The standard approach described above where a pair of prisms with a fixed separation is used to give sufficient negative dispersion, and where applicable, higher-order dispersion compensation is not suitable for highly compact laser designs. For example, a Cr^{3+}:LiSAF laser with dispersion compensation provided by a pair of LaK31 prisms requires a prism separation of ~200mm to achieve the required second-order and third-order dispersion compensation. One approach to solving this is to move to a situation where dispersion compensation is provided by a single prism (Kopf *et al.* 1996, Ramaswamypaye and Fujimoto 1994). Simple geometrical arguments can be used to demonstrate that in an appropriately designed cavity, the single prism can have the same effect as that first described by Fork *et al.* (1984) for multiple prisms. The single prism also reduces the number of cavity components and hence the optical loss in the cavity thereby affording a lower threshold.

Following the method described by Kopf *et al.* (1997), Hopkins *et al.* (1999) constructed a simplified laser based on diode-pumping of Cr^{3+}:LiSAF. The single-prism dispersion technique allowed the construction of a laser system where all of the components including the pump lasers, associated electronics and power supply fitted comfortably onto a 28×22cm optical breadboard. This laser was modelocked using a SESAM device and repetition rates as high as 450MHz with 250fs pulses were observed with an average output power of 6.4mW for only 72mW of input pump power. At lower repetition rates (350MHz), pulses as short as 133fs were obtained with an output power of 5mW for the same incident pump power. This system represented the first simplified femtosecond-laser system that also demonstrated the potential for low pump power operation in a very compact geometry suitable for taking femtosecond-laser sources out of laboratory based environments.

Ramaswamypaye and Fujimoto (1994) described a highly simple and compact laser system based on KLM and a single prism for dispersion compensation (see Figure 10). By making use of a gain medium with plane/Brewster facets, with a high reflector (pump transparent) mirror coated on the plane endface and providing dispersion compensation using a Littrow prism with an output coupler at the laser wavelength coated on the plane face of the prism, Ti:Sapphire lasers with 111fs pulses operating at 1 GHz have been obtained. Tomaru and Petek (2000) have adapted this resonator for Cr^{4+}:YAG lasers and obtained a 55fs pulse with a repetition rate of 1.2GHz. A further development of this oscillator is illustrated in Figure 11. This system, comprising only two cavity elements, has given 115fs pulses at a pulse repetition rate of 2.6GHz (Tomaru 2001).

It is also possible to modelock Cr^{4+}:YAG using SBR/SESAM based technologies although the fabrication of the saturable absorber elements is considerably more difficult than at the shorter wavelengths associated with the Ti:Sapphire and Cr^{4+}:LiSAF lasers discussed above. Recently Lagatsky *et al.* (2002) have demonstrated a Cr^{4+}:YAG laser giving 120fs at a repetition frequency of 205MHz pulses using an SBR. This laser also made use of a highly compact 10W fibre-laser-based pump source that further enhanced

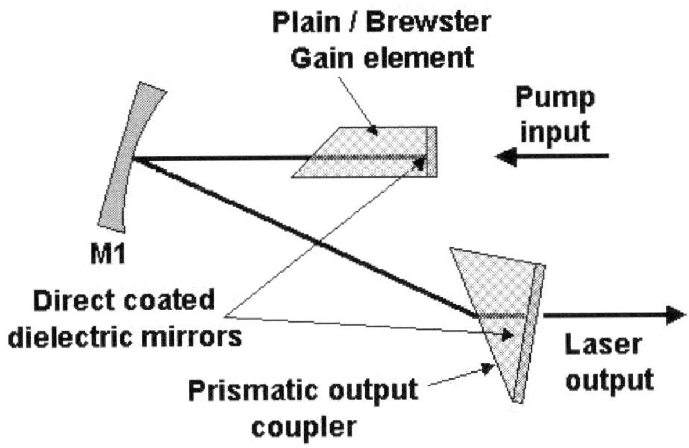

Figure 10. *Simplified resonator for femtosecond operation: dispersion compensation is provided by the prismatic output coupler which is a Littrow prism with an output mirror coated on the rear face. Note also that the cavity has been further simplified by the coating of the input mirror directly onto the gain element (Ramaswamypaye and Fujimoto (1994)).*

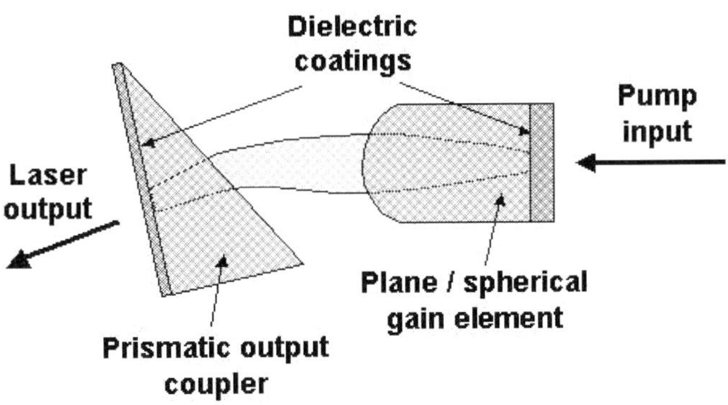

Figure 11. *Two element cavity for high repetition rate operation of Cr^{4+}:YAG. The curved polish on the endface of the gain medium provides a stable cavity suitable for obtaining KLM (Tomaru 2001).*

its simplicity and practicability.

Whilst Cr^{4+}:YAG systems may not be designed specifically for low threshold operation, they m suggest a route for the development of diode-pumped ultra-high repetition rate femtosecond sources. Cr^{4+}:YAG is a particularly attractive gain medium in this regime because the output is well matched to the third telecommunications window around 1.5μm and the material can be pumped using high power diodes around 980nm more normally associated with Er-doped fibre amplifiers or Raman amplifiers (Sorokin *et al.* 2001).

To simplify cavity design still further and reduce the component count and parasitic loss within the cavity, it is also possible to remove the single remaining prism. The dispersion compensation can now be provided using one of the mirror-based schemes described

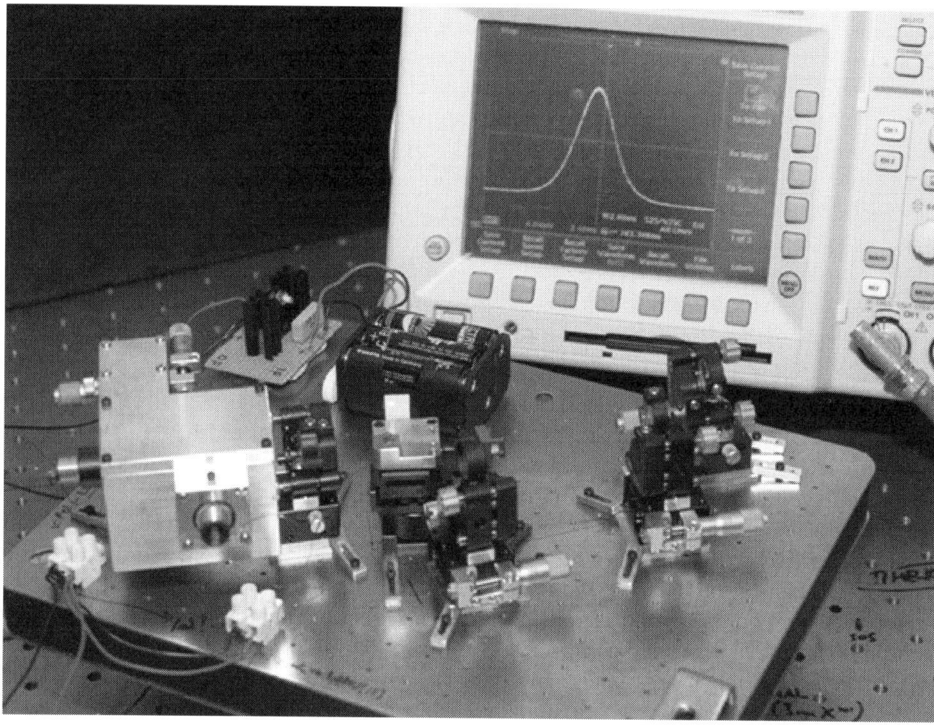

Figure 12. *Photograph of a compact, battery-powered, femtosecond Cr^{3+}:LiSAF laser cavity. Dispersion compensation is provided by GTI mirrors coated onto the cavity-fold optics and modelocking is initiated using a SESAM. This system has demonstrated electrical-optical efficiencies of 4% when generating fs pulses (Agate et al. 2002).*

above. In this case, however, care must be taken to ensure that the excess loss from the relatively complicated mirror coating should be minimised. By using a GTI-coated on the cavity fold mirrors, Agate *et al.* (2002a) have demonstrated a Cr^{3+}:LiSAF laser with a modelocking threshold < 100mW with modelocking provided by a SESAM. The threshold of this system is similar to that described by Hopkins *et al.* and thus it has been possible to use laser diodes powered only by AA electric batteries. A photograph of this laser system is shown in Figure 12 and it can be seen that there are no additional elements present in the cavity other than those required to form the basic laser resonator. This source is very compact, has good mechanical stability and electrical to femtosecond-pulse efficiencies of 4% have been demonstrated. This represents a five order-of-magnitude improvement in power conversion efficiencies since the first femtosecond lasers were demonstrated in organic dye-based lasers of the 1980's. The modular pumping arrangement described by Agate *et al.* also allows higher peak powers to be obtained in a very similar laser cavity. The utility of such sources has been demonstrated by the construction of a simple blue femtosecond source (Agate *et al.* 2002b) with electrical-to-blue-pulse efficiencies of 1%. It is believed that such sources may have significant applications in the photo-biology industry and may also have potential to provide direct optical clock distribution to silicon integrated circuits.

3.2 High repetition rate operation of SESAM-based lasers

Whilst this chapter is mainly concerned with the development of femtosecond-lasers based on crystal gain materials, it is worth taking a brief diversion to consider the world-leading work being conducted by Krainer *et al.* at ETH on the development of extremely high repetition rate picosecond lasers that were modelocked using SESAM mirrors (Krainer *et al.* 1999a, 1999b, 2000, 2002). The performance of these lasers has shown that it is possible for crystal laser media to meet, and in some cases exceed, some of the performance characteristics of semiconductor lasers. Indeed pulse repetition rates as high as 157GHz with pulse durations of 2.7ps have been observed from the monolithic laser structure illustrated in Figure 13. The studies carried out in the course of this work have direct implications in the use of these SESAM structures to produce femtosecond pulses at high repetition rates and a brief summary of the work is presented below.

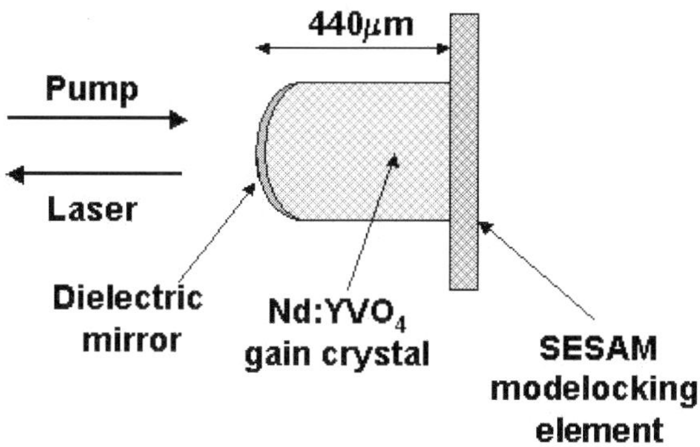

Figure 13. *Design for a 157GHz, monolithic, laser capable of generating 2.7ps pulses (Krainer et al. 2002)*

.

Krainer *et al.* (1999a) and Honninger *et al.* (1999) have shown that when the SESAM is being used as a fast saturable absorber, the following inequality concerning the laser output power, P_{out}, must be satisfied to obtain stable mode locked operation and to avoid Q-switching instabilities that can damage the modelocking element:

$$P_{\text{out}} > \sqrt{F_{L,\text{sat}}.F_{A,\text{sat}}.\Delta R.A.f.T}, \tag{4}$$

where $F_{A,\text{sat}}$ is the fluence on the saturable absorber, ΔR is the absorber modulation depth, A is the mode area, f is the repetition rate and T is the fractional output coupling.

$F_{L,\text{sat}}$ is the laser saturation fluence obtained from the familiar formula relating the frequency, ν and emission cross-section, σ:

$$F_{L,\text{sat}} = \frac{h\nu}{2\sigma}. \tag{5}$$

Thus to obtain a laser that will modelock successfully at high repetition rates the following properties have been shown to be desirable (Kemp 2000):

- Laser medium with large cross-section.
- Small mode area
- SESAM with low saturation fluence
- Modulation depth no larger than required to start and maintain modelocking.
- Low loss cavity including the output coupling.
- High brightness pump for good efficiency in small laser mode.
- Good quality crystal with short absorption depth to maintain efficiency in high repetition rate operation.

Whilst these conditions do provide some constraint in laser design, they also provide a route to search for suitable gain materials. Placing realistic values for the SESAM parameters (see eg. Krainer *et al.* (2000), one can readily check available gain materials. From such an analysis Nd:YVO_4 stands out and has been shown to be successful at the highest possible repetition rates by Keller *et al.* (1996) but the bandwidth of this crystal is not suitable for femtosecond operation. Studying likely broad-bandwidth gain materials for femtosecond-operation shows that both Cr^{4+}:YAG and Ti:Sapphire are promising candidates for high repetition rate operation. It should also be noted that when operating in a soliton modelocked regime, Honninger *et al.* (1999) have shown that the constraints for Q-switch modelocking are somewhat relaxed over the fast saturable absorber case thus offering extra flexibility for femtosecond-sources based on this technology.

4 The future

In this chapter we have concentrated on the development of sources of high-quality femtosecond pulses using vibronic crystal media to provide the broad bandwidth gain necessary to support the pulses. It has shown that while it is possible to build miniaturised sources of high-repetition rate pulses with relatively low threshold powers, some questions remain to be answered for such sources.

To be practical, we require femtosecond pulses to be available ‘on-demand’ and for the output from a laser to be encoded with data. This would allow the user to perform optical time division multiplexing (OTDM) on the laser output. Combining this technology with the potential for wavelength division multiplexing (WDM) available from the inherent bandwidth of the femtosecond-pulse could provide a single source capable of encoding vast amounts of data. The problem remains that the femtosecond-sources discussed above are based on passive modelocking technologies. This means that the output of the laser is normally a regular train of pulses with the repetition rate of the pulse train being determined by the laser cavity design. Whilst this may offer the possibility of very high-quality all-optical clock recovery, the applicability to OTDM is limited. Thus, to create pulses on demand, we need to think of novel methods for the generation of pulses. It may be possible to include an electrical modulator in a structure similar to a SESAM/SBR. Using such a structure, we may be able to provide a coded electronic drive to the modelocking element that would in turn provide a coded output from the laser. In order to achieve this, it is also important that we fully understand and characterise the mechanisms by which femtosecond-pulses build up in a passively mode locked laser system.

The pulse repetition rate for these laser systems is governed by the cavity length, which is ultimately controlled by the length of the gain medium. A 5mm Ti:sapphire crystal with a doping level of 0.25wt% ($\alpha_{abs} \approx 4cm^{-1}$) has a pump absorption of 87.5% and a crystal length limited repetition rate of 30GHz. Halving the crystal length would give a crystal length repetition rate of 60GHz, but would reduce the pump absorption to only 63% thereby placing stringent requirements on the available pump power. It is also difficult to envisage fully monolithic cavities which satisfy the cavity design requirements for passive modelocking but the work conducted by Anderson *et al.* (1997) and Hickey *et al.* (1996) on the fabrication of crystal waveguide lasers in broadband vibronic gain media may provide a way forward.

Another possibility for increasing the pulse repetition rate may be to consider harmonic modelocking of such lasers. In this case a small external Fabry-Perot cavity is placed after the laser output coupler and the output at a multiple of the laser cavity length is fed back into the femtosecond laser. This allows multiple pulses to circulate within the cavity and hence increases the repetition rate in the integer ratio of the cavity length divided by the secondary cavity length. This approach has been demonstrated by Wessel *et al.* (2000) using integrated $LiNbO_3$ and has shown that pulse repetition rate multiplication factors of $4\times$ can be achieved using this method for picosecond pulses created by active modelocking. A further option may be to use the output from a picosecond-diode laser to drive a femtosecond-source into mode locked operation. This intriguing hybrid arrangement could allow 3R regeneration of the picosecond output into a femtosecond-pulse stream giving the allowing the full electronic control of the semiconductor source to be transferred to the femtosecond laser.

As mentioned in the introduction, the ultimate goal must be the development of a femtosecond-laser based on semiconductor technologies. In the chapter by Penty (this volume), the development of short pulse sources in semiconductor materials is more fully discussed but it is worth highlighting some of the issues here from the point of view of the existing vibronic systems. Two chief challenges are present for the initial demonstration of <500fs pulses; firstly generating a broad bandwidth spectrum and secondly maintaining the spectrum during lasing to allow short pulse operation. It is possible to obtain the broad gain bandwidth (>100nm) using both asymmetric quantum wells and ensemble of quantum dot materials with a distribution of dot sizes. The problem comes that as the devices lase, gain narrowing takes place limiting the available bandwidth and serious issues must be overcome before these laser systems can support the coherent gain bandwidth required. Much research is currently being carried out into these processes with the aim of achieving a gain material with the correct balance of homogeneous and inhomogeneous broadening. The next step will be in the laser cavity design. Recent developments in external cavity VCSEL lasers have pointed to a geometry where the gain medium and saturable absorber can be combined with a more traditional based laser cavity to produce short pulse output (Garnache *et al.* 2002). The final step will be full integration of the laser components perhaps using photonic bandgap technologies for mirror fabrication and dispersion compensation if required, to produce electrically pumped femtosecond-lasers of the same size and simplicity as today's semiconductor picosecond lasers.

5 Conclusion

In this chapter we have considered the requirements for femtosecond pulse generation from lasers that incorporate vibronic crystals. It has been shown that femtosecond pulses can be produced using relatively simple passive modelocking effects. Careful cavity design has proved that such laser systems can offer compact and low threshold laser systems with the potential for battery powered operation where required. High-repetition rates are also possible although much work needs to be conducted before repetition rates comparable with existing diode lasers are possible with short femtosecond-pulses.

A future vision has also been presented where the ultimate goal is the development of fully integrated semiconductor sources of femtosecond pulses. Such lasers offer the prospect of greatly enhanced data carrying capacities in data communications and thus represent exciting and key components in the design and realisation of future femtosecond networks.

References

Agate B, Stormont B, Kemp A J, Brown C T A, Keller U and Sibbett W, 2002a, "Simplified cavity designs for efficient and compact femtosecond Cr : LiSAF lasers." *Optics Communications* **205** 207.

Agate B, Kemp A J, Brown C T A and Sibbett W, 2002b, "Efficient, high repetition-rate femtosecond blue source using a compact Cr : LiSAF laser." *Optics Express* **10** 824.

Alfrey A J, 1989, "Modeling of Longitudinally Pumped Cw Ti - Sapphire Laser- Oscillators." *IEEE Journal of Quantum Electronics* **25** 760.

Anderson A A, Eason R W, Hickey L M B, Jelinek M, Grivas C, Gill D S and Vainos N A, 1997, "Ti:sapphire planar waveguide laser grown by pulsed laser deposition." *Optics Letters* **22** 1556.

Burns D, Critten M P and Sibbett W, 1996, "Low-threshold diode-pumped femtosecond Cr^{3+}:$LiSrAlF_6$ laser." *Optics Letters* **21** 477.

Cautaerts V, Richardson D J, Paschotta R and Hanna D C, 1997, "Stretched pulse Yb^{3+}:silica fiber laser." *Optics Letters* **22** 316.

DeMaria A J, Stester D A and Heynau H, 1966, "Self Mode-Locking of Lasers with Saturable Absorbers." *Applied Physics Letters* **8** 174.

Finch A, Chen G, Sleat W and Sibbett W, 1988, "Pulse Asymmetry in the Colliding-Pulse Mode-Locked Dye-Laser." *Journal of Modern Optics* **35** 345.

Fork R L, Martinez O E and Gordon J P, 1984, "Negative Dispersion Using Pairs of Prisms." *Optics Letters* **9** 150.

Garnache A, Hoogland S, Tropper A C, Sagnes I, Saint-Girons G and Roberts J S, 2002, "Sub-500-fs soliton-like pulse in a passively mode-locked broadband surface-emitting laser with 100 mW average power." *Applied Physics Letters* **80** 3892.

Hickey L M B, Martins E, Roman J E, Brocklesby W S and Wilkinson J S, 1996, "Fluorescence of Ti^{3+} ions thermally diffused into sapphire." *Optics Letters* **21** 597.

Honninger C, Paschotta R, Morier-Genoud F, Moser M and Keller U, 1999, "Q-switching stability limits of continuous-wave passive mode locking." *Journal of the Optical Society of America B-Optical Physics* **16** 46.

Hopkins J M, Valentine G J, Sibbett W and Keller U, 1999, "CTuA4 Highly Compact Femtosecond Cr^{3+}:LiSAF Laser with a Simplified Cavity Geometry". *Conference on Lasers and Electro-Optics, Baltimore, Optical Society of America.*

Jung I D, Kartner F X, Matuschek N, Sutter D H, MorierGenoud F, Zhang G, Keller U, Scheuer V, Tilsch M and Tschudi T, 1997, "Self-starting 6.5-fs pulses from a Ti:Sapphire laser." *Optics Letters* **22** 1009.

Keller U, 1994, "Ultrafast All-Solid-State Laser Technology." *Applied Physics B-Lasers and Optics* **58** 347.

Keller U, Weingarten K J, Kartner F X, Kopf D, Braun B, Jung I D, Fluck R, Honninger C, Matuschek N and der Au J A, 1996, "Semiconductor saturable absorber mirrors (SESAM's) for femtosecond to nanosecond pulse generation in solid-state lasers." *IEEE Journal of Selected Topics in Quantum Electronics* **2** 435.

Kemp A J, 2000, Personal Communication.

Khrushchev I Y, White I H and Penty R V, 1998, "High-quality laser diode pulse compression in dispersion- imbalanced loop mirror." *Electronics Letters* **34** 1009.

Kopf D, Weingarten K J, Zhang G, Moser M, Emanuel M A, Beach R J, Skidmore J A and Keller U, 1997, "High-average-power diode-pumped femtosecond Cr:LiSAF lasers." *Applied Physics B-Lasers and Optics* **65** 235.

Kopf D, Zhang G, Fluck R, Moser M and Keller U, 1996, "All-in-one dispersion-compensating saturable absorber mirror for compact femtosecond laser sources." *Optics Letters* **21** 486.

Krainer L, Paschotta R, der Au J A, Honninger C, Keller U, Moser M, Kopf D and Weingarten K J, 1999a, "Passively mode-locked Nd : YVO4 laser with up to 13 GHz repetition rate." *Applied Physics B-Lasers and Optics* **69** 245.

Krainer L, Paschotta R, Spuhler G, Moser M and Keller U, 1999b, "29GHz modelocked miniature Nd : YVO_4 laser." *Electronics Letters* **35** 1160.

Krainer L, Paschotta R, Moser M and Keller U, 2000, "Passively mode-locked picosecond lasers with up to 59 GHz repetition rate." *Applied Physics Letters* **77** 2104.

Krainer L, Paschotta R, Lecomte S, Moser M, Weingarten K J and Keller U, 2002 "Compact Nd : YVO_4 lasers with pulse repetition rates up to 160 GHz." *IEEE Journal of Quantum Electronics* **38** 1331.

Lagatsky A A, Brown C T A and Sibbett W, 2002, CWA54 "Self-starting Passively Mode-locked Femtosecond Cr:YAG Laser Diode Pumped by a Yb-Fiber"". Conference on Lasers and Electro-Optics, Long Beach, Optical Society of America.

Lefort L, Albert A, Couderc V and Barthelemy A, 2002, "Highly stable 68fs pulse generation from a stretched-pulse Yb^{3+}-doped fiber laser with frequency shifted feedback." *IEEE Photonics Technology Letters* **14** 1674.

Mocker H W and Collins R J, 1965, "Mode Competition and Self-Locking Effects in a Q-Switched Ruby Laser." *Applied Physics Letters* **7** 270.

Paschotta R and Keller U, 2001, "Passive mode locking with slow saturable absorbers" *Applied Physics B-Lasers and Optics* **73** 653.

Ramaswamypaye M and Fujimoto J G, 1994, "Compact Dispersion-Compensating Geometry for Kerr-Lens Mode-Locked Femtosecond Lasers." *Optics Letters* **19** 1756.

Shank C V and Ippen E P, 1974, *Applied Physics Letters* **24** 373.

Smith P W, 1970, "Mode-locking of Lasers." *Proceedings of the Institute of Electrical and Electronic Engineers* **58** 1342.

Sorokin E, Sorokina I T and Wintner E, 2001, "Diode-pumped ultra-short-pulse solid-state lasers." *Applied Physics B-Lasers and Optics* **72** 3.

Spence D E, Kean P N and Sibbett W, 1991, "60-Fsec Pulse Generation from a Self-Mode-Locked Ti-Sapphire Laser" *Optics Letters* **16** 42.

Sutter D H, Steinmeyer G, Gallmann L, Matuschek N, Morier-Genoud F, Keller U, Scheuer V, Angelow G and Tschudi T, 1999, "Semiconductor saturable-absorber mirror-assisted Kerr-lens mode-locked Ti : sapphire laser producing pulses in the two- cycle regime" *Optics Letters* **24** 631.

Szipocs R and KohaziKis A, 1997, "Theory and design of chirped dielectric laser mirrors. *Applied Physics B-Lasers and Optics* **65** 115.

Szipocs R, Kohazi-Kis A, Lako S, Apai P, Kovacs A P, DeBell G, Mott L, Louderback A W, Tikhonravov A V and Trubetskov M K, (2000), "Negative dispersion mirrors for dispersion control in femtosecond lasers: chirped dielectric mirrors and multi-cavity Gires-Tournois interferometers." *Applied Physics B-Lasers and Optics* **70** S51.

Tamura K, Doerr C R, Nelson L E, Haus H A and Ippen E P, 1994a, "Technique for Obtaining High-Energy Ultrashort Pulses from an Additive-Pulse Mode-Locked Erbium-Doped Fiber Ring Laser." *Optics Letters* **19** 46.

Tamura K, Nelson L E, Haus H A and Ippen E P, 1994b, "Soliton Versus Nonsoliton Operation of Fiber Ring Lasers." *Applied Physics Letters* **64** 149.

Tomaru T, 2001, "Two-element-cavity femtosecond Cr^{4+}: YAG laser operating at a 2.6-GHz repetition rate." *Optics Letters* **26** 1439.

Tomaru T and Petek H, 2000, "Femtosecond Cr^{4+}: YAG laser with an L-fold cavity operating at a 1.2-GHz repetition rate." *Optics Letters* **25** 584.

Tsuchiya M, Igarashi K, Yatsu R, Taira K, Koay K Y and Kishi M, 2001, "Sub-100 fs SDPF optical soliton compressor for diode laser pulses." *Optical and Quantum Electronics* **33** 751.

Tsuda S, Knox W H, Cundiff S T, Jan W Y and Cunningham J E, 1996, "Mode-Locking Ultrafast Solid-State Lasers with Saturable Bragg Reflectors." *IEEE Journal of Selected Topics in Quantum Electronics* **2** 454.

Valdmanis J A and Fork R L, 1986, "Design Considerations for a Femtosecond Pulse Laser Balancing Self Phase Modulation, Group Velocity Dispersion, Saturable Absorbtion and Saturable Gain." *IEEE Journal of Quantum Electronics* **12** 112.

Valentine G J, Critten M P, Kennedy G T, J-M., H., Loza-Alvarez P and Sibbett W, 1997, "CMI1 Novel Resonator Designs for Low-Threshold Self Modelocking". Conference on Lasers and Electro-Optics, Baltimore, Optical Society of America.

Valentine G J and Sibbett W, 1999, "Compact, low-threshold femtosecond lasers." *Recent Research Developments in Quantum Electronics* **1** 43.

Walmsley I W L, and Dorrer C, 2001, "The Role of Dispersion in Ultrafast Optics." *Review of Scientific Instruments* **72** 1.

Wessel R, Ricken R, Rochhausen K, Suche H and Sohler W, 2000, "Supermode stabilized coupled-cavity 5-and 10-GHz mode-locked Ti : Er : $LiNbO_3$ waveguide lasers." *IEEE Journal of Quantum Electronics* **36** 394.

Short pulse generation using semiconductor lasers

Richard Penty

Cambridge University, England

1 Introduction

Semiconductor lasers are key components for optical communications systems as they are able to work at extremely high data rates (10Gb/s and above), operate over a wide range of wavelengths (defined by the material systems) and can be designed to be extremely wavelength stable with low chirp under modulation. Compared to solid state, gas and dye lasers, they have the advantage of being extremely small, typically a few hundred microns in length, have high efficiency and low threshold, are driven directly by an electrical current and can be extremely cheap, the manufacturing cost being of the order of a dollar.

In typical communications systems, diode lasers are either used as a CW light source, which is followed by a modulator, or are directly driven by a data modulated current which acts to imprint the data modulation on the laser output light. The predominant modulation format is non return to zero (NRZ). Recently however there has been interest in more advanced modulation formats, such as multilevel amplitude shift keying (ASK) or quadrature amplitude modulation (QAM), in order to reduce the system fibre dispersion and improve the spectral efficiency. There has also been much interest in using pulsed return to zero (RZ) modulation formats. These can be used to provide (i) improved performance in transmission, for example, via solitons, (ii) to increase the channel data rate via optical time division multiplexed (OTDM) bit interleaving and (iii) to allow bit level processing of an optical signal using all-optical switches.

The purpose of this chapter is to describe how semiconductor lasers can be used to generate high quality pulse trains at high frequency. It will begin with a brief description of semiconductor lasers and will then describe the three main techniques for generating pulse trains from laser diodes, namely gain switching, Q-switching and mode-locking. It will then describe techniques to improve the quality of the pulse train via pulse compression.

For reasons of space, it is only possible to provide a brief overview of semiconductor lasers but for a more detailed description the reader is directed to the excellent books of Thompson (1980), Agrawal and Dutta (1993) and Petermann (1988) and to that of Vasil'ev (1995) for a first-rate review of short pulse generation using diode lasers.

2 Semiconductor laser diodes

Semiconductor lasers are all variations of basic p-n junctions constructed from direct band-gap material systems such as AlGaAs or InGaAsP. Applying a forward bias to the junction allows charge carriers (electrons and holes) to be transported to a narrow depletion region (in the case of the p-n junction) or to the confinement region of a p-i-n diode. Here the carriers can recombine, releasing a photon with an energy equivalent to the band-gap energy of the laser material. Sufficient current results in population inversion and hence gain.

The gain is not sufficient to cause lasing action and optical feedback is also required. The simplest way of providing this is to form a cavity by cleaving the laser along the appropriate crystal planes such that it has two facets which act as partially reflecting mirrors. This type of laser is known as a Fabry-Perot (or FP) laser. Alternatively weak Bragg gratings can be incorporated into the laser structure to provide distributed feedback (hence this type of device is known as the DFB laser) at a given wavelength. DFB lasers have the advantage that they emit at only one frequency and are therefore ideal devices for optical communications, where a narrow linewidth results in low fibre dispersion.

2.1 Material considerations

As mentioned above, the output photon energy is approximately defined by the band-gap energy of the semiconductor material making up the laser diode. Typically diode lasers are constructed from III–V semiconductor materials such as GaAs and InP. It is, however, extremely desirable to be able to tune the laser wavelength to, for example, 1300nm, the zero dispersion wavelength for conventional single mode fibre, or to 1550nm, the wavelength of minimum fibre loss. This is achieved by "mixing" two materials with similar lattice constants, such as GaAs, which has a band-gap energy of 1.43eV, and AlAs, with 2.16eV. Al and Ga have lattice constants which differ by less than 0.1% and so an Al atom can substitute for a Ga atom in GaAs without any significant strain to form the compound semiconductor $Al_xGa_{1-x}As$. The Al fraction of this ternary semiconductor can be used to define the band-gap energy of the material to be anywhere between 1.43 and 2.16eV, equating to a band-gap wavelength of 870 to 575nm respectively.

The other important compound material system for semiconductor lasers is the quaternary system which consists of InP and the GaInAs ternary compound which is lattice matched to this. If these are mixed together, they form InGaAsP which is an extremely technologically important system as it covers both the 1300 and 1550nm wavelength regimes. The wavelength dependence of the two material systems is shown in Figure 1.

2.2 Laser structures

In order to obtain efficient laser operation, it is necessary to confine both the carriers and the photons in the same region of the laser. Carriers may be effectively confined in the plane perpendicular to the junction by the use of hetero-junctions. Here an active region is sandwiched between two cladding layers which have a higher band-gap. When current is injected into the laser, the potential barriers at the boundaries between the layers confine the charge carriers to the active layer, allowing gain to be achieved at much lower current densities than possible with homo-junction lasers.

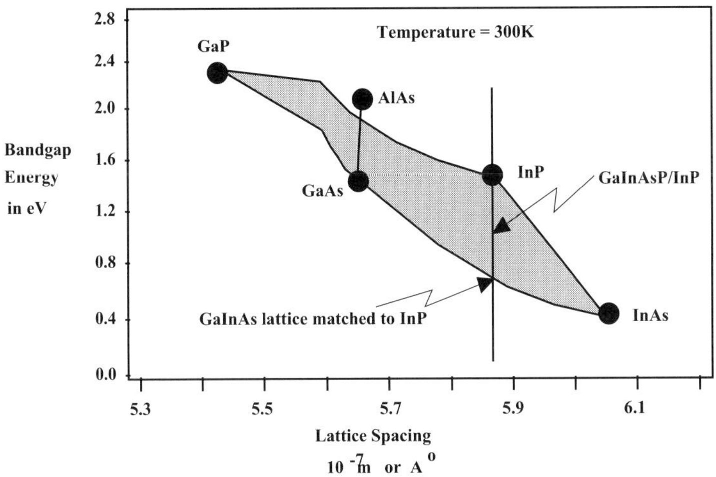

Figure 1. *Bandgap energy for the AlGaAs and InGaAsP material systems*

Lateral confinement of the injected carriers is also very important. The simplest way of achieving this is to open up a narrow stripe in the insulator on the top of the laser (see Figure 2, top). Whilst the carriers are not confined, the active region is only of the order of a micron below the top surface of the laser and hence the charge carriers can only diffuse by a few microns. A more sophisticated approach is one in which lateral confinement is achieved via the use of a buried heterostructure (Figure 2, bottom). Whilst this results in high efficiency, such devices require regrowth and thus are more complicated and expensive than conventional diode lasers.

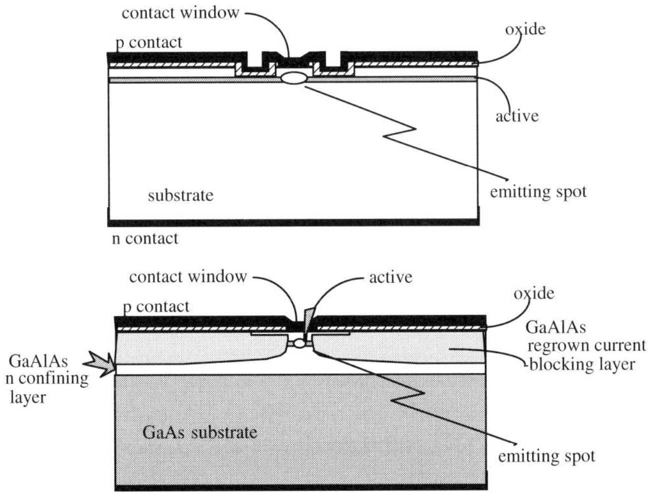

Figure 2. *Two types of diode laser with strong transverse carrier confinement: (top) ridge laser and (bottom) buried hetero-structure laser which has the further advantage of lateral current confinement.*

Photons can be confined in the same region as the electrons via the use of optical waveguides. One advantage of the heterojunction for lateral current confinement, is that it also acts as lateral slab waveguide as the refractive index of the active region is higher than the cladding regions. Transverse optical confinement can be achieved either via the use of a ridge waveguide (Figure 2a), which provides an effective index step, or via the real index step provided by the buried heterostructure (Figure 2b).

2.3 Rate equations

In order to understand short pulse operation of laser diodes, it is important to understand the dynamics of these devices. A simple analysis of the basic dynamic phenomena that describe the generation of short pulses can be carried out by the nonlinear rate equations which connect the photon density, P, in the cavity with the carrier density in the laser active region n (Adams 1973). They can be written in several forms but one of the most common is

$$\frac{dn}{dt} = \frac{j(t)}{ed} - g_0 (n - n_t) P - \frac{n}{\tau_s}, \quad (1)$$

$$\frac{dP}{dt} = \Gamma g_0 (n - n_t) P - \frac{P}{\tau_{ph}} + \frac{\beta \Gamma n}{\tau_s}, \quad (2)$$

where $j(t)$ is the current density waveform, e is the electronic charge, d is the active layer thickness, n_t is the transparency carrier density, g_0 is the differential gain coefficient, Γ is the optical confinement factor and τ_s and τ_{ph} are the carrier and photon lifetimes respectively. The carrier lifetime is a measure of how quickly the electrons in the active region recombine in such a manner as not to emit a stimulated photon. This recombination can occur in several different ways, namely by non-radiative recombination, spontaneous emission and Auger recombination. The photon lifetime is a measure of the length of time a photon remains within the cavity and is a function of the cavity loss its mirror reflectivities. Whilst it is possible to include sophisticated physical effects, such as gain saturation, in these rate equations, these are sufficient to gain physical understanding of the short pulse generation techniques described in this chapter. For further details, the reader is directed to Adams (1973) and Agrawal and Dutt (1993).

2.4 Light-current characteristic and relaxation oscillations

The laser rate equations, described above, may be solved to describe a laser's operation. It is relatively simple to solve the equations for CW operation where the rate of change of carrier density and photon density are zero and j takes a constant value. The solution of the equation is shown (Figure 3) in the light–current characteristic which describes how the light output varies as a function of the applied current. As can be seen, there is very little light generated until the laser reaches a threshold current, I_{th}, when the light varies linearly with the current above I_{th}.

When the current is varied, it is extremely difficult to solve the rate equations analytically, since they are coupled nonlinear differential equations, and hence numerical solutions are necessary. This is not a difficult procedure though a reasonable level of computing power and memory are necessary to solve them accurately in a timely fashion.

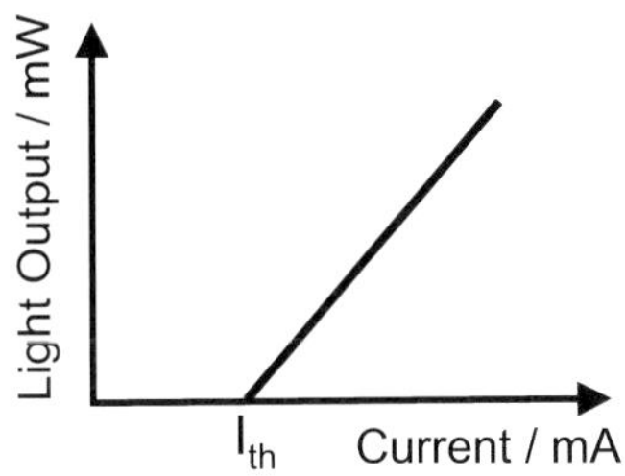

Figure 3. *Light Current Characteristic of Laser Diode*

Even a simple situation, such as the response of the laser to a current step function, results in complicated behaviour that requires numerical solution. Figure 4 shows this response schematically. As soon as the current turns on, the carrier density begins to grow. Once it has reached the transparency density, it is possible for stimulated photons to be generated (see the photon rate equation) but the laser does not begin to lase until

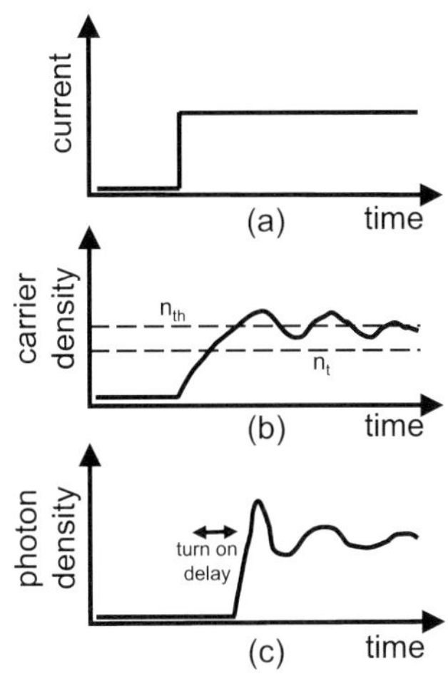

Figure 4. *Schematic of the response of a diode laser to a current step function (a) current waveform and evolution of (b) carrier and (c) photon density waveforms*

the carrier density reaches a higher value, the threshold carrier density, n_{th}. The time between the current pulse being applied and the laser turning on is known as the turn on delay and typically is of the order of a few nanoseconds if the laser is turned on from zero current. Once the laser does begin to lase, the coupling between the rate equations results in a damped, coupled oscillation in both the carrier and photon densities. The oscillation in the photon density is known as the relaxation oscillation and it possesses a characteristic frequency, known as the relaxation oscillation frequency, that is a function of the laser design and the value of the current above threshold (Agrawal and Dutta 1993). It is possible to design diode lasers with very high relaxations oscillation frequencies of up to 40GHz but it is more usual for values in the range 5–20GHz to be obtained.

3 Gain switched semiconductor lasers

Gain switching is probably the simplest technique for generating short pulses from laser diodes. As the name implies, it relies on the fast switching of optical gain of the device through the modulation of the driving current. Unlike Q-switching or mode-locking, gain switching has the advantage that no external cavity or sophisticated drive electronics are required. The principle behind gain switching is to excite the first relaxation oscillation, as described in Section 2.4, as the laser is turned on but to rapidly terminate the electrical drive before the second oscillation occurs. This requires that the electrical pulse source should generate short pulses, typically of the order of a nanosecond or less. Popular pulse sources are comb generators and avalanche transistor generators. Alternatively gigahertz frequency sinusoidal current sources driving a sub threshold laser also generate gain switched pulse trains.

Figure 5 shows the dynamics of gain switched operation of a laser diode. A below threshold d.c. bias is applied to the laser, resulting in the steady state carrier density being below the threshold carrier density. As the electrical pulse is applied to the laser, the carrier density rises above its threshold value. However, since it takes a finite time for the photon density to build up in the cavity, the laser does not begin to lase immediately and hence the carrier density builds up so that the gain is significantly in excess of the threshold gain. Once the laser begins to lase, the high gain means that the photon density rises very quickly. This rapidly depletes the carrier density, and hence the gain, and the laser switches off rapidly.

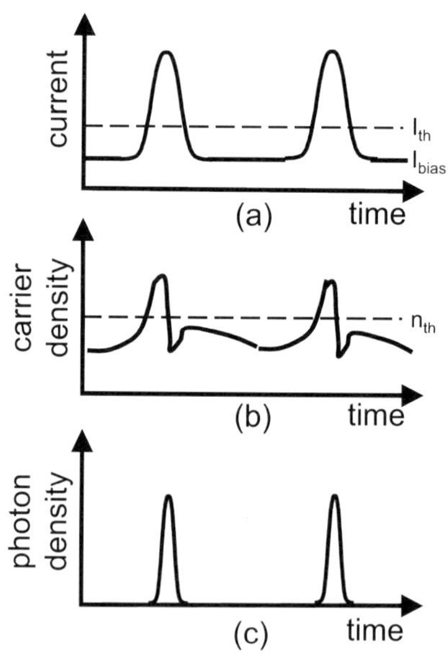

Figure 5. *Gain switched laser dynamics (a) applied current, (b) carrier density and (c) optical power.*

3.1 Gain switched laser performance

Gain switched laser pulse widths are typically of the order of several multiples of the cavity round trip time. Aspin *et al.* (1981) showed that there was a linear relationship between pulse width and laser length in AlGaAs lasers where a 15ps pulse was generated by a 60μm long device whilst a 500μm device resulted in 60ps pulses. It has also been established that the pulse width decreases with increased RF modulation power until an unwanted secondary optical pulse appears (AuYeung 1981). By optimising the laser construction and drive conditions 10ps pulses have been generated at a wavelength of 850nm (Elliot *et al.* 1983) whilst an 11ps width has been obtained at 1550nm (White *et al.* 1985). However more representative pulse widths are 15–20ps for AlGaAs lasers and 25–30ps using InGaAsP devices.

Typical peak powers from gain-switched lasers lie in the range 0.1–1W for AlGaAs (AuYeung 1981) and 10–100mW for InGaAsP (White *et al.* 1985) devices respectively. The lower value at the longer wavelength is usually ascribed to increased nonlinear gain compression in this material system (Petermann 1988). The spectrum of gain switched pulses from diode lasers is usually far from transform limited. Even in the case of a DFB laser, which supports a single wavelength, the wavelength chirp is significant. As can be seen in Figure 5, the carrier concentration, and hence the gain, changes rapidly throughout the optical pulse. Any change in the imaginary part of the susceptibility is accompanied by a change in the real part as described by the Kramers-Kronig relations (Agrawal and Dutta 1993). Consequently as the gain changes, the refractive index, and hence the device wavelength, also changes throughout the pulse. DFB lasers typically exhibit time bandwidth products for gain switched pulses in the range 1–3. This frequency chirp may be used positively in pulse compression schemes as described in Section 6.

4 Q-switched semiconductor lasers

As the name implies, Q-switched pulses are generated in semiconductor lasers by the modulation of the cavity Q. This is usually achieved by changing the cavity loss. This can either be done actively where the loss is externally modulated or passively via the use of a saturable absorber. Usually Q-switched laser diodes are fabricated specially and consist of more than one section to provide the required gain and loss functions. Whilst this leads to a more complicated device which requires more sophisticated drive control than gain switched lasers, the main advantage of the Q-switching technique is that it allows the generation of pulse with peak powers at least an order of magnitude greater than those available from the other techniques described in this chapter.

4.1 Active Q-switching

An actively Q-switched laser diode consists of two, or possibly more, sections as shown in Figure 6. One section is driven with a CW forward current to provide optical gain. The other section is the loss section which is modulated electrically. Since the active region of the gain and loss sections are monolithically integrated, the two sections will contain the same active region material, unless techniques such as selective area growth or quantum

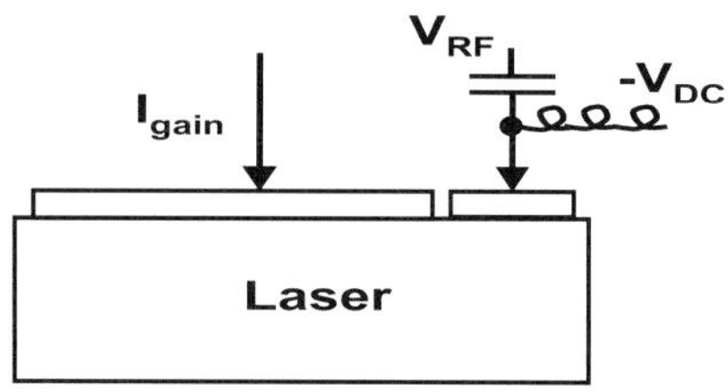

Figure 6. *Schematic of actively Q-switched laser diode showing electrical driving scheme.*

well intermixing are used. Consequently the band-gap in the absorber section must be shifted to lower energies to increase the modulation efficiency. This is often done using a DC bias via the Quantum Confined Stark Effect (QCSE).

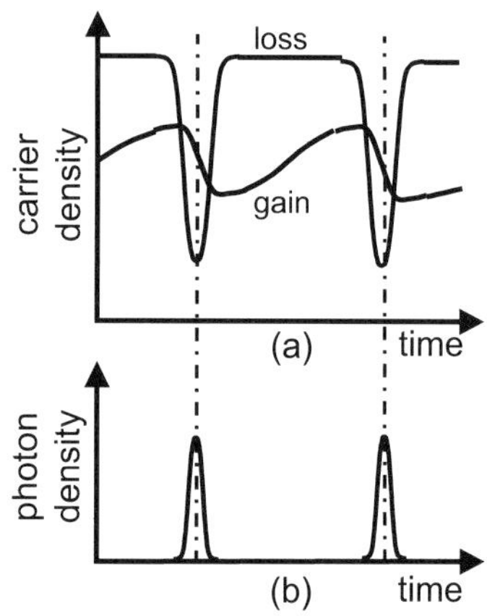

Figure 7. *Time evolution of (a) loss, gain and (b) optical power in a Q-switched laser*

Figure 7 shows schematically the operation of a Q-switched laser. The loss of the absorber section is initially set to be high via for example, QCSE. The initial low Q of the cavity allows the carrier density in the gain section to build up to a very high value and hence a very high gain. The voltage to the loss section is then reduced resulting in higher cavity Q and the laser is then allowed to lase. Since the gain has built up to a high value, the optical pulse increases in energy very rapidly as the stored electrical energy in the gain section is dumped into photons. As photons are generated, the gain section carrier density drops rapidly below threshold and the laser turns off rapidly.

It can be shown from the rate equations that the pulse width and peak power are strongly dependent on the inversion just before the pulse is generated and on the photon lifetime (Vasil'ev 1995). In order to obtain a short pulse, the inversion should be high and the photon lifetime low. Williams *et al.* (1993) demonstrated active Q-switching in a two section InGaAsP device to achieve 26ps long pulse with 80mW peak power at a 1.25GHz repetition rate. Q-switched pulses with up to 6W peak power and 20–30ps width were generated from a two section AlGaAs laser, though here the loss was modulated by illuminating a gap in the contacts with short pulses from an Nd:glass laser (Thedrez *et al.* 1993).

4.2 Passive Q-switching

In many ways passive Q-switching is similar to active Q-switching. The laser again usually consists of two sections, one of which provides gain and one of which provides loss. However, in this case, the loss section consists of a saturable absorber. Figure 8 shows the operating principle of passive Q-switching. Initially the loss is high and the gain is low. However as current is injected into the gain section, the gain builds up. This leads to an increase in the amount of spontaneous emission, some of which is coupled into the loss section. As these spontaneous photons are absorbed, the loss is reduced. Eventually the loss is reduced to such an extent that the Q of the cavity is high enough for lasing to occur. Consequently the high electrical energy is converted rapidly into photons, leading to a rapid reduction in optical gain and the laser turns off. If the absorber carrier lifetime is short enough, the carrier density in the absorber will reduce after the pulse has been generated and the loss will recover. The process then begins anew.

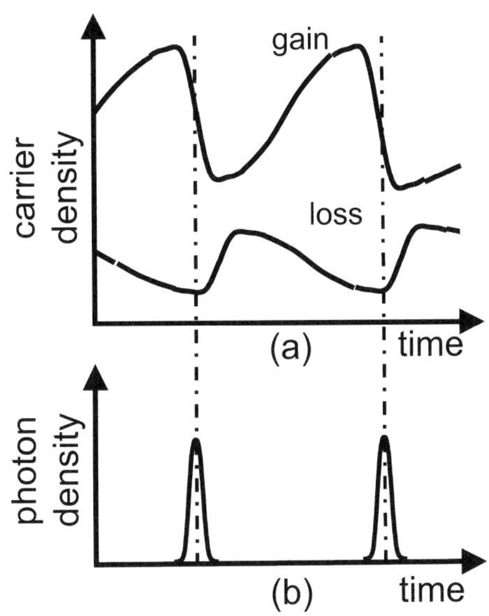

Figure 8. *Evolution of (a) loss, gain and (b) optical power in a passively Q-switched laser diode.*

It can be shown that in order to generate passively Q-switched pulses the ratio of the differential loss and differential gain should be greater than unity, $(da/dn)/(dg/dn) > 1$, and the loss carrier lifetime should be shorter than the gain carrier lifetime (i.e. $\tau_a < \tau_g$.) As can be seen from the schematic of the dependence of loss/gain on carrier density, shown in Figure 9, the values of dg/dn and da/dn can easily be controlled using different current inputs to the two sections. To a lesser extent, the carrier lifetimes can also be controlled. In particular τ_a may be made very short relative to τ_g by reverse biasing the loss section. In this way the pulse width, peak power and repetition frequency can be controlled by optimising the currents and voltages to the Q-switched laser contacts.

Using a proton bombarded absorber section, very short absorber carrier lifetimes

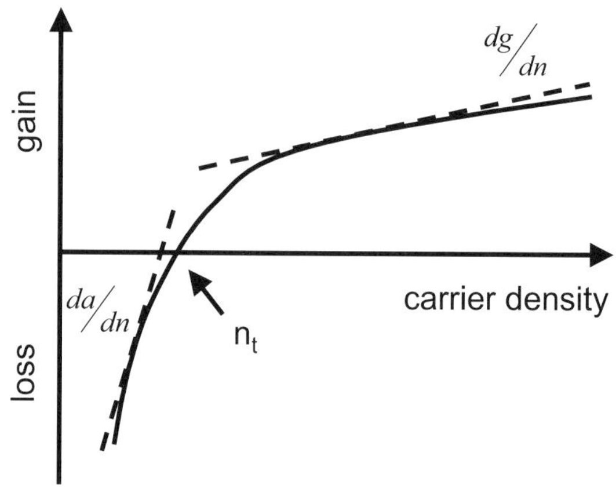

Figure 9. *Schematic of the dependence of loss and gain on carrier density for the laser active region.*

(<100ps) were generated which allowed 12ps pulses with repetition rates up to 5GHz to be generated (Alferov *et al.* 1986). Using a reverse biased saturable absorber on a three section device, Vasil'ev *et al.* (1993) generated pulses with peak powers in excess of 10W, with 1.6ps pulse duration and repetition rates up to 18.5GHz.

5 Mode-locked semiconductor lasers

A third technique to generate short pulses from semiconductor lasers is to lock the longitudinal modes of the laser together. In general the phase relationship between the cavity modes is a random one. If the electric field of the modes can be expressed as

$$E(t) = \sum A_m \exp i\left[(\omega_0 + m\delta\omega)t + \phi_m\right],$$

where A_m and ϕ_m describe the amplitude and phase of the m-th mode and $\delta\omega$ is the frequency difference between two adjacent modes. If the modes can be phase locked then the output of the laser will be a periodic function of time with a repetition rate of the round trip frequency and can be described as

$$E(t) = A_0 \frac{\sin\left[(k+1)\delta\omega t/2\right]}{\sin\left[\delta\omega t/2\right]} \exp(i\omega_0 t).$$

The main advantage of mode-locking over either gain-switching or Q-switching is that it can generate much shorter pulses at much higher repetition rates, often with low timing jitter. Mode-locked pulses can be generated using four main techniques, namely (i) passive mode-locking using a saturable absorber, (ii) active mode-locking where gain or loss modulation at the round trip frequency is used to lock the modes of a single contact laser, (iii) hybrid mode-locking where the gain/loss modulation is applied to a multiple section or external cavity laser and (iv) colliding pulse mode-locking where a centrally placed absorber is used to shorten two counter propagating pulses.

5.1 Passive mode-locking

A saturable absorber is a vital element in any passive mode-locked scheme. As the pulse arrives at the absorber, the front edge of the pulse experiences loss. During the pulse the loss saturates faster than the gain but towards the end of the pulse, the gain becomes saturated, causing the trailing edge to experience loss. If the recovery time of the absorber is faster than the gain, the loss will be greater than the gain everywhere but the peak of the pulse. Hence the broadening of the pulse through dispersion as it propagates through the laser is corrected via the operation of the saturable absorber. This principle is illustrated schematically in Figure 10.

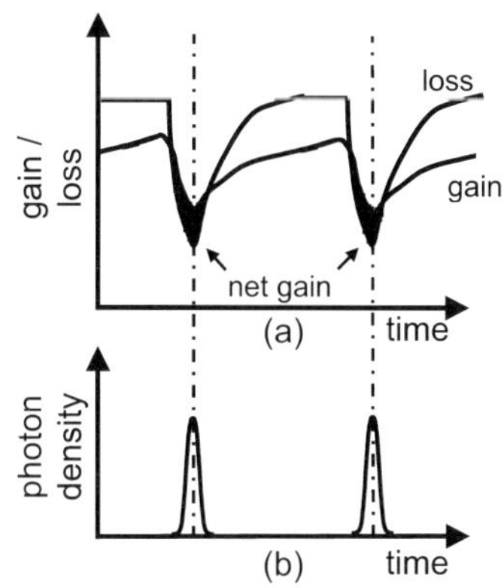

Figure 10. *Schematic of passive mode-locking operation using a saturable absorber. Top: gain and loss dynamics, bottom: optical output.*

In practice the saturable absorber in monolithic diode lasers is usually achieve via proton or ion bombardment of a short section of the device or by the use of QCSE via reverse biasing a section. Using the latter technique, Arahira *et al.* (1996) demonstrated 40GHz active mode-locking in an integrated device with 25ps pulse widths.

5.2 Active mode-locking

Active mode-locking is realised by modulating the gain of a diode laser at a frequency equal to the round trip frequency of the device. Since diode lasers are typically rather short, of the order of $\sim$300μm, their intrinsic round trip frequency is rather high, of the order of 100GHz. As these frequencies are too high to be of practical interest and are difficult to generate electrically, the length of the laser system is usually extended either by placing the laser in an external cavity or by monolithically integrating a passive extended cavity. In this way, total cavity lengths from 1mm to 1cm will give a range of frequency of operation from $\sim$40GHz to 1GHz respectively which is the typical range for many of the applications for mode-locked lasers around today. An advantage of active mode-locking over passive mode-locking is that the operating frequency is set by an electrical clock generator rather than being allowed to free run. This is particularly important in optical communications when the optical pulse rate must be very accurately controlled.

The principle of operation of current driven active mode-locking is shown schematically in Figure 11. A sinusoidal current modulation at the round trip frequency of the cavity is superimposed on a DC bias. The resultant gain modulation acts to lock the cavity modes together in phase and the result is a repetitive pulse train at the frequency of the electrical modulation.

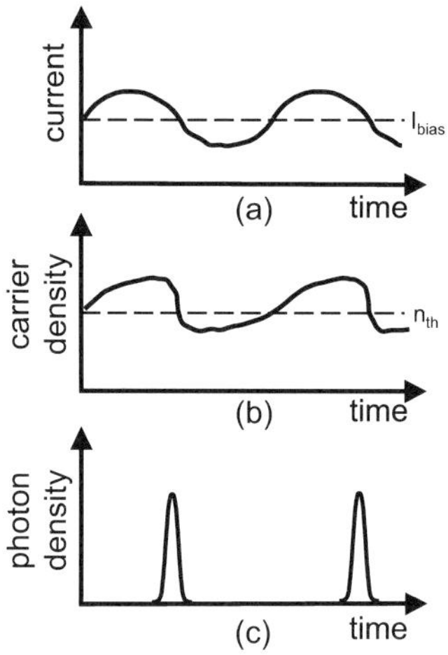

Figure 11. *Schematic of active mode-locking operation showing (a) drive current, (b) resultant carrier density and (c) output pulse train.*

An example of external cavity active mode-locking was demonstrated by Morton *et al.* (1992) who coupled the output of a semiconductor laser into an optical fibre which contained a fibre Bragg grating to define the optical cavity and the wavelength of operation. The laser operated at a frequency of 2.4GHz with 20ps transform limited pulses having a peak power of ~50mW in the fibre.

5.3 Hybrid mode-locking

Hybrid mode-locking is very similar to active mode-locking except that the laser is a multisection device which has separate gain and modulation sections, sometimes in an external cavity. Figure 12 shows a generic hybrid mode-locking scheme. Here a DC bias to one contact provides gain for the mode-locked laser whilst the DC bias to the other contact is below transparency to provide an adjustable saturable loss. The RF current is at the round trip frequency and is used to lock the cavity modes.

Weber *et al.* (1992) demonstrated hybrid mode-locking at a frequency of 1GHz in an external cavity using an InGaAsP semiconductor laser which generated high quality pulses as short as 730fs. Zielinksi *et al.* (1997) fabricated a monolithic laser which contained a gain section, a modulated absorber section and a passive section incorporating a DBR grating. The different material bandgaps of the active and passive sections were achieved by selective area growth. The 2mm long device operated at a frequency of 40GHz with 9ps pulses.

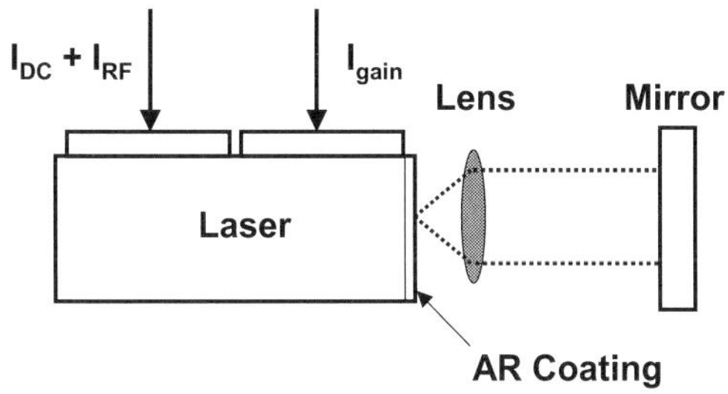

Figure 12. *Hybrid mode-locking scheme.*

5.4 Colliding pulse mode-locking

The colliding pulse mode-locking (CPM) technique is very effective in generating ultra-short pulses at very high repetition rates. The method is based on the interaction of two counter-propagating pulses inside the saturable absorber of a laser. The effectiveness of the pulse shortening mechanism of a saturable absorber, described in Section 5.1, is determined by the ratio of the differential loss of the absorber to the differential gain of the amplifying region of the laser; in other words, the easier the absorber is saturated relative to the gain, the better will be the mode-locking performance.

This optimisation is usually achieved by varying the material properties of the device, but the colliding pulse mode-locking technique achieves an improvement by design of the laser cavity. In the simplest configurations the saturable absorber is placed precisely at the centre of the cavity. If two counter-propagating pulses are excited in the cavity, they will overlap in the absorber but will pass through the gain regions separately, as shown in Figure 13. The absorber is thus subjected to twice as much energy as the gain regions as the pulses pass through it and thus will be more easily bleached than the case of a single pulse in the cavity. Naturally, as there are two pulses present in the cavity, the repetition rate of a CPM laser is twice that of a conventional mode-locked laser for a given cavity length.

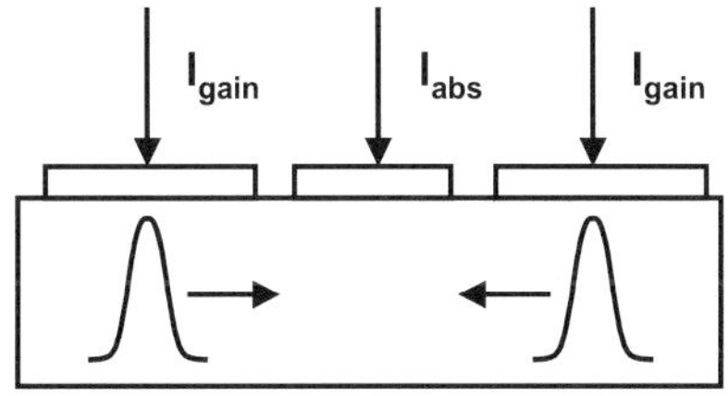

Figure 13. *Schematic of colliding pulse mode-locking operation.*

CPM was first demonstrated using a GaAs laser placed in an external ring cavity and pulses with widths of 0.6ps and peak power of 80mW were generated (Van der Ziel *et al.* 1981). Subsequently monolithic CPM lasers have been fabricated. Lee *et al.* (2002) demonstrated 36GHz CPM where low phase noise 2.5ps pulses, with a time-bandwidth product of 0.45, were generated.

6 Pulse compression

In general, the pulses generated by semiconductor lasers are not as short as those it is possible to obtain from solid state sources such as vibronic lasers. It is possible, however, to compress the diode laser pulses to reduce the pulse width. In general the pulse generation schemes described in Sections 3–5 result in carrier density variations in the semiconductor material as the pulse is produced. This, of course, means that the refractive index, and by extension the effective cavity length, varies during the pulse. Hence the output wavelength "chirps", typically to longer wavelengths, as the pulse is generated (see Figure 14), leading to time-bandwidth products in excess of the transform limit.

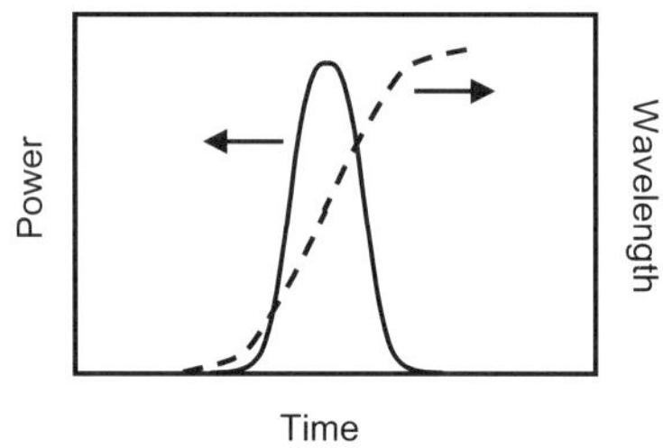

Figure 14. *Time evolution of wavelength chirp for short pulse generated by laser diode.*

It is possible to make use of this property to compress the pulse by chirp compensation techniques. These include the use of bulk grating compressors, negative dispersion fibre and chirped fibre gratings. Delfyett *et al.* (1992) showed that it was possible to generate 200fs pulses with a peak power of 165W using the bulk grating technique on pulses from an amplified mode-locked laser. In addition, Hou *et al.* (1990) showed that it was possible to use the dispersion characteristics of an optical fibre to compress chirped pulses. Here pulses from an actively mode-locked laser diode were compressed from 15.5 to 4.1ps using standard single mode fibre in the normal (positive) group velocity dispersion regime at an operating wavelength of 1286nm. New fibre types, such as dispersion compensating fibre (DCF) have now been developed to allow a similar technique to be applied to pulses within the more technologically important 1550nm wavelength range.

7 Conclusions

Ultrashort pulse generation from laser diodes is a fast growing field of modern optoelectronics. The success of this field has lead to many applications in the areas of optical communications, optical signal processing, optoelectronic measurement and sampling and optical sensing. The advantages of laser diodes, namely the wide range of operating wavelength, the widely tunable repetition frequency (from a few Megahertz to hundreds of Gigahertz), their small size and low power consumption, have meant that they are being widely adopted within the photonics industry.

Table 1 gives an overview of the performance of the main techniques for ultrashort pulse generation using laser diodes. Gain switching is the simplest technique, using only current modulation of a single contact laser, and can give relatively short 5ps pulses with

Technique	Pulse Width	Repetition Rate	Pulse Energy	Comments
Gain Switching	5ps	<10GHz	10pJ	Good control of repetition rate via electrical drive
Passive Q-Switching	1ps	<20GHz	100pJ	Limited control of repetition rate and poor jitter performance
Active Q-Switching	5ps	<10GHz	10pJ	Repetition rate set by electrical drive
Passive Mode-Locking	50fs	<2THz	1pJ	Repetition rate set by cavity - medium jitter performance
Active Mode-Locking	1ps	<50GHz	1pJ	Repetition rate set by cavity and good jitter performance

Table 1. *Performance summary of the main types of short pulse generation techniques for semiconductor lasers.*

repetition rates up to approximately 10GHz. Passive Q-switching requires a slightly more complicated two-electrode structure but only requires low frequency electrical drives. It can give pulses as short as 1ps with high pulse energy but suffers from limited control of the repetition rate and poor jitter performance. Active Q-switching requires a high frequency drive but this allows the repetition rate to be set though at the expense of achievable pulse energy. Mode-locking is the best method for achieving very short pulses at very high repetition rates but this rate is set by the cavity length, leading to inflexibility in performance. Nevertheless passive mode-locking can give pulses as short as 50fs with Terahertz repetition rates. Active mode-locking allows some limited control of the repetition rate, particularly important in communications where the system has to operate with a synchronous clock, and improved jitter performance when compared to passive mode-locking. Typical pulse widths are of the order of 1ps with repetition rates up to 50GHz achievable.

The continued research into ultrashort pulse generation using diode lasers will enable a new generation of photonic devices and systems to be developed. It is very possible that they will be the key to overcoming the inevitable limitations of electronic signal processing that will occur within the next ten to twenty years.

References

Adams M J, 1973, "Rate equations and transient phenomena in semiconductor lasers", *IEE Optoelectronics*, **5** 201.

Agrawal G P and Dutta N K, 1993, *Semiconductor lasers*, (Van Nostrand Reinhold).

Alferov Zh I, Zhuravlev A B, Portnoi E L and Stel'makh N M, 1987, "Generation of picosecond pulses in hetero-lasers with a modulated durability", *Soviet J. of Quantum Electron.*, **17**, 835.

Arahiri S, Matsui Y and Ogawa Y, 1996, "Mode-locking at very high repetition rates more than terahertz in passively mode-locked distributed Bragg reflector laser diodes", *IEEE J. of Quantum Electron.*, **32** 1211.

Aspin G J, Carroll J E and Plumb R G S, 1981, "The effect of cavity length on picosecond pulse generation with highly RF modulated AlGaAs double heterostructure lasers", *Appl. Phys. Lett.*,**39**, 860.

AuYeung J, 1981, "Picosecond optical pulse generation at gigahertz rate by direct modulation of a semiconductor laser", *Appl. Phys. Lett.* **38**, 309.

Delfyett P J, Thirion Florez L, Stoffel N , Gmitter T, Andreadakis N C and Silberberg Y, 1992, "High power ultrafast laser diodes", *IEEE J. Quantum Electron.*, **28**, 2203.

Elliott R A, DeXiu H, DeFreez R K, Hunt J M and Rickman P G, 1983, "Picosecond optical pulse generation by impulse train current modulation of a semiconductor laser", *Appl. Phys. Lett.*, **42**, 1012.

Hou A S, Tucker R S and Eisenstein G, 1990, "Pulse compression of an actively modelocked diode laser using linear dispersion in fibre", *IEEE Photon. Tech. Lett.*, **2**, 322

Lee H K, Loyo-Maldonado V, Qiu B C, Lee K L, Shu C, Pinches S, Thayne I G, Bryce A C and Marsh J H, 2002, "Efficient Direct Locking of Colliding Pulse Mode-Locked Lasers on Semi-Insulating Substrate at 1.5μm", *IEEE Photon. Tech. Lett.*, **14**, 1049.

Morton P A, Mizrahi V, Kosinski S G, Mollenauer L F, Tanbun-Ek T, Logan R A, Coblentz D L, Sergent A M and Wecht K W, 1992, "Hybrid soliton pulse source with fibre external cavity and Bragg reflector", *Electron. Lett.*, **28** 561.

Petermann K, 1998, *Laser diode modulation and noise*, (Kluwer Academic Publishers).

Thedrez B J, Saddow S G, Liu Y Q, Wood C, Wilson R and Lees C M, 1993, "Experimental and theoretical investigation of large output power Q-switched AlGaAs semiconductor lasers", *IEEE Photon.Tech.Lett.*, **5** 19.

Thomson G H B, 1980, *Physics of Semiconductor Laser Devices*, (John Wiley and Sons).

Van der Ziel J P, Tsang W T, Logain R A, Mikulyak R M and Augustyniak W M, 1981, "Generation of sub-picosecond pulses from an actively mode locked GaAs laser in an external ring cavity", *Appl. Phys. Lett.*, **39** 867.

Vasil'ev P P, White I H, Burns D and Sibbett W, 1993, "Direct high bit rate encoding of a high power picosecond pulse train generated by a self-Q-switched multicontact laser", *Paper CThC8, Conference on Lasers and Electro-Optics, Baltimore.*

Vasil'ev P, 1995, *Ultrafast diode lasers: fundamentals and applications*, (Artech House).

Weber A G, Schell M, Fishbeck G and Bimberg D, 1992, "Generation of single femtosecond pulses by hybrid mode-locking of a semiconductor laser", *IEEE J. of Quantum Electron.*, **28**, 2220.

White I H, Gallagher D F G, Osinski M and Bowley D, 1985, "Direct streak camera observation of picosecond gain-switched optical pulses from a 1.5μm semiconductor laser", *Electron. Lett.*, **21**, 197.

Williams K A, Burns D, White I H, Sibbett W and Fice M J , 1993, "Picosecond pulse generation with ultra-low jitter 1.5μm multicontact MQW lasers using Q-switching", *IEEE Photon. Tech. Lett.*, **5**, 867.

Zielinski E, Lach E, Bouayad-Amine J, Haisch H, Kuhn E, Schilling M and Weber J, 1997, "Monolithic multisegment mode-locked DBR laser for wavelength tunable picosecond pulse generation", *IEEE J. Selected Topics in Quantum Electron.*, **3**, 230.

Measuring ultrafast laser pulses

Derryck T Reid

Heriot-Watt University, Scotland

1 Introduction

The last decade has witnessed a revolution in the practice of ultrashort laser pulse measurement. Since the mid-1990s a spectrum of techniques has been developed which makes it possible to accurately record the intensity and phase profiles of even the most complex pulses from almost any laser. This chapter will attempt to describe in broad terms the more established methods, giving attention to their relative strengths and weaknesses and ultimate limitations. Conventional measurements will be contrasted with the more recent exact techniques and the chapter will conclude by describing some examples of experiments only made possible by using complete pulse characterisation methods.

2 Mathematical description of ultrafast laser pulses

Ultrafast laser pulses can almost always be accurately described in the time domain by treating the pulses as a carrier wave of constant frequency ω modulated by a complex envelope function $A(t)\exp(i\phi(t))$ which is responsible for the pulse duration and any frequency chirp content. The electric field amplitude of the pulse can be represented as

$$E(t) = A(t)\exp\big(i\phi(t)\big)\exp\big(i\omega t\big)\,, \tag{1}$$

where $E(t)$ is assumed to consist of a real amplitude $A(t)$ and a complex phase $\phi(t)$. The frequency chirp of the pulse is therefore contained entirely in the phase $\phi(t)$ which can be differentiated to give the instantaneous frequency of the pulse at any given time:

$$\Omega(t) = \omega + \frac{d\phi(t)}{dt}\,. \tag{2}$$

Common functions which are often measured are the pulse intensity

$$I(t) = E(t)E^*(t) = |E(t)|^2 , \qquad (3)$$

and the power spectrum

$$I(\omega) = E(\omega)E^*(\omega) = |E(\omega)|^2 , \qquad (4)$$

where $E(\omega)$ is the Fourier transform of $E(t)$, namely

$$E(\omega) = \frac{1}{\sqrt{2\pi}} \int_{-\infty}^{\infty} E(t) \exp(-i\omega t) dt . \qquad (5)$$

It is interesting to consider for a moment the importance of the pulse spectral phase (the phase of $E(\omega)$). Controlling the spectral phase of a laser pulse is the key to controlling its temporal shape. To illustrate this, consider the following exercise. Take two images (Figures 1a and 1b) and Fourier transform them into their frequency-domain representations, combine them by retaining the phase of one (Figure 1a) and the amplitude of the other (Figure 1b) and finally Fourier transform this hybrid "spectrum" (Figure 1c) back into the spatial domain (Figure 1d). The intensity of the final image is largely determined not by the image whose spectral amplitude was retained but by the image which contributed its spectral phase. A similar argument holds for ultrashort laser pulses and explains why, as more effective methods were developed for dispersion cancellation in laser cavities, ever shorter laser pulses were produced.

3 Autocorrelation techniques

The fastest photodiodes available today have rise times measured in many hundreds of femtoseconds and therefore direct electronic detection is not fast enough to characterise the shape of an ultrashort pulse or even to estimate its duration. For this reason it is necessary to employ all-optical methods to make the measurement. Because most sources of ultrashort pulses have a very regular pulse repetition frequency the ideal measurement device resembles a super-fast mechanical shutter or *gate* which can be opened periodically to sample the energy in a particular time-slice of every pulse. An *autocorrelator* is an optical analogue of this system in which the pulse is overlapped in a nonlinear medium with its time-delayed replica and generates a mixing signal.

In the idealised measurement device, a fast gate would be opened repeatedly at a time τ before or after the pulse and would record samples of the pulse intensity as this delay was scanned. In autocorrelation measurements, the gate pulse is identical to the measurement pulse and the shutter mechanism is achieved by focusing the two pulses in a nonlinear medium to produce a mixing signal whose size depends on the degree of mutual overlap. Because autocorrelation fails to mimic the idealised case where the gate is much shorter than the measurement pulse the result is a broadened and smoothed signal which, although not a direct map of the pulse envelope, gives a useful estimate of its duration.

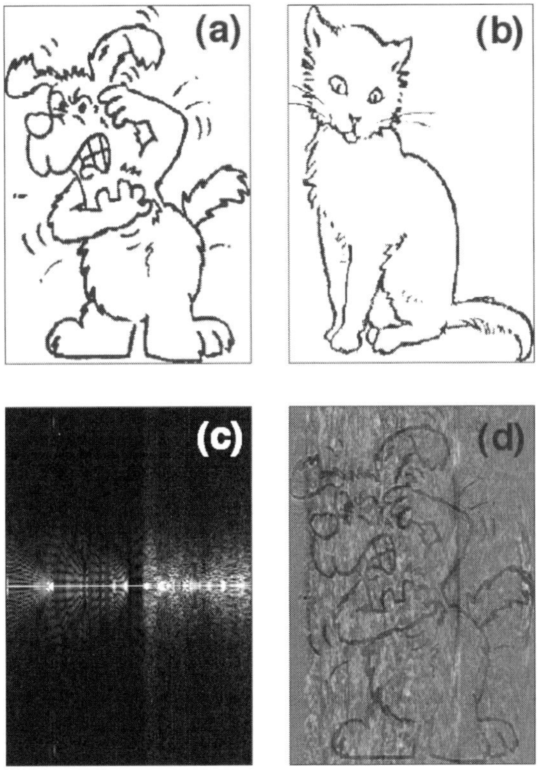

Figure 1. *Combining two images by taking the spectral phase of one (a) and the spectral intensity of the other (b) results in a hybrid spectrum (c) which when transformed back into the spatial domain (d) has a shape strongly determined by the image which contributed its spectral phase. By analogy the spectral phase of an ultrashort pulse is what chiefly determines the shape of the pulse in time.*

Autocorrelation is commonly implemented using second-harmonic generation (SHG) in a nonlinear crystal to produce a mixing signal which has the form

$$g_2(\tau) = \frac{\int\limits_{-\infty}^{\infty} \left| \left\{ E(t) + E(t-\tau) \right\}^2 \right|^2 dt}{2 \int\limits_{-\infty}^{\infty} \left| E(t)^2 \right|^2 dt}. \tag{6}$$

The interference term in Equation 6 leads to $g_2(\tau)$ being known as the *interferometric autocorrelation* profile. The interferometric autocorrelation is a useful diagnostic for detecting the presence of chirp on the pulse but unambiguous determination of the sign and shape of the chirp is not possible from this measurement alone. Pulses with a linear frequency chirp have an instantaneous frequency which is higher in the leading edge of the pulse and lower at the trailing edge, or vice versa. This means that when two identical pulses are combined interferometrically as required by Equation 6 the resulting interferogram only shows fringes in the region near $\tau = 0$ because elsewhere the local pulse frequencies are different. When $g_2(\tau)$ is recorded using a measurement system which has insufficient bandwidth to resolve the interference fringes, a modified signal is recorded which is known as the *intensity autocorrelation* profile and is given by

$$G_2(\tau) = 1 + \frac{2\int\limits_{-\infty}^{\infty} I(t)I(t-\tau)\,dt}{\int\limits_{-\infty}^{\infty} I(t)^2\,dt}. \tag{7}$$

Because the pulse intensity, not the electric field, is the important quantity in Equation 7, the intensity autocorrelation is totally insensitive to the temporal chirp present on the pulse. The situation described by Equation 7 commonly occurs in scanning autocorrelators in which the resistive-capacitive time constant is comparable to the time taken to scan through a single interference fringe.

Regardless of pulse shape, the autocorrelation functions described above always exhibit a characteristic peak-to-background contrast ratio which for the interferometric signal is 8:1 and for the intensity signal is 3:1. A special situation exists when the second-harmonic light generated by each separate pulse (mixing with itself) is not detected, and in this case a background-free intensity autocorrelation is recorded which is equal to the last term of Equation 7. Mathematically the background-free autocorrelation can be conveniently calculated by taking the convolution ($\otimes$) of the pulse intensity and its time-reversed replica:

$$G_2(\tau) = \int\limits_{-\infty}^{\infty} I(t)I(t-\tau)dt = I(t) \otimes I(-t). \tag{8}$$

Separate optical implementations are generally used for autocorrelating amplifier and oscillator sources because of the differences in pulse energy and repetition rate. To characterise pulses from an oscillator, the output is split using a scanning Michelson interferometer (see Figure 2a) which then recombines the two beams by focusing into a second-harmonic crystal and detecting the frequency-doubled light using a photomultiplier or other linear detector. The same functionality can be obtained by mixing the pulses in a semiconductor whose bandgap energy is related to the pulse photon energy by

$$E_g > E_{\text{photon}} > E_g/2, \tag{9}$$

and detecting the photocurrent due to two-photon absorption (Laughton *et al.* 1992, Reid *et al.* 1997). Scanning autocorrelators are most suitable for high-repetition-frequency (~MHz) pulses because each part of the autocorrelation trace results from mixing between distinct pairs of pulses and the measurement is averaged over a time equal to many millions of pulse repetition periods. Mechanical scanning is inconvenient and too slow for lower repetition-rate sources such as kHz ultrafast amplifiers, and for these systems a single-shot autocorrelator is used. Single-shot methods measure the background-free autocorrelation by using non-collinear SHG between two pulses, often by mixing copies of a single beam as shown in Figure 2b. When a sufficiently large mixing angle is used the spatial intensity distribution of the frequency-doubled light is proportional to the

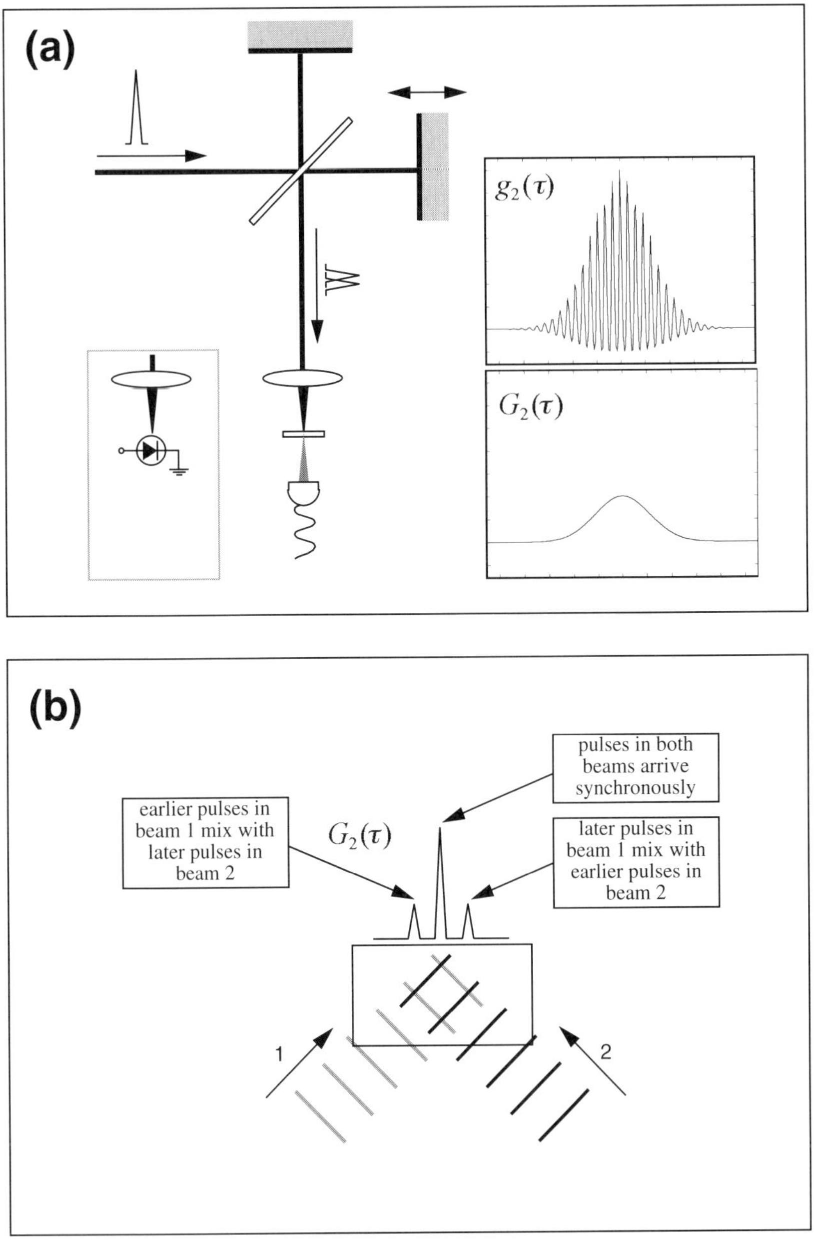

Figure 2. *A typical scanning autocorrelator using either second-harmonic generation or two-photon absorption (dashed inset). As the path difference is scanned the output signal, depending on the bandwidth of the detection system, is either the interferometric (top) or the intensity (bottom) autocorrelation. (b) Illustration of the principle of single-shot autocorrelation in a second-harmonic crystal.*

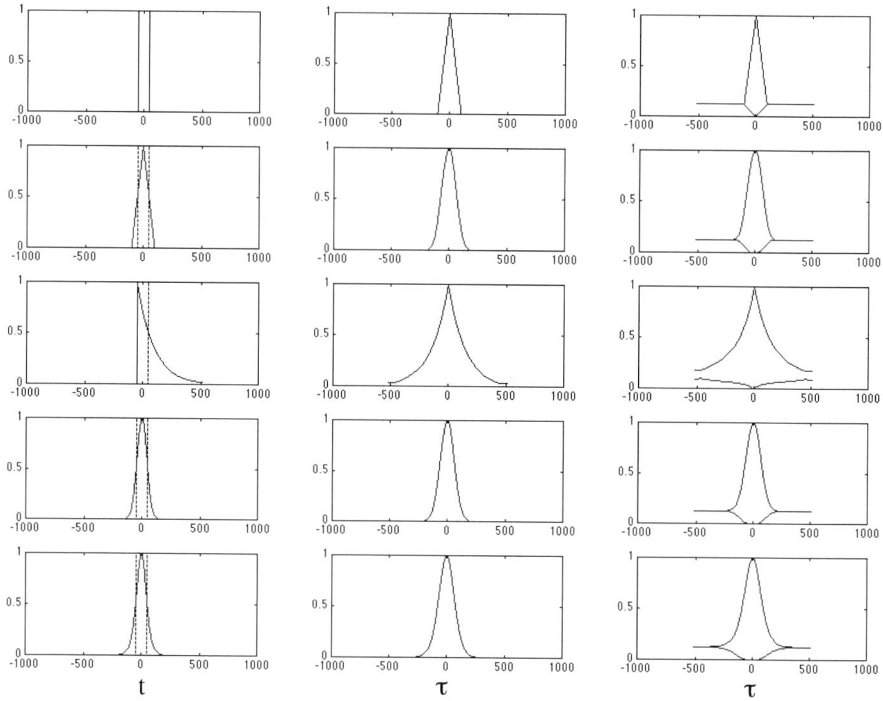

Figure 3. *Illustration of the sensitivity of the second-order autocorrelation profile to pulse shape. Rows: pulses of different shapes but equal FWHM (dashed lines) durations; (a) square, (b) triangular, (c) single-side exponential, (d) Gaussian, (e) sech*$^2(t)$. *Columns (left to right): temporal pulse intensity, background-free intensity autocorrelation, envelope of the interferometric autocorrelation.*

intensity autocorrelation because the transverse phase difference across the beam focus is more than 2π and interference fringes are averaged out.

Autocorrelation measurements do not allow the pulse duration to be determined exactly because the width of the autocorrelation function, although directly proportional to the pulse width, is also dependent on the shape of the pulses. Figure 3 reproduces the intensity and interferometric autocorrelations of five representative pulses with identical FWHM durations and demonstrates the difficulty of trying to infer pulse shape information from autocorrelation measurements. Practically, autocorrelation can be used to estimate the pulse duration from a laser by assuming a pulse shape—normally a Gaussian or $\mathrm{sech}^2 t$ intensity profile—and using a known *deconvolution factor* to infer the pulse duration, $\Delta\tau$, from the autocorrelation width $\Delta\tau_a$ using

$$\begin{aligned} c\Delta\tau_a/\Delta\tau &= 1.414 \qquad \text{for Gaussian pulses,} \\ c\Delta\tau_a/\Delta\tau &= 1.543 \qquad \text{for } \mathrm{sech}^2 t \text{ pulses.} \end{aligned}$$

4 Exact pulse measurement methods

In many experiments, precise quantitative information about the intensity and phase profile of a pulse is invaluable. Exact pulse measurement methods have been used to tailor pulse chirp to a pre-determined value for quantum-control studies; to test the validity of competing laser theories (Taft *et al.* 1995); and to obtain pulse data for numerical simulations (Loza-Alvarez *et al.* 1999). Throughout the 1990s, various new methods were developed which were able to provide exact shape and chirp information about ultrashort pulses from high-repetition-rate modelocked oscillators or single-shot laser amplifiers. The approaches used can be classified either as two-dimensional methods which measure the pulse spectrogram or sonogram signal in a hybrid time-frequency domain, or one-dimensional methods which measure one or more signals exclusively in either time or frequency. In this section, the principal implementations of each type of technique will be explained and their advantages and limitations discussed.

4.1 Two-dimensional 'time-frequency' measurements

Although approximate methods for determining the spectral phase of ultrashort laser pulses have existed for some time (Treacy 1971), the first truly exact method was frequency-resolved optical gating (FROG) which was reported by Kane and Trebino (1993). Based on measuring a spectrogram, FROG records the frequency-spectrum of different time-components of the pulse and uses a 2-D phase-retrieval algorithm (for which a unique solution exists) to obtain from this the pulse characteristics. Figure 4 illustrates this for the particular case of second-harmonic generation (SHG) FROG and the insets show a SHG-FROG trace and the corresponding double-pulse which produced it. FROG has been developed extensively since its original demonstration, and a wide variety of implementations has been introduced which allow the measurement of pulses from sources covering a diverse range of wavelengths, pulse durations and energies.

The mixing of two pulses,the 'probe' and the 'gate', in a nonlinear medium to obtain a cross- (or auto-) correlation signal is the basis of FROG. The FROG spectrogram is built up by measuring the spectrum of the correlation signal for a continuous range of positive and negative delays between the probe and gate pulses. The specific form of the FROG spectrogram depends on the nonlinearity used to obtain the mixing signal and, in addition to SHG-FROG, common FROG implementations include Kerr-effect polarisation-gating (PG-FROG), third-harmonic generation (THG-FROG), difference-frequency generation (DFG-XFROG), and sum-frequency generation (SFG-XFROG). Table 1 lists the fields of the nonlinear mixing signal, $E_{\mathrm{sig}}(t,\tau)$, associated with the more popular FROG imple-

Second-harmonic FROG	$E_{\mathrm{sig}}(t,\tau)=E(t)E(t-\tau)$
Third-harmonic FROG	$E_{\mathrm{sig}}(t,\tau)=E^2(t)E(t-\tau)$
Polarisation gating FROG	$E_{\mathrm{sig}}(t,\tau)=E(t)\lvert E(t-\tau)\rvert^2$

Table 1. *Nonlinear mixing field associated with common FROG geometries.*

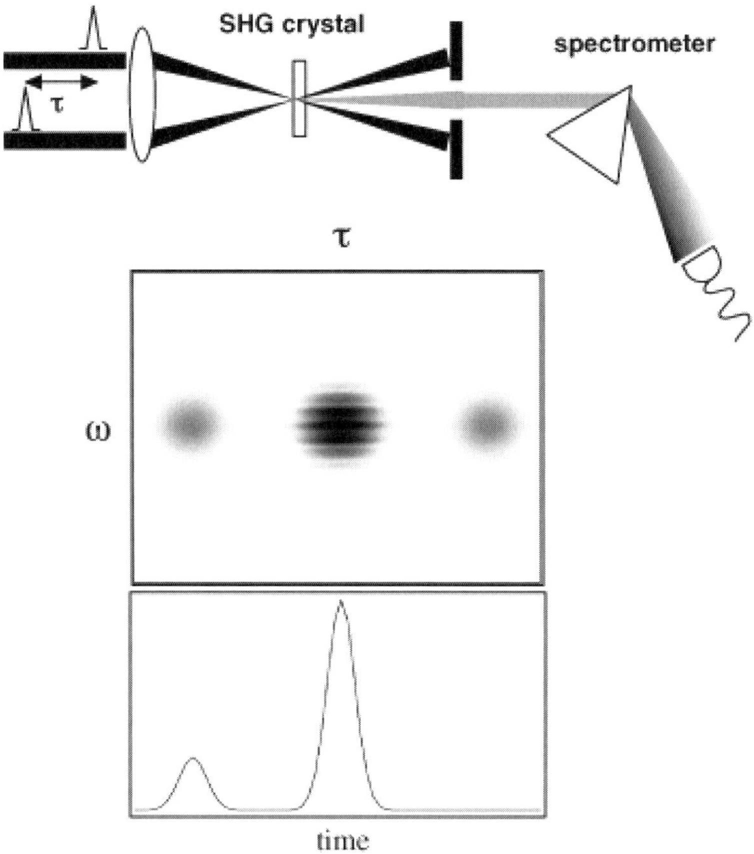

Figure 4. *Schematic of a typical SHG-FROG configuration (top) showing an example of a SHG-FROG trace (middle) and the pulse which produced it (bottom).*

mentations. The FROG spectrogram is related to the mixing signal by

$$S_{\mathrm{FROG}}(\omega, \tau) = \left| \int_{-\infty}^{\infty} E_{\mathrm{sig}}(t, \tau) \exp(-i\omega t) dt \right|^2 . \tag{10}$$

In some FROG geometries, notably SHG-FROG, the roles of the probe and gate pulses are interchangeable and this leads to an ambiguity in the direction of time of the inferred pulse. Once the FROG spectrogram has been measured and calibrated, the pulse intensity and phase can be obtained using an iterative *retrieval algorithm* based on an initial guess pulse whose shape is then refined by applying prior knowledge of the nonlinearity and by constraining the FROG trace of the retrieved pulse to match the experimental data. The details of FROG algorithms are beyond the scope of this chapter but have been described fully by several authors (Trebino and Kane 1993, Delong *et al.* 1994, Kane 1998). With the availability of fast computers, algorithms retrieving in less than one second have been

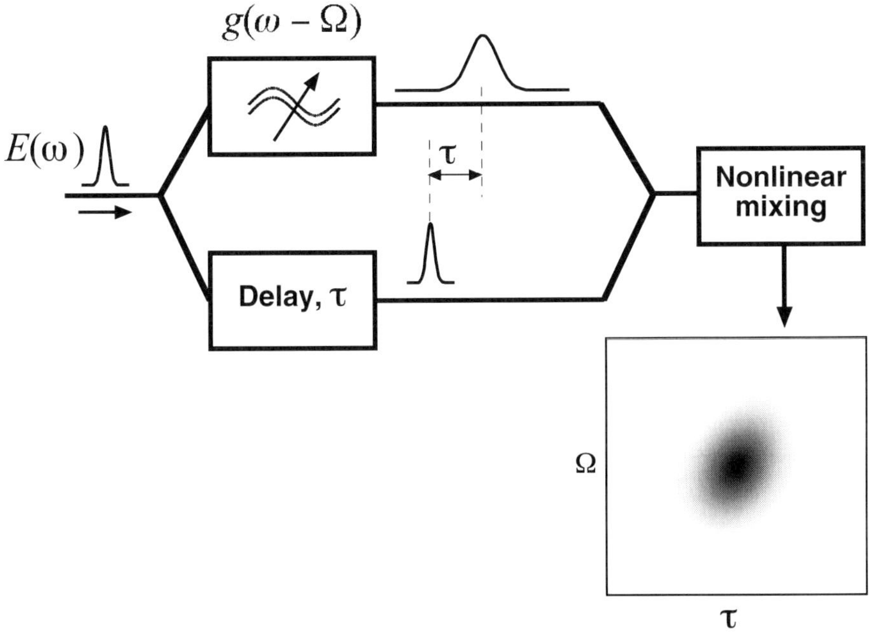

Figure 5. *Illustration of the sonogram measurement in which a pulse with a spectral amplitude $E(\omega)$ is split into two replicas, one which remains unchanged except for experiencing a variable delay, τ, and the other which is bandpass filtered by a tunable filter described by an amplitude transmission function $g(\omega)$. Nonlinear mixing using either sum-frequency mixing or two-photon absorption yields a trace whose slope depends on the pulse chirp.*

reported, allowing real-time implementations of FROG to be demonstrated (Kane 1998).

A pulse *sonogram* can also be used for characterisation purposes and is recorded by measuring the cross-correlation signal between different frequency-filtered components of the pulse and a shorter (uncharacterised) reference (see Figure 5). Commonly the reference is the original pulse itself (Wong and Walmsley 1997, Reid 1999) but any available pulse which is synchronous with the pulse to be measured can be used. The sonogram obtained using a zero-dispersion bandpass filter $g(\omega-\Omega)$ with a variable centre frequency Ω is

$$S_{\mathrm{SONO}}(\Omega, t) = \left| \int_{-\infty}^{\infty} E(\omega) g(\omega - \Omega) \exp(i\omega t)\, d\omega \right|^2 , \tag{11}$$

and the trace is distinctive from SHG-FROG because it displays a tilt proportional to the magnitude and sign of the linear chirp on the pulse. Sonogram measurements can be made using two-photon-absorption cross-correlation and, using waveguide detectors, can be both sensitive and broadband. The requirement for a dispersion-free frequency-filter makes the application of the sonogram technique to sub-50fs pulses difficult but, unlike

SHG-FROG, sonogram measurements have no time ambiguity (Reid 1999). Recently, another method known as dispersive-propagation time-resolved optical gating (DP-TROG) has used multiple autocorrelation measurements to record a sonogram using a dispersive filter (Koumans and Yariv 2000). In general, two-dimensional techniques have accessible to them error-checking methods based on comparing the trace marginals (the trace integral over delay or frequency) with independently measurable quantities. The form and application of the trace *marginals* has been reported for the FROG, sonogram and DP-TROG techniques and these are reproduced in Table 2.

	Delay marginal	Frequency marginal
SHG-FROG	$M_\tau(\tau)=A^{(2)}(\tau)=\int_{-\infty}^{\infty} I(t)I(t-\tau)dt$	$M_\omega(\omega)=I(\omega)\otimes I(\omega)$
SONOGRAM	$M_\tau(\tau)=A^{(2)}(\tau)\otimes\left\|F\{g(\omega)\}\right\|^2$	$M_\Omega(\Omega)=I(\Omega)\otimes\left\|g(\Omega)\right\|^2$
DP-TROG	$M_\tau(\tau)=A^{(2)}(\tau)=\int_{-\infty}^{\infty} I(t)I(t-\tau)dt$	$M_\Omega(\Omega)=\mathrm{A}^{(2)}(\Omega)=\int_{-\infty}^{\infty} I(\omega)I(\omega-\Omega)d\omega$

Table 2. *Trace marginals for the FROG, sonogram and DP-TROG techniques.*

4.2 One-dimensional measurements

Two-dimensional methods are characterised by a high redundancy factor because the measurement of a pulse described by N intensity and N phase samples requires N^2 data points. Such oversampling is desirable because it makes the measurement more resistant to corruption by noise, but the acquisition of 2D datasets often necessitates mechanical scanning and complicated data handling algorithms. It is well known that when a previously-characterised reference pulse exists at the same wavelength as an unknown pulse, the interference signal between the pulses is sufficient for characterisation. The use of 2D methods was originally adopted because no well-characterised reference pulse generally exists, but in 1998 a 1D interferometric method was reported (Iaconis and Walmsley 1998) which used a form of spectral shearing interferometry to solve this problem. Known as spectral phase interferometry for direct electric-field reconstruction (SPIDER), the method eliminated the need for a reference pulse by interfering two time-delayed pulses with a relative frequency-shift. The SPIDER signal is the interference spectrum between the two pulses,

$$S_{\mathrm{SPIDER}}(\omega)=\left|E(\omega)\right|^2+\left|E(\omega+\Omega)\right|^2+2\left|E(\omega)E(\omega+\Omega)\right|\cos\left[\varphi(\omega+\Omega)-\varphi(\omega)+\omega\tau\right] \quad (12)$$

and, unlike 2D methods, both the time-delay, τ, and the frequency-shift, Ω, are fixed. In practice, the time-delay is achieved using an etalon or a fixed Michelson interferometer and the frequency-shift by sum-frequency mixing the delayed pulses with a quasi-monochromatic wave which is a highly-chirped copy of the original pulse. SPIDER is attractive because it requires no moving parts and its 1D nature means than spatially-resolved measurements of chirp can be recorded (Gallmann *et al.* 2001), but, for optimum performance, the technique requires careful selection of time and frequency shifts. Using SPIDER, pulses as short as 6fs have been measured.

The practicality of measuring pulses using non-phasematched $\chi^{(3)}$ two-photon processes in semiconductors instead of phasematched $\chi^{(2)}$ frequency-conversion in bulk nonlinear crystals has motivated efforts to extract intensity and phase information from spectral and auto-/cross-correlation measurements alone. This approach is essentially one-dimensional because measurements are made in either the time or frequency domain and the resulting datasets are vectors, not arrays. Various algorithms which use autocorrelation measurements for exact characterisation have been reported but complex pulse shapes are known to present problems for such techniques. A variation based on the cross-correlation between the measurement pulse and its frequency-chirped replica has been shown to cope successfully with asymmetric and double pulses (Nicholson *et al.* 2000). The approach uses the spectrum with an initial random phase to estimate the complex spectral field amplitude $E(\omega)$ from which the temporal intensities of the pulse before and after a dispersive element are calculated. By minimising the RMS error between the measured and calculated cross-correlations of the chirped and unchirped pulses, the algorithm converges to the correct pulse shape. The simplicity of techniques which require only measurements of the spectrum and correlation of the pulse is attractive but the minimisation algorithms can be sensitive to the initial guess phase and their convergence can be slow in comparison to iterative spectrogram or sonogram algorithms.

5 Applications of exact pulse measurement

The ability to exactly characterise an optical pulse provides a basis for controlling experiments in which the outcome is sensitive to the precise pulse shape, frequency content and chirp. A contemporary theme in photo-chemistry uses femtosecond pulses to drive reactions into particular product channels or prepare molecules in precise quantum states. Work by Kent Wilson and others (Kohler *et al.* 1995) applied the FROG pulse measurement technique to prepare negatively-chirped femtosecond pulses with the appropriate duration required to excite diatomic iodine molecules so that, some time after excitation, their vibrational wavepackets had evolved to a state of minimum uncertainty ($\Delta p \Delta x = \hbar/2$) corresponding to a known atomic velocity, separation and momentum. The equivalent positively-chirped pulse was shown, in agreement with theory, to fail to produce localised wavepackets. A related strategy has been used by Gerber and others to optimise the branching ratio of products resulting from the photodissociation of $CpFe(CO)_2Cl$ (Assion *et al.* 1998). The experiment used a genetic algorithm controlling a pulse-shaper to maximise or minimise a feedback signal based on the ratio of $Fe(CO)_5^+$ to Fe^+ ions by creating a pulse with an optimised phase profile.

The FROG method has been used successfully in conjunction with pulse propagation theory to develop models of pulse evolution in lasers and nonlinear materials. One early example of this was the application of FROG to determine the mechanism responsible for limiting the durations of the shortest pulses which could be produced by a self-modelocked Ti:sapphire laser. Two contradictory but viable theories were proposed, one which stated that uncompensated high-order cavity dispersion was responsible and the other that attributed the limit to coherent ringing in the oscillator (Taft *at al.* 1995). Both theories predicted autocorrelation profiles and spectra which agreed with those measured experimentally, although the actual pulse shapes were different in the two cases. By comparing the experimentally measured FROG trace with those calculated on the basis of each theory it was possible to unambiguously show that uncorrected high-order dispersion was in fact responsible for limiting the shortest pulse durations which could be generated.

Another example of the power of exact pulse measurements was demonstrated in experiments which used FROG to characterise the infrared pulses incident on a quasi-phasematched nonlinear crystal containing an aperiodic grating structure. When chirped input pulses are incident on an aperiodically-poled second-harmonic crystal the output pulses are either compressed or stretched, depending on the orientation of the crystal grating. By using FROG to provide a mathematically formal description of the input pulses we were able to run a pulse propagation model which accurately predicted the fields of the stretched or compressed output pulses (Loza-Alvarez *et al.* 1999).

6 Perspective and conclusions

Developments in pulse characterisation have transformed the practice of ultrafast laser science and this chapter has provided only a brief review of contemporary themes in the area. Several directions remain open for investigation and these include improved real-time (Kane 1999), and single-shot (O'Shea 2000, Reid and Cormack 2002) measurement methods together with techniques for making spatio-temporal measurements of the output beams from ultrafast lasers (Gallmann *et al.* 2001). Very recently the question of how to measure the absolute phase of pulses from a modelocked laser has been raised because of its relevance to optical comb-generation and frequency metrology. Some experiments are sensitive to the relative phase between the carrier field and the pulse envelope (Drescher *et al.* 2001) and these provide, in principle, a way to extract pulses from a periodic sequence which have perfect carrier-envelope phase synchronism. The ability to characterise and control laser pulses to this degree may indeed only be the beginning of a new and even more precise approach to optical pulse measurement.

References

Assion A, Baumert T, Bergt M, Brixner T, Kiefer B, Seyfried V, Strehle M, Gerber G, 1998, 'Control of chemical reactions by feedback-optimized phase-shaped femtosecond laser pulses', *Science* **282** 919.

Cormack I G, Sibbett W, Ortega-Martinez R and Reid D T, 2001, *Rev. Sci. Instrum.* **72** 4071.

Delong K W, Fittinghoff D N, Trebino R, Kohler B and Wilson K, 1994, 'Pulse Retrieval In Frequency- Resolved Optical Gating Based On The Method Of Generalized Projections', *Opt. Lett.* **19** 2152.

Drescher M, Hentschel M, Kienberger R, Tempea G, Spielmann C, Reider G A, Corkum P B and Krausz F, 2001, 'X-ray pulses approaching the attosecond frontier', *Science* **291** 1923.

Gallmann L, Steinmeyer G, Sutter D H, Rupp T, Iaconis C, Walmsley I A and Keller U, 2001, 'Spatially Resolved Amplitude And Phase Characterization Of Femtosecond Optical Pulses', *Opt. Lett.* **26** 96.

Iaconis C and Walmsley I A, 1998, 'Spectral phase interferometry for direct electric- field reconstruction of ultrashort optical pulses', *Optics Letters* **23** 792.

Kane D J, 1998, 'Real-Time Measurement Of Ultrashort Laser Pulses Using Principal Component Generalized Projections', *IEEE J. Sel. Top. Quant.* **4** 278.

Kane D J and Trebino R J, 1993, 'Single-Shot Measurement Of The Intensity And Phase Of An Arbitrary Ultrashort Pulse By Using Frequency-Resolved Optical Gating', *Optics Letters* **18** 823.

Kohler B, Yakovlev V V, Che J W, Krause J L, Messina M, Wilson K R, Schwentner N, Whitnell R M and Yan Y J, 1995, 'Quantum control of wave-packet evolution with tailored femtosecond pulses', *Phys. Rev. Lett.* **74** 3360.

Koumans R G M P and Yariv A, 2000, 'Pulse Characterization At 1.5m Using Time- Resolved Optical Gating Based On Dispersive Propagation', *IEEE Photonic Tech Lett.* **12** 666.

Laughton F R, Marsh J H and Kean A H, 1992, 'Very Sensitive 2-Photon Absorption GaAs/AlGaAs Wave-Guide Detector For an Autocorrelator', *Electron Lett.* **28** 1663.

Loza-Alvarez P, Reid D T, Faller P, Ebrahimzadeh M, Sibbett W, Karlsson H and Laurell F, 1999, 'Simultaneous femtosecond pulse compression and second-harmonic generation in aperiodically poled $KTiOPO_4$', *Optics Letters* **24** 1071.

Nicholson J W, Mero M, Jasapara J and Rudolph W, 2000, 'Unbalanced Third-Order Correlations For Full Characterization Of Femtosecond Pulses', *Optics Letters* **25** 1801.

OShea P, Kimmel M, Gu X and Trebino R, 2000, *Optics Letters* **26 932**

Reid D T and Cormack I G, 2002, *Optics Letters* **27** 658.

Reid D T, Padgett M, McGowan C, Sleat W E and Sibbett W, 1997 'Light-Emitting Diodes As Measurement Devices For Femtosecond Laser Pulses', *Opt. Lett.* **22** 233.

Reid D T, 1999, 'Algorithm For Complete And Rapid Retrieval Of Ultrashort Pulse Amplitude And Phase From A Sonogram', *IEEE J. Quantum Elect.* **35** 1584.

Taft G, Rundquist A, Murnane M M, Kapteyn H C, DeLong K W, Trebino R and Christov I P, 1995, *Opt. Lett.* **20** 743.

Taft G, Rundquist A, Murnane M M, Kapteyn H C, DeLong K W, Trebino R and Christov I P, 1995, 'Ultrashort optical waveform measurements using frequency resolved optical gating', *Opt. Lett.* **20** 743.

Treacy E B, 1971, 'Measurement and interpretation of dynamic spectrograms of picosecond light pulses', *J. Appl. Phys.* **42** 3848.

Trebino R and Kane D J, 1993, 'Using Phase Retrieval To Measure The Intensity And Phase Of Ultrashort Pulses - Frequency-Resolved Optical Gating', *J. Opt. Soc. Am.* **A 10** 1101.

Wong V and Walmsley I A, 1997, 'Ultrashort-Pulse Characterization From Dynamic Spectrograms By Iterative Phase Retrieval', *J. Opt. Soc. Am.* **B 14** 944.

Cascaded nonlinear optical processes

Solomon M Saltiel

University of Sofia, Bulgaria

1 Introduction

Cascaded nonlinear optical processes (CNOP) have played an important role in the development of nonlinear optics during the last decade. Many of the experiments for the generation of extreme ultraviolet and X-ray radiation are accompanied by the generation of low number harmonics; this is a sign of the involvement of CNOP (Misoguti *et al.* 2001). The latest experimental achievements in the attempt to form stable 2D and 3D solitary waves in quadratic crystals are based on the great progress made in the understanding of cascaded processes (Stegeman and Torner 2001). The devices based on frequency shifting in quadratic crystals have reached commercial applications in the optical communication industry (Chou *et al.* 1998, Gallo and Assanto 1999) and are an excellent example of the importance of CNOP.

Cascading in nonlinear optics is an effect noticed during the first steps of nonlinear optics (Akhmanov and Khokhlov 1964, Ostrovskii 1967), but only recently has it received the attention adequate to its importance. In fact CNOP are accompanying effects for all known nonlinear optical processes. As an example, the well known depletion of the fundamental wave during the process of phase-matched second harmonic generation is a cascaded process. Indeed, depletion of the fundamental wave is based on two steps: (i) generation of a second harmonic wave and (ii) reconstruction of the fundamental wave by difference frequency mixing. Another very simple experiment demonstrating cascaded processes can be done with a quadratic crystal tuned to phase-matched second harmonic generation. If we block fundamental and second harmonic waves after passing through the crystal and allow the shorter wavelengths to go through, we can record a sizeable third harmonic signal and also fourth and fifth harmonics with decreasing efficiencies. This generation of high-order harmonics is a result of CNOP.

Cascaded nonlinear optical processes can be separated into two big groups. In the first one, due to simultaneous action of two or more sub-processes, the fundamental (input) waves change their parameters such as amplitude, phase, polarization state, beam size, temporal shape. The second group includes those cascaded interactions in which new waves are generated. The majority of the cascaded interactions simulate direct $\chi^{(3)}$ or direct $\chi^{(5)}$ processes. The advantage of CNOP is its higher efficiency and extremely short time constant. Cascading effects can be used for the construction of nonlinear optical devices for efficient high order harmonic generation, for the construction of devices for optical communications, all optical switching devices, mode locking devices, etc.

The other possible classification of nonlinear optical cascaded processes is based on the number of sub-processes involved in the cascading. There are two possibilities: one in which all sub-processes are governed by the same mismatch parameter and the other in which several mismatch parameters control the whole cascaded process. An example of a cascaded process that belongs to the first type is the cascading connected with the process of second harmonic generation. The efficiencies of the two cascading steps $\omega + \omega = 2\omega$ and $2\omega - \omega = \omega$ depend on *single* phase-matching parameter $\Delta k_2 = k_2 - 2k_1$ where k_1, k_2 are the wave vectors of the fundamental and second harmonic wave, respectively. The effect of third harmonic generation in a single quadratic crystal, where the two cascading steps are second harmonic generation and sum frequency mixing, is an example of the cascaded process that is controlled by *two* phase-matching parameters: $\Delta k_2 = k_2 - 2k_1$ and $\Delta k_3 = k_3 - k_2 - k_1$ (k_3 is the wave vector of the third harmonic wave).

Above we gave examples of cascading of second-order processes but, as will be shown below, cascading of third-order effects is also an important field in the nonlinear optics of cascaded processes. In a similar way, third-order cascading can be observed inside a single third-order interaction or can be a result of cascading different third order processes. We will consider in this chapter both second and third-order cascading as well.

2 Control by a single phase mismatched parameter

2.1 Cascading with Type I second harmonic generation (SHG)

Let us consider the cascading effects connected with the most popular nonlinear interaction: second harmonic generation, Type I. The cascading steps are two: $\omega + \omega = 2\omega$ and $2\omega - \omega = \omega$. In both steps the second-harmonic generation (SHG) and the difference frequency mixing involve only two waves, namely the fundamental and second harmonic (SH) waves. So, both sub-processes depend only on one phase-mismatch parameter $\Delta k_2 = k_2 - 2k_1$. We will show that, due to cascading, the fundamental wave accumulates a nonlinear phase shift (NPS) during the interaction. This effect was indeed realized last decade (Stegeman *et al.* 1996) in spite of some early works.

The plane wave equations describing the process of SHG are:

$$\frac{dA_1}{dz} = -i\sigma A_2 A_1^* \exp(-i\Delta k_2 z), \tag{1a}$$

$$\frac{dA_2}{dz} = -i\sigma A_1^2 \exp(i\Delta k_2 z)\,. \tag{1b}$$

By representing the complex amplitudes A_1 and A_2 in the form

$$A_1 = a_1 \exp(i\varphi_1),$$
$$A_2 = a_2 \exp(i\varphi_2),$$

and introducing the phase $\Phi = \varphi_2 - 2\varphi_1 - \Delta k_2 z$ we can express the phase of the fundamental wave and the modulus of the SH wave by the following differential equations

$$\frac{d\varphi_1}{dz} = -\sigma a_2 \cos\Phi, \tag{2a}$$
$$\frac{da_2}{dz} = -\sigma a_1^2 \sin\Phi\,. \tag{2b}$$

With the help of two invariants of the system

$$\sigma a_1^2 a_2 \cos\Phi + \frac{\Delta k_2}{2} a_2^2 = \Gamma = 0, \tag{3a}$$
$$a_1^2 + a_2^2 = a_{10}^2 + a_{20}^2 = u^2, \tag{3b}$$

we obtain

$$\frac{da_2^2}{dz} = -2\sigma u^3 \sqrt{\left(\frac{a_2}{u}\right)^2 - \frac{1}{M}\left(\frac{a_2}{u}\right)^4 + \left(\frac{a_2}{u}\right)^6}\,. \tag{4}$$

Neglecting the last term in the square root in (4) and after integration of (4) we get the expressions for the conversion efficiency of the SH and fundamental waves:

$$\left(\frac{a_1}{u}\right)^2 = 1 - M\sin^2(Qz), \tag{5a}$$
$$\left(\frac{a_2}{u}\right)^2 = M\sin^2(Qz), \tag{5b}$$

where $M = (\sigma u/Q)^2$, $Q = \sqrt{2\sigma^2 u^2 + \Delta k_2^2/4}$. If a_1 and a_2 are known, Equation 2a can be integrated to result in

$$\Delta\varphi_{\mathrm{NPS}} = \frac{\Delta k_2 L}{2}\left(\frac{\arctan\left[\sqrt{1-M}\tan(QL)\right]}{QL\sqrt{1-M}} - 1\right). \tag{6}$$

Expressions (5) and (6) describe the cascading effects for moderate conversion efficiencies (<30%) and the depletion and accumulation of a NPS by the fundamental wave. These dependencies are shown in Figure 1.

For bigger input fundamental intensities the adequate description of the phase and the amplitudes requires a numerical solution of Equation 1. We see from the curves shown that at phase-mismatches $\Delta k_2 L = \pm\pi$ the NPS accumulated by the fundamental wave has a maximum. The remarkable point is that the sign of the NPS can be changed by the simple change of sign of the phase-mismatch parameter. In the real experiments the change of Δk_2 can be done by changing for example, the angular position of the crystal or the temperature. The higher the fundamental input intensity, the bigger is the accumulated NPS.

In order to demonstrate that this type of cascading is in fact a simulation of a third order effect and can be described in terms of $\chi^{(3)}_{\mathrm{casc}}$ let us present (6) in a different form. Expressing $\cos\Phi$ from (3a) and substituting it in (2a) and using the amplitude of the SH wave defined in (5b) after integration of (2) we obtain

$$\Delta\varphi_{\mathrm{NPS}} = \Delta k_2 L \left(\frac{\sigma u}{2Q}\right)^2 \left[1 - \frac{\sin(2QL)}{2QL}\right],$$

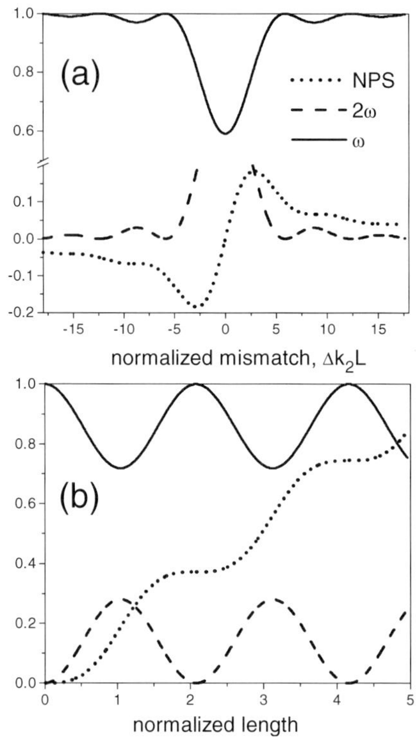

Figure 1. *(a) Fundamental depletion, second harmonic efficiency and nonlinear phase shift versus normalized mismatch $\Delta k_2 L$, ($\sigma u L = 0.8$) and (b) the same parameters versus the normalized crystal length $\sigma u L$, ($\Delta k_2/\sigma u = 2.5$)*

which for very small input intensities ($\sigma u L \ll 1$) has the form

$$\Delta\varphi_{\rm NPS} = \left(\frac{\sigma\, u}{\Delta k}\right)^2 [\Delta k_2 L - \sin(\Delta k_2 L)] \ . \tag{7}$$

For $\Delta k_2 L = \pi$ and keeping in mind that $\sigma = 2\pi d^{(2)}/\lambda n = \pi\chi^{(2)}/\lambda n$, we find that the NPS accumulated by the fundamental wave is proportional to the square of the second-order susceptibility

$$\Delta\varphi_{\rm NPS} = \frac{\pi u^2 L^2}{\lambda^2 n^2}\left[\chi^{(2)}\chi^{(2)}\right] . \tag{8}$$

The last expression illustrates the origin of the $\chi^{(2)}$:$\chi^{(2)}$ cascading terminology.

In centrosymmetric media NPS is a result of self-phase modulation due to the inherent $\chi^{(3)}$ nonlinearity:

$$\Delta\varphi_{\rm NPS} = \frac{3\pi u^2 L}{4\lambda n}\chi^{(3)} . \tag{9}$$

Comparing Equations (8) and (9) we find

$$\chi^{(3)}_{\rm casc} = \frac{4L}{3\lambda n}\left[\chi^{(2)}\chi^{(2)}\right] , \tag{10}$$

$$n_{2,\text{casc}} = \frac{L}{2\lambda n^2}\left[\chi^{(2)}\chi^{(2)}\right] . \tag{11}$$

The main conclusion at this point is that cascading in quadratic media can simulate cubic processes and we can expect many of the known $\chi^{(3)}$ processes to be observed on the basis of $\chi^{(2)}$:$\chi^{(2)}$ cascading. Since $\chi^{(2)}$ effects are instantaneous, not connected with the real transitions in the media, the $\chi^{(3)}_{\text{casc}}$ is also instantaneous.

2.2 Cascading with Type II SHG

With this type of SHG the input fundamental wave splits into two input waves, namely ordinary and extraordinary. When the goal is maximum SH efficiency, the ordinary and extraordinary waves are usually chosen to have equal intensities. To obtain the maximum effect from second order cascading with a Type II interaction the ordinary and extraordinary waves are required to have *unequal* intensities. The plane wave equations describing Type II SHG are:

$$\frac{dA_{1o}}{dz} = -i\sigma A_2 A_{1e}^* \exp(-i\Delta kz), \tag{12a}$$

$$\frac{dA_{1e}}{dz} = -i\sigma A_2 A_{1o}^* \exp(-i\Delta kz), \tag{12b}$$

$$\frac{dA_2}{dz} = -i2\sigma A_{1e} A_{1o} \exp(i\Delta kz) . \tag{12c}$$

Three waves are involved in this quadratic interaction: ordinary (A_{1o}), extraordinary (A_{1e}) and SH (A_2)waves. The three cascading steps are:

$$\omega_o + \omega_e = \omega_{2e}; \qquad \omega_{2e} - \omega_o = \omega_e; \qquad \omega_{2e} - \omega_e = \omega_o .$$

The single phase-matching parameter $\Delta k = k_{2e} - k_o - k_e$ controls each cascading step.

Similarly, as in Section 2.1, we derive approximate "low intensity" expressions for the NPS collected by both the ordinary and extraordinary waves (Saltiel *et al* 1995, 1996a):

$$\Delta\varphi_{1o} = \left(\frac{\sigma u_{1e}}{\Delta k}\right)^2 [\Delta kL - \sin(\Delta kL)] , \tag{13a}$$

$$\Delta\varphi_{1e} = \left(\frac{\sigma u_{1o}}{\Delta k}\right)^2 [\Delta kL - \sin(\Delta kL)] . \tag{13b}$$

It is important to note that the intensity of the ordinary wave at the input of the crystal defines the NPS of the extraordinary wave and vice versa.

Figure 2 (obtained by numerical integration of Equations 12) shows the NPS collected by both the ordinary and extraordinary waves for several ratios of their input intensities. The less intense wave accumulates the bigger NPS. This effect can be used for the construction of all-optical transistors (Assanto *et al.* 1995) and for many polarization devices as described below.

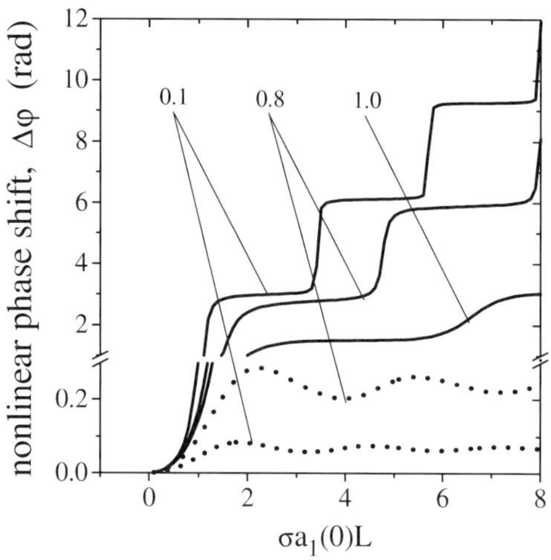

Figure 2. *Nonlinear phase shifts collected by the less intense (solid lines) and more intense (dotted lines) fundamental waves during the process of Type II SHG as a function of normalized input amplitude of the more intense fundamental wave. The parameter is the ratio of the intensities of the two fundamental waves. The phase-matching parameter is* $\Delta kL = 0.3$.

2.3 Evolution of the polarization state as a result of $\chi^{(2)}$:$\chi^{(2)}$ cascading in Type II SHG.

As an interesting example of the capability of second order cascading to simulate third order processes we consider change of the polarization state of waves involved in Type II SHG (Buchvarov *et al.* 1997). The cubic effect simulated is the induced polarization rotation and induced ellipticity observed along four-fold axes in cubic centro-symmetric crystals. In this cubic effect the change of the polarization is a result of the anisotropy of the third order susceptibility. We show here that the same effect can be observed in quadratic crystals due to cascaded effects. The idea of the effect is shown in Figure 3. The change of the polarization state of the input fundamental wave involved in Type II SHG is a

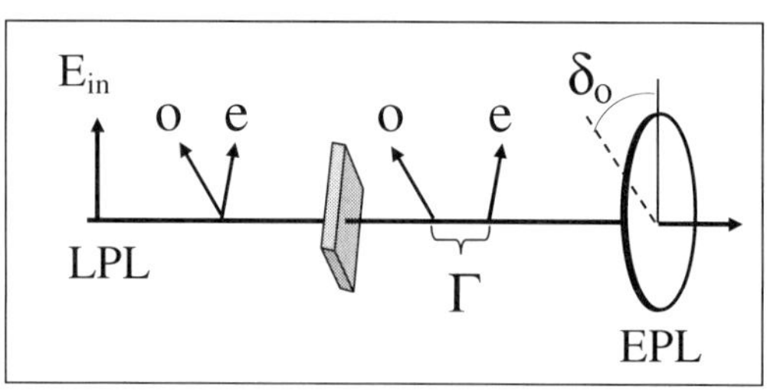

Figure 3. *Induced change of the polarization state as a result of second order cascading in a Type II SHG crystal. LPL, linearly polarized light; EPL, elliptically polarized light;* Γ *is the accumulated nonlinear phase shift difference between ordinary/extraordinary wave.*

result of the effect of the asymmetry on the accumulation of NPS with respect to the ordinary and extraordinary waves as considered in the previous paragraph. Let us first note that linear birefringence has to be compensated by a proper wave plate (not shown). Then, if the NPS collected by the two waves is different, this will lead to a change of the polarization state at the output of the quadratic crystal. We have to remember also that the depletion of the two eigen-waves (ordinary and extraordinary) will also influence the output polarization state.

The output complex amplitudes of the two fundamental waves as a result of cascading becomes

$$E_{\text{out}}^{(o)} = E_{\text{in}}^{(o)} \rho_o \exp\left(i\Delta\varphi_{\text{NPS}}^{(o)}\right), \tag{14a}$$

$$E_{\text{out}}^{(e)} = E_{\text{in}}^{(e)} \rho_e \exp\left(i\Delta\varphi_{\text{NPS}}^{(e)}\right), \tag{14b}$$

where ρ_o and ρ_e are the amplitude transmission coefficients of the two eigen-waves.

If we introduce the complex ratio $R = E_{\text{out}}^{(e)}/E_{\text{out}}^{(o)}$ then the induced ellipticity ϵ and induced azimuth rotation δ_o can be calculated using the following relations (Azzam and Bashara 1977)

$$\tan(2\delta_o) = \frac{2\text{Re}\,(R)}{\left(1 - |R|^2\right)} = \frac{\rho_o\rho_e \cos(\Gamma)\sin(2\alpha)}{\rho_e^2 \sin^2(\alpha) - \rho_o^2\cos^2(\alpha)}, \tag{15a}$$

$$\sin(2\epsilon) = \frac{2\text{Im}\,(R)}{\left(1 + |R|^2\right)} = \frac{\rho_o\rho_e \sin(\Gamma)\sin(2\alpha)}{\rho_e^2 \sin^2(\alpha) + \rho_o^2\cos^2(\alpha)}. \tag{15b}$$

In Equations 15, α is the angle between the input polarization direction and the polarization plane of the ordinary wave and $\Gamma = \Delta\varphi_{\text{NPS}}^{(e)} - \Delta\varphi_{\text{NPS}}^{(o)}$. Figure 4 shows these two parameters and the transmission of the fundamental wave as a function of the input amplitude. The important factor for the observation of these induced effects is that the input amplitudes for the ordinary and extraordinary waves be unequal ($\alpha \neq 45°$).

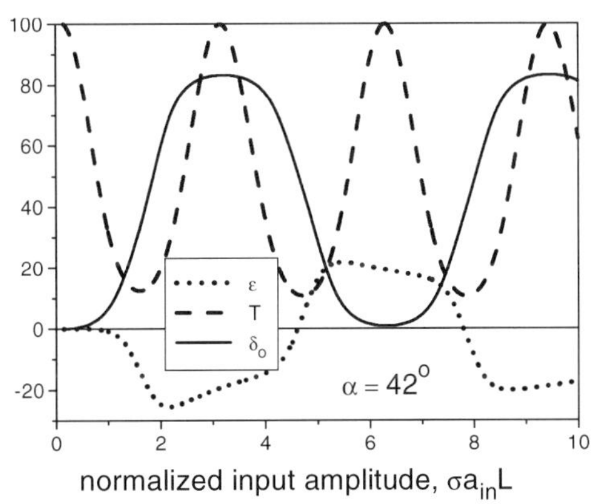

Figure 4. *The angle of ellipticity ϵ , the rotation angle δ_o and the transmission of the fundamental wave T (in %) as a function of the normalized input amplitude for the indicated input angle α. The normalized phase-mismatch $\Delta kL = 0.3$.*

Similar effects can also be observed with Type I SHG. If a linearly polarized wave is falling on the crystal, the latter should be rotated round the laser beam axis in such a way that the input wave is split into ordinary and extraordinary waves with equal intensities. If the interaction is $o_1 o_1 – e_2$ only the ordinary wave accumulates NPS and experiences depletion and the extraordinary wave is not involved in the process of SHG then $\rho_e = 1$ and $\Delta\varphi_{\text{NPS}}^{(e)} = 0$. Then we can use again (14) and (15) to calculate the induced change in the polarization state.

2.4 Polarization interferometer based on $\chi^{(2)}$:$\chi^{(2)}$ cascading.

The effect of intensity induced polarization rotation can be used to construct devices in which the frequency doubling crystal(s) is sandwiched between two polarizers. The schematic diagram of such a device, called by us a frequency doubling polarization interferometer (Saltiel *et al.* 1996b, Buchvarov *et al.* 1997) is shown in Figure 5. Both Type I and Type II frequency doubling schemes can be used; the device contains also a phase

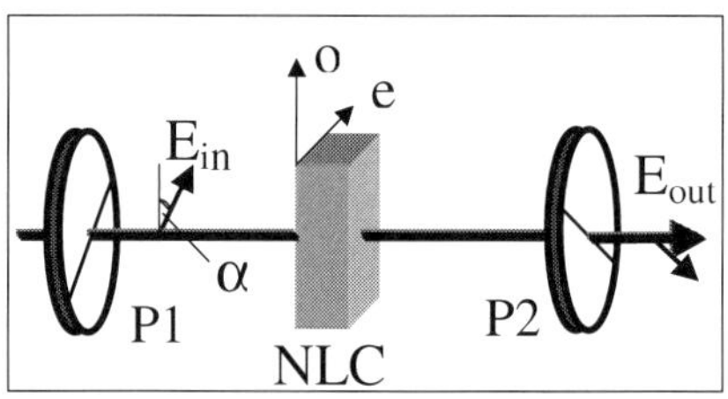

Figure 5. *Frequency doubling polarization interferometer. NLC, nonlinear crystal for SHG; P1, polarizer; P2, analyzer. α is the angle between the input polarization and the polarization plane of the ordinary wave.*

corrector (not shown). The angle α is the angle between the input polarization plane and the direction of the ordinary wave polarization vector. Depending on the proposed application the polarization planes of the two polarizers can be perpendicular, parallel or at an angle 2α. The phase corrector is used to compensate the linear phase shift that the two waves obtain in the crystal. Depending on the mutual orientation of the polarizer, analyzer and the frequency doubling crystal, the nonlinear polarization interferometer exhibits self induced transparency or self induced darkening. The throughput of the frequency doubling polarization interferometer (FDPI) for the three different arrangements of the FDPI is given by

$$T_{\perp} = \frac{1}{4}\left(\rho_o^2 + \rho_e^2 - 2\rho_o\rho_e \cos\Gamma\right)\sin^2(2\alpha)\,, \tag{16a}$$

$$T_{\parallel} = \rho_o^2 \cos^4\alpha + \rho_e^2 \sin^4\alpha + \frac{1}{2}\rho_o\rho_e \cos\Gamma \sin^2(2\alpha)\,, \tag{16b}$$

$$T_{2\alpha} = \rho_o^2 \cos^4\alpha + \rho_e^2 \sin^4\alpha - \frac{1}{2}\rho_o\rho_e \cos\Gamma \sin^2(2\alpha)\,, \tag{16c}$$

where, as before, $\Gamma = \Delta\varphi_{\text{NPS}}^{(e)} - \Delta\varphi_{\text{NPS}}^{(o)}$ and ρ_o and ρ_e are the amplitude transmission coefficients for the ordinary and extraordinary waves, respectively.

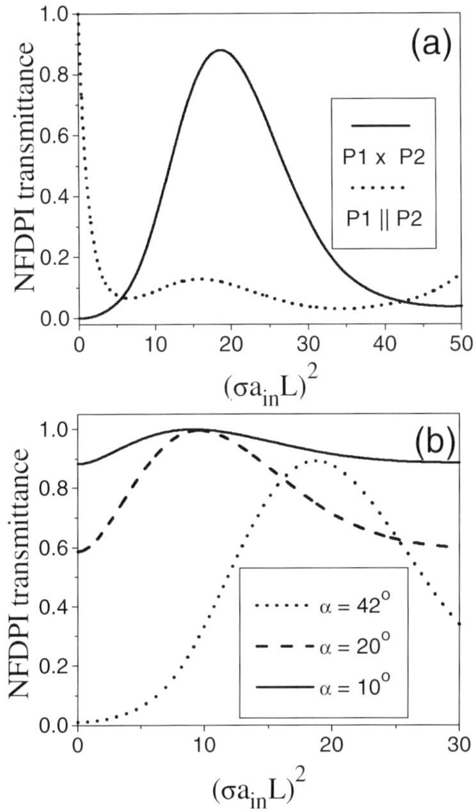

Figure 6. *Transmittance of the FDPI as a function of normalized input intensity for three different arrangements: a) for parallel and crossed polarizers and b) for polarizers that constitute angle 2α. The phase-mismatch is $\Delta kL = 0.3$*

The third expression describes the "2α " version of the interferometer, that is the situation when the analyzer makes an angle 2α with the polarizer.The transmission curves for the three different versions of the FDPI, shown in Figure 5 are presented in Figure 6. When the polarizer and the analyzer are crossed, strong shortening of the pulses incident on the FDPI can be obtained (Buchvarov *et al.* 1997).

It is clear that the two capabilities of the FDPI: pulse shortening and self induced transparency effect at relatively low power (we calculate 60MW/cm^2 for a 10mm long KTP crystal) make this device suitable for mode locking of lasers. Another possible application could be all optical switching, and sensor protection.

Experimentally FDPI has been realized with bulk phase-matched KTP (Lefort and Barthelemy 1995, Louis *et al.* 2001). As seen from Figure 6b, by varying the angle α between 0 and $\pi/4$ with the " 2α " configuration of the interferometer one can achieve any contrast ratio in between the open and closed working points. Flexibility to change the contrast ratio is very important when this effect is used to mode-lock lasers. Successful mode-locking with such an interferometer was reported by Louis *et al.* (2001)

2.5 Cascading in centro-symmetric media

As we demonstrate in the previous paragraphs $\chi^{(2)}$:$\chi^{(2)}$ cascading simulates cubic effects and causes self-interaction effects known as "n_2" effects typical for cubic interactions, such as accumulation of nonlinear phase shift, self-focusing and self defocusing, change of polarization stage, solitons etc. The question is whether it is possible to observe cascading effects in centrosymmetric media as a result of cascading of third order processes? Let us consider centrosymmetric media with the possibility for third harmonic generation $\omega + \omega + \omega = 3\omega$ with a very small phase-mismatched parameter $\Delta k = k_3 - 3k_1$. The generated third harmonic wave interacts again with the fundamental wave through a four wave mixing down-conversion process $3\omega - \omega - \omega = \omega$. The fundamental wave, reconstructed through this two step cascading process is phase shifted by an amount that depends on the size of the phase-mismatch parameter Δk. Such a fundamental wave at the output experiences a NPS. The resulting phase shift depends on the square of the input intensity indicating that this cascading effect has "n_4" behaviour simulating artificially the role of $\chi^{(5)}$ in the process of NPS collection in centrosymmetric media.

The plane wave equations describing the effect of third harmonic generation are:

$$\frac{dA}{dz} = -i\left(\gamma_1 |A|^2 + \gamma_2 |T|^2 + \delta|A|^4\right) A - i\gamma_{13} T \left(A^*\right)^2 \exp(-i\Delta k z), \tag{17a}$$

$$\frac{dT}{dz} = -i\left(\gamma_4 |A|^2 + \gamma_5 |T|^2\right) T - i\gamma_{31} A^3 \exp(i\Delta k z), \tag{17b}$$

where

$$\gamma_{13} = \frac{3\omega}{8c}\frac{\chi^{(3)}}{n_1}, \qquad \gamma_{31} = \frac{3\omega}{8c}\frac{\chi^{(3)}}{n_3},$$

and $\gamma_1, \gamma_2, \gamma_3, \gamma_4$ and δ are nonlinear coupling coefficients responsible for self- and cross-phase modulation.

To obtain the effect of third order cascading, however, one has to solve them (applying the same procedure as in Section 2.1) with respect to the phase of the fundamental beam (Saltiel *et al.* 1997). In the approximation of a small conversion efficiency into the third harmonic wave the output phase of the fundamental wave is

$$\Delta\varphi_1^{\rm NPS} = \Delta\varphi_{1,\rm dir}^{\rm NPS} + \Delta\varphi_{1,\rm casc}^{\rm NPS}, \tag{18a}$$

$$\Delta\varphi_{1,\rm dir}^{\rm NPS} = \left[\gamma_1 a_{10}^2 + \delta a_{10}^4\right] L, \tag{18b}$$

with

$$\Delta\varphi_{1,\rm casc}^{\rm NPS} \approx -\gamma_{13}\gamma_{31}\frac{a_{10}^4}{(\Delta k)^2}\left[\Delta k L - \sin(\Delta k L)\right]. \tag{19}$$

Two important conclusions can be deduced from the result for the NPS given by (19). The NPS has a quadratic dependence on the input intensity $I_1 \propto a_{10}^2$. The product $\gamma_{13}\gamma_{31} \propto \chi^{(3)}$:$\chi^{(3)}$ directly indicates that the NPS is a result of third order cascading, called also $\chi^{(3)}$:$\chi^{(3)}$ cascading. As in second-order cascading the sign of the NPS due to third-order cascading depends on the sign of the phase-mismatch parameter Δk. By comparing with Equation 7 we see the dependence of the nonlinear phase shift on the phase mismatch parameter is the same as for $\chi^{(2)}$:$\chi^{(2)}$ cascading. In fact, Equation 19 can be considered as an additional term $n_4^{\text{casc}} E^4$ in the expansion of the index of refraction with the power of the electric field.

$$n = n_0 + n_2 E^2 + n_4 E^4 + n_4^{\text{casc}} E^4 + \cdots , \quad (20)$$

where n_4^{casc} is defined as

$$n_4^{\text{casc}}(\max) \approx [-\text{sign}(\Delta k)]\, 2n_2^2 \frac{L}{\lambda} . \quad (21)$$

This new kind of saturating nonlinearities, non-Kerr nonlinearities, arising from a phase-matched interaction between the fundamental beam and its third-harmonic can have a dramatic effect on the propagation of spatial solitary waves. Conditions for propagation of solitary waves (both bright and dark) in centro-symmetric media in the presence of third harmonic waves are described by Kivshar (1998).

3 Control by several phase-mismatch parameters.

In this part we consider the second type of nonlinear cascaded processes that involve several processes, each of them with its own phase mismatch parameter. Here we show that these processes play an important role in high order harmonic generation, frequency shifting processes, nonlinear light scattering, the generation of cross polarized waves and nonlinear polarization rotation. Cascaded interactions controlled by several phase-mismatch parameters have been observed in both centro-symmetric and non-centro-symmetric media.

In quadratic media the most efficient are those cascaded interactions that start with the SHG process. Indeed recently the single crystal cascaded interactions listed in the table below have attracted great interest.

Cascaded interaction	First $\chi^{(2)}$ step	Second $\chi^{(2)}$ step	Equivalent process
Third harmonic generation	$\omega + \omega = 2\omega$	$\omega + 2\omega = 3\omega$	$\omega + \omega + \omega = 3\omega$
Fourth harmonic generation	$\omega + \omega = 2\omega$	$2\omega + 2\omega = 4\omega$	$\omega + \omega + \omega + \omega = 4\omega$
Generation of cross polarised wave	$\omega + \omega = 2\omega$	$2\omega - \omega = \omega_\perp$	$\omega + \omega - \omega = \omega_\perp$
Frequency shifting (frequency conversion)	$\omega + \omega = 2\omega$	$2\omega - \omega_a = \omega_b$	$\omega + \omega - \omega_a = \omega_b$

Table 1. *Two step cascaded nonlinear optical interactions starting with second harmonic generation.*

Third harmonic generation in a single quadratic crystal is the most studied cascaded interaction controlled by two phase-mismatching parameters and we will start with this interaction.

3.1 Third harmonic generation in a single quadratic crystal

Third harmonic generation in a single non-centro-symmetric crystal can arise from the direct process $\omega + \omega + \omega = 3\omega$, or can result from the cascading of two second order processes $\omega + \omega = 2\omega$ and $2\omega + \omega = 3\omega$. When the quadratic processes are close to their exact phase matching the direct process is relatively weak and can be neglected and the efficiency of the generated third harmonic is a result mostly of second order cascading. Third harmonic generation in a single $\chi^{(2)}$ crystal can occur under four different phase matching conditions. They can be easily found from plane wave equations describing this process:

$$\frac{dA_1}{dz} = -i\sigma_1 A_2 A_1^* \exp(-i\Delta k_{\mathrm{SHG}} z) - i\sigma_2 A_3 A_2^* \exp(-i\Delta k_{\mathrm{SFM}} z), \tag{22a}$$

$$\frac{dA_2}{dz} = -i\sigma_1 A_1^2 \exp(i\Delta k_{\mathrm{SHG}} z) - i2\sigma_2 A_3 A_1^* \exp(-i\Delta k_{\mathrm{SFM}} z), \tag{22b}$$

$$\frac{dA_3}{dz} = -i3\sigma_2 A_2 A_1 \exp(i\Delta k_{\mathrm{SFM}} z), \tag{22c}$$

where $\Delta k_{\mathrm{SFG}} = k_3 - k_2 - k_1; \qquad \Delta k_{\mathrm{SHG}} = k_2 - 2k_1.$

Indeed if we neglect the depletion of the fundamental and the second-harmonic waves Equations 22 simplify to Equations 23 which can be easily solved analytically.

$$\frac{dA_1}{dz} = 0\,, \tag{23a}$$

$$\frac{dA_2}{dz} = -i\sigma_1 A_1^2 \exp(i\Delta k_{\mathrm{SHG}} z)\,, \tag{23b}$$

$$\frac{dA_3}{dz} = -i3\sigma_2 A_1 A_2 \exp(i\Delta k_{\mathrm{SFG}} z)\,. \tag{23c}$$

The are four situations which permit the efficient generation of the third-harmonic A_3. They are as follows.

- **Case 1.** Phase matching for SHG: $\Delta k_{\mathrm{SHG}} = k_2 - 2k_1 \to 0$.

$$A_3(z) = \frac{i3\sigma_1\sigma_2 A_1^3 z}{\Delta k_{\mathrm{SFG}}} \exp\left(i\Delta k_{\mathrm{SFG}} z\right)\,. \tag{24}$$

- **Case 2.** Phase matching for sum frequency mixing: $\Delta k_{\mathrm{SFM}} = k_3 - k_2 - k_1 \to 0$.

$$A_3(z) = -\frac{i3\sigma_1\sigma_2 A_1^3 z}{\Delta k_{\mathrm{SHG}}}\,. \tag{25}$$

- **Case 3.** Phase matching for the direct third harmonic generation (ω+ω+ω=3ω): $\Delta k_{\mathrm{SFM}} + \Delta k_{\mathrm{SHG}} = k_3 - 3k_1 \to 0$.

  This is the only condition when A_3 will be the result of interference between the cascaded contribution and the direct contribution based on the inherent $\chi^{(3)}$ nonlinearity $\left(\gamma \propto \chi^{(3)}\right)$ in the media.

$$A_3(z) = i\left(\frac{3\sigma_1\sigma_2}{\Delta k_{\mathrm{SHG}}} + \gamma\right) A_1^3 z\,. \tag{26}$$

For the three cases considered above the generated third harmonic intensity depends on the square of the length of the nonlinear media and on the cube of the input intensity. In the literature the biggest efficiency reported is 6% measured in a single BBO crystal for the case (iii) (Banks *et al.* 1999).

The most efficient and interesting case is when both quadratic steps are simultaneously phase matched

- **Case 4:** $\Delta k_{\mathrm{SFM}} = k_3 - k_2 - k_1 \to 0$ and $\Delta k_{\mathrm{SHG}} = k_2 - 2k_1 \to 0$

  In this case the intensity of the third harmonic wave will depend on the fourth power of the length of the nonlinear media!

$$|A_3(z)|^2 = (3\sigma_1\sigma_2)^2 |A_1|^6 z^4 . \tag{27}$$

  For realization of this fourth case, simultaneous phase matching of two processes is required. The methods to do that will be discussed below in Section 4. The highest reported efficiency of third harmonic generation in condition of double phase matching in $LiTaO_3$ is as high as 27% (Zhang *et al* 2001).

For higher conversion efficiencies it is no longer correct to neglect the depletion of the fundamental and second harmonic waves. In this case direct numerical integration of (22) is required. In general the solution gives a periodic change of the third-harmonic efficiency with the increase in the length of the media or input intensity. However, for a properly chosen ratio of the nonlinear coupling coefficients σ_1 and σ_2, the theoretically predicted efficiency is 100% (Egorov and Sukhorukov 1998) as illustrated in Figure 7.

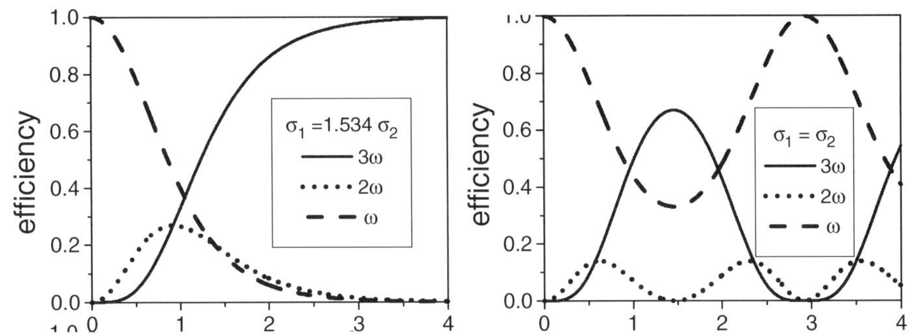

Figure 7. *Results of numerical integration of Equations 22: efficiency of third and second-harmonic waves and the depletion of the fundamental wave as a function of the normalized length of the crystal* $\sigma_1 a_{10} L$ *for the case of exact phase-matching for the two cascaded steps and different ratios for* σ_1/σ_2*, as indicated.*

The effect of high order harmonic generation (and in particular third harmonic generation) in single quadratic crystals is promising for construction of compact frequency converters for diode-pumped solid-state lasers. The second application of these processes is for measurement of the magnitude of cubic nonlinearities. Indeed if the tensors of quadratic nonlinearities are known and we compare the intensity at phase matching conditions (i), (ii) and (iii) we can express the cubic coupling coefficient in terms of second-order coupling coefficients. Cascaded effects allow us to build a calibration link between high order nonlinearities and low order nonlinearities. In this way first estimations of $\chi^{(5)}$ in calcite (Akhmanov *et al.* 1975) and $\chi^{(4)}$ in Lithium formiate crystal (Akhmanov *et al.* 1974) have been done. The same way, by using different phase matching conditions for single crystal third harmonic generation, unknown $\chi^{(2)}$ components can be expressed in terms of known $\chi^{(2)}$ components. This was demonstrated in the work of Petrov *et al.* (1999), where, by measuring the polarization anisotropy of cascaded third-harmonic generation in LBO, unknown $\chi^{(2)}$ components in this crystal have been measured.

3.2 Cross-polarized wave generation in a quadratic crystal.

This is another example of a second-order cascaded interaction controlled by two phase matching parameters listed in Table 1. Indeed (see Figure 8), let us consider a quadratic medium that simultaneously supports Type I (AA–S) and Type II interaction (AB–S); here A and B denote two orthogonal polarizations of the fundamental waves with and S marks the second harmonic field. In the first step with the process AA–S a second harmonic wave is generated; with the second step, the generated second harmonic wave mixed with the fundamental wave is down-converted by the process SA–B to a wave at the fundamental frequency, but with a polarization plane perpendicular to the input

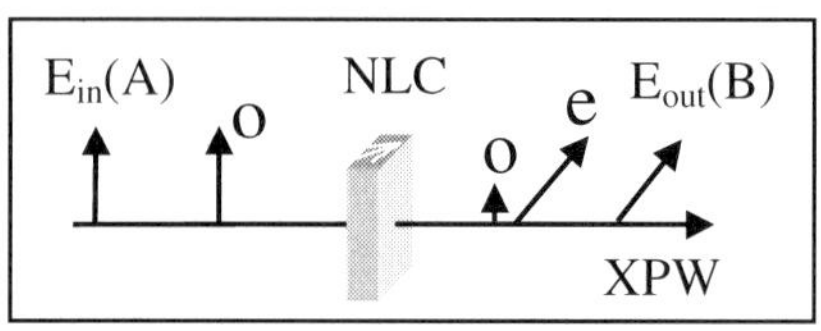

Figure 8. *The idea for cross-polarized wave (XPW) generation via second order cascaded processes. NLC - nonlinear crystal for SHG; o - ordinary; e - extraordinary.*

one. If both processes are exactly phase matched the theoretical prediction is that one can achieve more than 60% conversion from ordinary to extraordinary wave (Saltiel and Deyanova 1999) or vice versa for equal nonlinearities for the two steps. Such a device can be called a “nonlinear polarization rotator”. As for most second-order cascaded processes the effect of cross-polarized wave (XPW) generation has an equivalent cubic process. With the above notation this is the process AAA*–B. In a way similar to that considered for cascaded third-harmonic generation in Section 3.1 the process of XPW generation will be efficient for three phase-matched conditions: (a) phase-matching for the first step—Type I SHG; (b) phase-matching for the second step—Type II SHG; (c) phase-matching for the direct process $\omega + \omega - \omega = \omega_\perp$. The direct process is automatically phase matched along the optic axis of the uniaxial quadratic crystals, where ordinary and extraordinary waves have the same phase velocities. In this direction however the cascaded process of XPW is interfering with the direct process governed by the inherent $\chi^{(3)}$ nonlinearity of the quadratic crystal.

Experimentally the process of XPW generation in a single quadratic crystal was demonstrated in BBO crystal (Petrov *et al.* 2001), which is one of the crystals that

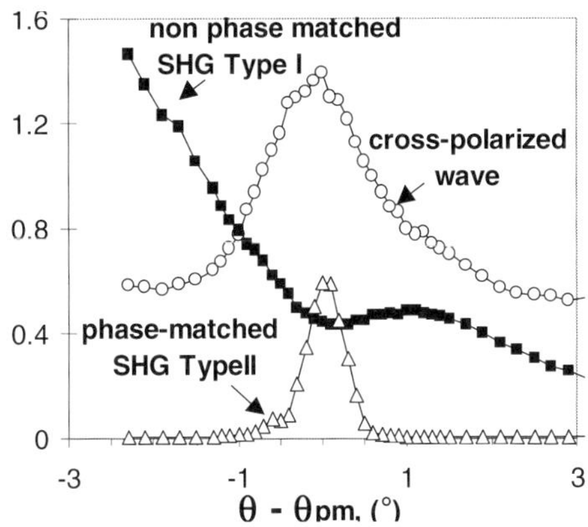

Figure 9. *Experimentally measured cross polarized wave and non-phase matched SH signals (in a.u.) as a function of the deviation of the crystal orientation from the phase-matching angle for Type II SHG. The bottom curve represents the phase-matched Type II SHG signal measured in a separate experiment; (the purpose of this recording is to mark the position of the phase-matching angle for Type II SHG).*

supports the co-existence of Type I and Type II SHG. One of the processes (AA–S) was not phase-matched and that is why the measured overall efficiency was small. The XPW signal was observed when the crystal was tuned to the exact phase matching condition for the second step $\Delta k_{\text{2-step}} = k_{\text{SH}} - k_A - k_B \to 0$. The experimental evidence of generated XPW is shown in Figure 9. The fact that the SH wave is depleted exactly at the point of maximum XPW wave efficiency confirms that this "cubic" process is a result of two-step second order cascading.

Third harmonic generation and XPW generation in a single quadratic crystal are only two of all possible cascaded interactions that lead to generation of new waves. Other important interactions are fourth harmonic generation in single crystals (Akhmanov *et al.* 1974, Kildal and Iseler 1978, Hooper *et al.* 1994) and cascaded single crystal frequency conversion (Chou *et al.* 1998, Gallo and Assanto 1999) which is the very efficient analogue of the cubic frequency shifting through non-degenerate four wave mixing.

Cascaded interactions that lead to generation of new waves can also be realized in centrosymmetric media. In this case we have cascading of two third-order processes. Typical examples are fifth harmonic generation as a result of cascading of :

- (a) third-harmonic generation $\omega + \omega + \omega = 3\omega$ with the phase-matching condition $\Delta k_{\text{1-step}} = k_3 - 3k_1$ and
- (b) four-wave mixing up-conversion $3\omega + \omega + \omega = 5\omega$ with the phase-matching condition $\Delta k_{\text{2-step}} = k_5 - k_3 - 2k_1$.

For the first time such an interaction was realized in $CaCO_3$ by Akhmanov *et al.* (1975). Later many experiments in gases confirmed the important role of cascaded processes in the efforts to generate coherent vacuum ultraviolet and soft X-ray radiation by high order harmonic generation (Misoguti *et al.* 2001). In the next paragraph we will consider another type of cascading of two third order processes.

3.3 Polarization rotation induced by two third-order processes.

In Section 2.3 we showed that in quadratic media cascaded nonlinear optical processes can cause cross-polarized wave generation and can change the polarization state of a wave. The same effect—XPW generation and respectively, induced polarization rotation and change of the polarization state—can also be caused by cascading two third-order processes in centro-symmetric media when the laser light is propagating along the z-axis. The two cascaded cubic sub-processes are self-phase modulation and four wave mixing.

To describe this effect let us consider, in the slowly varying envelope approximation, plane wave propagation equations, neglecting linear absorption and choosing a coordinate system defined by the directions of the crossed polarizer and analyser. The equations for self-phase modulation of the co-polarized component A and generation of XPW with amplitude B, under the condition where $|B| \ll |A|$ (i.e. neglecting the possible effect on cross-phase modulation of wave A caused by the presence of wave B and self-phase modulation of wave B), are (Minkovsky *et al.* 2002):

$$\frac{dA}{dz} = i\gamma_{\parallel} |A|^2 A \qquad \text{with} \quad \gamma_{\parallel} = \gamma_0 \left[1 - (\sigma/2)\sin^2(2\beta)\right], \tag{28a}$$

$$\frac{dB}{dz} = i\gamma_{\perp} |A|^2 A \qquad \text{with} \quad \gamma_{\perp} = -\gamma_0 (\sigma/4)\sin(4\beta), \tag{28b}$$

where β is the angle of the input polarization plane with respect to the [100] axis of the crystal, $\gamma_0 = (6\pi/8\lambda n)\,\chi^{(3)}_{xxxx}$ and σ, the anisotropy of the $\chi^{(3)}$ tensor, is given by

$$\sigma = \frac{\chi^{(3)}_{xxxx} - \left(2\chi^{(3)}_{xyxy} + \chi^{(3)}_{xxyy}\right)}{\chi^{(3)}_{xxxx}}.$$

In the general case $\gamma_{\parallel}$ and $\gamma_{\perp}$ are complex: $\gamma_{\parallel} = \gamma'_{\parallel} + i\gamma''_{\parallel}$ and $\gamma_{\perp} = \gamma'_{\perp} + i\gamma''_{\perp}$. The solution to Equations 28 with initial conditions $B(0) = 0$ and $A(0) = A_0$ (A_0 real) is:

$$A = A_0 \exp\left(i\gamma_{\parallel} A_0^2 L\right), \tag{29a}$$

$$B = A_0 \left(\gamma_{\perp}/\gamma_{\parallel}\right)\left[\exp(i\gamma_{\parallel} A_0^2 L) - 1\right]. \tag{29b}$$

The complex amplitudes B and A define the ratio $R = B/A$. The angle position δ_0 of the main ellipse axis and the ellipticity angle ϵ can then be found using Equations (15a,b). Assuming that δ_0, $\gamma'_{\parallel} A_0^2 L$, $\gamma''_{\parallel} A_0^2 L$, $\gamma'_{\perp} A_0^2 L$ and $\gamma''_{\perp} A_0^2 L$ are all small we get:

$$\delta_0 \approx \frac{1}{2}\gamma'_{\perp}\gamma'_{\parallel} A_0^4 L^2, \tag{30}$$

$$\epsilon^2 \approx \left(\gamma'_{\perp} A_0^2 L\right)^2. \tag{31}$$

Equation 31 indicates that the magnitude of the XPW efficiency is proportional to the square of the input intensity in accordance with the previous understanding of this effect. However Equation 30 predicts that the polarization rotation depends on the square of the input intensity and on the real part of the cubic nonlinearity. This is in contrast to the previously reported theoretical and experimental investigations obtained at relatively lower intensity and a bigger contribution of two photon absorption effects.

The experiment of Minkovsky *et al.* (2002) with the cubic crystal yttrium vanadate (YVO_4) studied this interesting cascaded effect. The output polarization state was defined

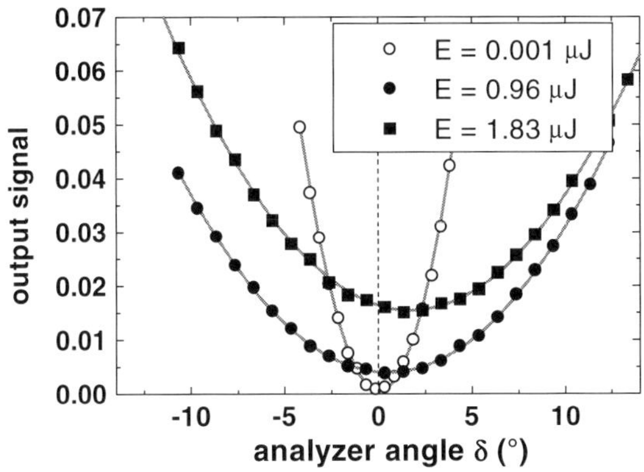

Figure 10. *Output signal (normalized to input intensity) with respect to angular detuning of the analyzer for three different input energies. Zero position corresponds to exactly crossed polarizer and analyzer. Solid lines are a parabolic fit.*

by extinction curve measurements, performed by introducing small angular deviations δ from the exact crossed position between the two polarizers. For analyzing the sign of the ellipticity, a $\lambda/4$ quartz plate with its c-axis parallel to the input polarization is introduced in front of the analyzer. The maximum XPW efficiency achieved is 1.2%.

Typical results of extinction curve measurements are shown in Figure 10 for two input intensities. The solid lines are parabolic fits of the type $Y = C_1(\delta - \delta_0)^2 + C_2$. The parameter δ_0 is a measure of the induced polarization rotation angle and $|\epsilon| = \sqrt{C_2/C_1}$ is a measure of the modulus of induced ellipticity. For the curve recorded at input energy 0.96μJ the fitting gives $\delta_0 = +8.7$mrad and $|\epsilon| = 64$mrad , and for the curve recorded at input energy 1.83μJ the fitting gives $\delta_0 = +30$mrad and $|\epsilon| = 122.5$mrad . Additional tests with a $\lambda/4$ plate proved that $\epsilon < 0$. As seen from Figure 10 the dependence of δ_0 on intensity measured for three values of input intensity is quadratic in good accordance with the theory described above. By measuring the amount of anisotropy of two-photon absorption it was estimated that induced rotation due to two-photon absorption cannot explain the recorded polarization rotation angles.

From Equation (30) we see the role of the cascaded processes. The expression for the induced polarization rotation angle δ_0 is proportional to the $\chi^{(3)}{:}\chi^{(3)}$ term. The reason for this "cascaded rotation" lies in the self-phase modulation effect experienced by the fundamental wave A. At high input intensities the phase shift between B and A will be different from $\pi/2$ and intensity dependent: $\arg(R) = (\pi - \gamma_{\parallel} A_0^2 L)/2$. This leads to a complex value of R and consequently to induced polarization rotation. It is the "cascading" of self-phase modulation in the fundamental wave and the generation of a cross-polarized wave by four-wave mixing that causes the induced polarization rotation. The effect described can be observed along crystallographic axes of cubic crystals or along the Z axis of tetragonal crystals. The condition for observing the effect is that the imaginary part of the coupling coefficient $\gamma_{\perp}''$ responsible for the four wave mixing process should be very small in combination with a relatively high intensity, so that $\gamma_{\perp}''/\left(\gamma_{\perp}'\gamma_{\parallel}' A_0^2 L\right) \ll 1$. This condition is easy to satisfy when the fundamental photon energy is below half the band gap energy. In our experiment the fundamental photon energy is near the edge of the half band gap energy but the above inequality is well fulfilled due to the high intensity used in the experiment described.

The effect of the $\chi^{(3)}:\chi^{(3)}$ cascaded induced rotation of the polarization is unique. This is not just a simple mimic of $\chi^{(5)}$ contribution—the effect of "cascaded" nonlinear polarization rotation has no $\chi^{(5)}$ analogue.

Knowledge of this effect is important for avoiding unwanted depolarization effects. For example, YVO_4 is known to be an optically active element for solid state lasers. Also this effect can be used for measuring the magnitude and sign of the components of the $\chi^{(3)}$ tensor and therefore the anisotropy of the $\chi^{(3)}$ tensor. The attractive aspect of such a method lies in the single beam technique with pumping and recording at a single wavelength and that the measurement is based on recording anglular rotation values. On the other hand, if the cubic nonlinearity is known, then the measurement of the induced polarization rotation angle can be used for calculating the intensity of the beam. Finally polarization-switching devices, can also be designed.

The common characteristic of the interactions considered in this section is that they consist of several steps, each with its own phase-matching condition. For best efficiency simultaneous phase matching of all steps is required. Some of the interactions (e.g. cascaded frequency conversion and the effect of cascaded polarization rotation in centro-symmetric media) do not require any special effort to fulfil double phase-matching because it is achieved automatically. For realization of the other interactions like, third-harmonic, fourth-harmonic and XPW-generation in a single quadratic crystal the nonlinear media must be prepared in a special way to support double or multiple phase-matching conditions. Methods for achieving double phase-matching are considered in the next section.

4 Methods of achieving simultaneous phase-matching

In this section we will present a brief overview of different methods for simultaneously phase-matching for several parametric optical processes in engineered structures with modulated second-order nonlinear susceptibility. In particular, we will discuss the possibility of double phase-matching (DPM) in both uniform and non-uniform quasi-phase-matched (QPM) periodic optical superlattices and also in the recently fabricated two-dimensional nonlinear photonic crystals.

4.1 DPM with uniform QPM structures—the method of two commensurable periods.

In a bulk homogeneous nonlinear crystal the quadratic nonlinearity is constant everywhere. Several methods (Fejer *et al.* 1992) have been suggested and employed in order to create a periodic change of the sign of the second-order nonlinear susceptibility $d^{(2)}$ in QPM structures as shown in Figure 11a. From the mathematical point of view, such a periodic sequence of two domains can be described by the periodic function

$$d(z) = d_0 \sum_{m \neq 0} g_m \exp\left(iG_m z\right), \tag{32}$$

$$g_m = (2/m\pi)\sin\left(\pi m D\right), \tag{33}$$

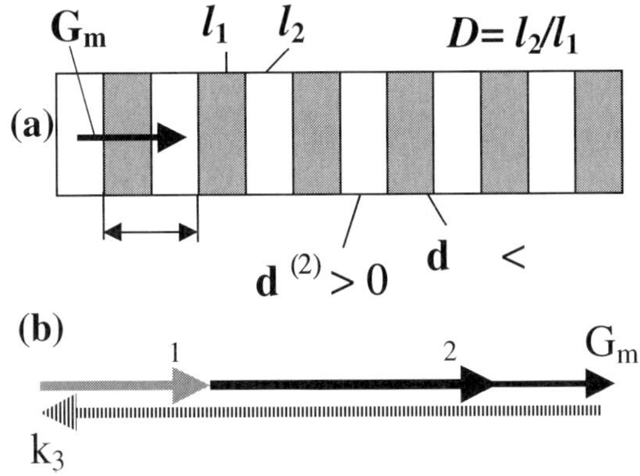

Figure 11. *QPM structure (a) and the principle (b) of phase-matching compensation with the help of the grating reversal vectors.*

where $G_m = 2\pi m/\Lambda$ is the m-th reciprocal vector of the QPM structure. The uniform QPM structure is characterized by a set of the reciprocal vectors: $\pm 2\pi/\Lambda$, $\pm 4\pi/\Lambda$, $\pm 6\pi/\Lambda$, $\pm 8\pi/\Lambda$, ..., which can be used to achieve the phase-matching conditions. The manner in which the reciprocal vector compensates the mismatch of a given nonlinear process is shown on Figure 11b, where the process of sum-frequency mixing is taken as an example. The integer number m (that can be both *positive* or *negative*) is called *the order of the QPM phase-matching*. According to Equation 33, the smaller the order of the QPM reciprocal wave vector, the larger is the effective nonlinearity. If the filling factor D = 0.5, the effective quadratic nonlinearities (proportional to the parameter $d_0 g_m$) that correspond to the even orders QPM vectors vanish. Importantly, such uniform QPM structures can be used for simultaneous phase-matching of two parametric processes (Pfister *et al.* 1997).

As an example, we consider cascaded third-harmonic generation in a single crystal under the condition that all interacting waves are *collinear* to the reciprocal wave vectors of the QPM structure. We denote the mismatches of the nonlinear material without modulation of the quadratic nonlinearity ("bulk mismatches") as Δb_1 and Δb_2 (i.e. we set $\Delta b_1 = k_2 - 2k_1$ and $\Delta b_2 = k_3 - k_2 - k_1$) and choose the period Λ of the QPM structure in order to satisfy the phase-matching conditions $G_{m_1} = -\Delta b_1$ and $G_{m_2} = -\Delta b_2$. Two processes, SHG and sum frequency mixing are characterized by the wave-vector mismatches $\Delta k_1 = k_2 - 2k_1 + G_{m_1} = 0$ and $\Delta k_2 = k_3 - k_2 - k_1 + G_{m_2} = 0$, respectively, and they become simultaneously phase-matched for this particular choice of the QPM period. A drawback of this method is that it can satisfy simultaneously two phase-matching conditions for *discrete values* of the optical wavelength only. The values of the fundamental wavelength λ for the double phase-matching condition can be found from the relation

$$\Delta b_2/m_2 - \Delta b_1/m_1 = 0, \tag{34}$$

since both Δb_1 and Δb_2 are functions of the wavelength. For a chosen pair of integer numbers (m_1, m_2) the QPM period Λ is found from the relation $\Lambda = 2\pi\,|m_1/\Delta b_1|$ or $\Lambda = 2\pi\,|m_2/\Delta b_2|$. Some tunability is possible due to the dependence of Δb_1 and Δb_2 on temperature.

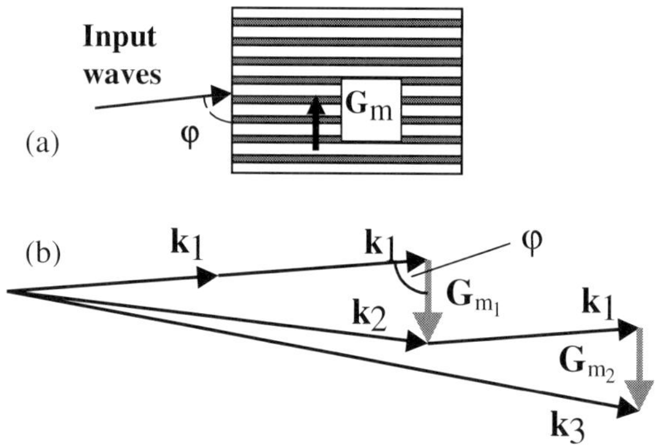

Figure 12. *Non-collinear interaction with QPM structure (a) and the principle (b) of double-phase-matching for the third harmonic generation via sum frequency mixing of fundamental and second harmonic waves.*

4.2 DPM with uniform QPM structures—the non-collinear case.

Phase-matching is possible even in the case when some of the waves propagate under a certain angle to the direction of the reciprocal vectors of the QPM structure, that is, for the non-collinear case. The advantage of this method is that double phase-matching can be realized in a broad spectral range. Such a type of non-collinear interaction will be efficient for the distances corresponding to the overlap of the interacting beams.

Let us consider the cascaded single crystal third-harmonic generation process in this type of geometry (see Figure 12). As has been discussed above, the simultaneous phase-matching of two parametric processes $\omega + \omega = 2\omega$ and $\omega + 2\omega = 3\omega$ is required in this case. We assume the fundamental wave at the input. As shown in Figure 12b, for the first process, SHG, the phase-matching is achieved by the reciprocal vector $\mathbf{G}_{m_1}$ and the generated SH wave has a wave-vector $\mathbf{k}_2$ which is not collinear to the fundamental wave: $2\mathbf{k}_1 + \mathbf{G}_{m_1} = \mathbf{k}_2$. The second process, sum frequency mixing, is phase-matched by the vector $\mathbf{G}_{m_2}$ such that $\mathbf{k}_1 + \mathbf{k}_2 + \mathbf{G}_{m_2} = \mathbf{k}_3$. From Figure 12 we can calculate the period of the QPM structure that allows double phase-matching of two processes—SHG and sum frequency mixing (Saltiel and Kivshar 2000a):

$$\Lambda = 2\pi \sqrt{\frac{2(k_3^2 - 9k_1^2)m_1 - 3(k_2^2 - 4k_1^2)(m_1 + m_2)}{(m_1 + m_2)(2m_2 - m_1)m_1}}. \qquad (35)$$

Because of the two degree of freedom: the angle of non-collinearity φ and the QPM period Λ , the double-phase matching can be realized in a wide spectral range.

4.3 DPM with non-uniform QPM structures

Several non-uniform QPM structures have been proposed and experimentally tested. One type is so called phase-reversed QPM structure. The idea of the phase-reversed QPM structures (Chou *et al.* 1999) is illustrated in Figure 13a. Such a structure can be

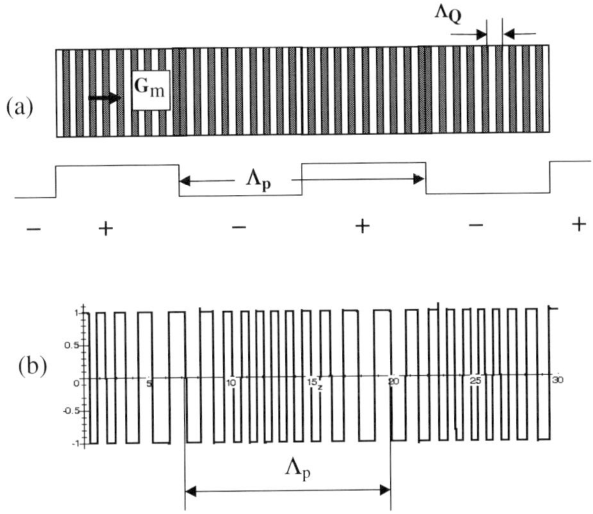

Figure 13. *Non-uniform QPM structures: (a) phase reversed QPM structure (Chou et al. 1999) and (b) periodically chirped QPM structure (Bang et al. 1999).*

described as a sequence of many equivalent uniform short QPM sub-structures with the length $\Lambda_p/2$ connected in such a way that at the join, the two end layers have the same sign of the quadratic nonlinearity. Any two neighbouring junctions have opposite sign of the $\chi^{(2)}$ nonlinearity. In other words, the phase-reversed QPM structure is a kind of uniform QPM structure with a change of the domain phase by π characterized by the other (larger) period Λ_p.

The modulation of the quadratic nonlinearity $d(z)$ in the phase-reversed QPM structure with the filling factor $D = 0.5$ can be described by the response function

$$d(z) = d_0 \, (-1)^{\mathrm{int}\left(2z/\Lambda_Q\right)} \, (-1)^{\mathrm{int}(2z/\Lambda_p)} \, , \tag{36}$$

and can be presented in the form of the Fourier series,

$$d(z) = d_0 \sum_{-\infty}^{+\infty} g_l e^{(iG_l z)} \sum_{-\infty}^{+\infty} g_m e^{(iF_m z)} = d_0 \sum_{-\infty}^{+\infty} \sum_{-\infty}^{+\infty} g_{l,m} e^{\left(iG_{l,m} z\right)}, \quad l, m \neq 0 \, , \tag{37}$$

where

$$g_m = \frac{2}{m\pi}, \quad g_{l,m} = g_l g_m, \quad G_l = \frac{2\pi l}{\Lambda_Q}, \quad F_m = \frac{2\pi m}{\Lambda_p}, \quad G_{l,m} = \frac{2\pi}{\Lambda_Q} l + \frac{2\pi}{\Lambda_p} m.$$

Another type non-uniform QPM structure is proposed by Bang *et al.* (1999). This structure is illustrated in Figure 13b and is characterized by the QPM period Λ_Q that is a periodic function of z:

$$\Lambda = \Lambda_Q + \epsilon_0 \cos\left(2\pi z/\Lambda_p\right) . \tag{38}$$

The formulae for double phase-matching are the same for both the phase reversed and periodically-chirped QPM structures shown in Figure 13.

$$\Lambda_Q = \left| \frac{2\pi\,(l_2 m_1 - l_1 m_2)}{m_2 \Delta b_1 - m_1 \Delta b_2} \right|, \tag{39}$$

$$\Lambda_p = \left| \frac{2\pi\,(l_2 m_1 - l_1 m_2)}{l_2 \Delta b_1 - l_1 \Delta b_2} \right|. \tag{40}$$

Equations (39,40) allow the two periods of these non-uniform QPM structures to be found. The structures designed in this way will support double phase matching of any two nonlinear optical processes with bulk phase mismatches Δb_1 and Δb_2. In very recent work Liu *et al.* (2001, 2002) demonstrated that phase-reversed structures are suitable for single-crystal third-harmonic generation.

4.4 DPM: quasi-periodic and aperiodic optical superlattices

Another method of double-phase-matching, investigated both theoretically and experimentally, is based on the use of the quasi-periodic optical superlattices (QPOS) (Zhu and Ming 1999) and a-periodic optical superlattices (Zhang *et al.* 2000). In most of the cases QPOSs are built with two-component blocks (A and B) aligned in a Fibonacci (or more general quasi-periodic) sequence (see Figure 14). Each of the blocks consists of two layers

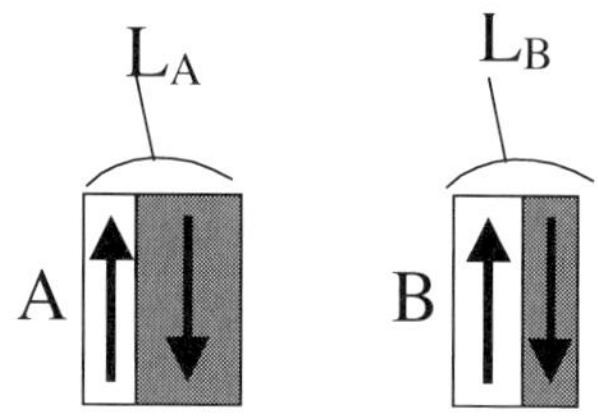

Figure 14. *The two basic blocks for the construction of quasi-periodic and aperiodic optical superlattices.*

with the opposite sign of the quadratic nonlinearity. Then the two blocks are aligned in Fibonacci (ABAABABA....) or different sequence. We illustrate the possibility of double-phase-matching in such a structure taking as an example, the structure that consists two blocks, aligned in a generalized sequence (Fradkin-Kashi *et al.* 1999, 2002). The modulation of the quadratic nonlinearity $d(z)$ can be described by the following Fourier expansion:

$$d(z) = d_0 \sum_{m,n} f_{m,n} \exp\left(i G_{m,n} z\right), \tag{41}$$

where the reciprocal vectors are defined as $G_{m,n} = 2\pi\,(m + n\tau)\,/S$ and $S = \tau L_A + L_B$.

For phase-matching of two processes with bulk mismatches Δb_1 and Δb_2, we solve the system of equations

$$G_{m_1,n_1} = 2\pi \left(\frac{m_1 + n_1\tau}{S} \right) = \Delta b_1, \tag{42a}$$

$$G_{m_2,n_2} = 2\pi \left(\frac{m_2 + n_2\tau}{S} \right) = \Delta b_2, \tag{42b}$$

to find the unknowns S and τ. Exact values for L, L_B and the lengths of the layers have to be found by maximizing f_{m_1,n_1} and f_{m_2,n_2}. This structure allows simultaneous phase-matching in a broad spectral range without constrains on the ratio $\Delta b_2/\Delta b_1$. Equations (42) are also valid for Fibonacci type QPOS but, since τ is fixed, double phase-matching can be satisfied for a limited number of the wavelengths.

In several recent papers a-periodic optical superlattices were used for simultaneous phase-matching of several processes. Presently the record efficiency in single crystal third harmonic generation with the use of generalized two-component QPOS is 27% (Zhang *et al.* 2001).

4.5 2D $\chi^{(2)}$ nonlinear photonic crystals and DPM

Nonlinear photonic crystals (NPC) with a homogeneous (linear) refractive index and containing a two-dimensional (2D) periodic lattice of volumes with reversed signs of the quadratic nonlinear susceptibility (Berger 1998) can be effectively employed as a host medium for many different types of CNOP. A schematic diagram of 2D NPC is shown in Figure 15. The simple way to obtain the phase-matching conditions for 2D NPC is

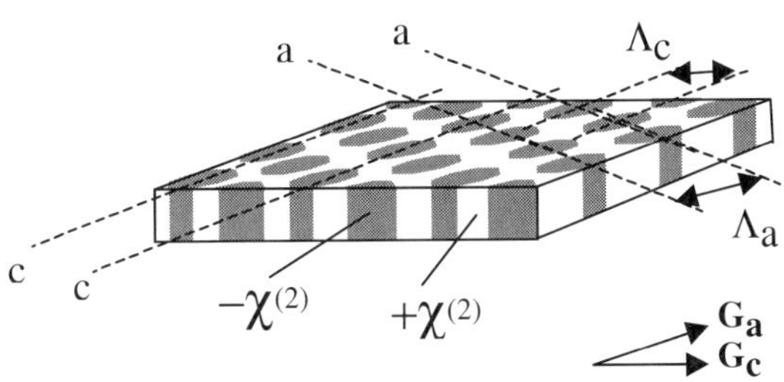

Figure 15. *A 2D Nonlinear photonic crystal is formed in a periodic lattice by volumes (marked in gray) with reversed sign of the* $\chi^{(2)}$. $\mathbf{G_a}$ *and* $\mathbf{G_c}$ *are the reciprocal vectors of the 2D structure.*

to use a reciprocal lattice formed by vectors $\mathbf{G_a}$ and $\mathbf{G_c}$, defined as $|G_a| = 2\pi/\Lambda_a$ and $|G_c| = 2\pi/\Lambda_c$ and directions—the normals to the lines **a** and **c**, respectively. Vectors $\mathbf{G_a}$ and $\mathbf{G_c}$ are used to construct the so called reciprocal lattice, that is very useful for analyzing phase matching conditions in these types of photonic structures.

For a hexagonal lattice, we find $\Lambda_a = \Lambda_c = a\sqrt{3}/2$, where a is the distance between the centres of two neighbouring volumes with inverted sign of $\chi^{(2)}$. For an asymmetric unit cell $\Lambda_a \neq \Lambda_c$ the two basic reciprocal vectors $\mathbf{G_a}$ and $\mathbf{G_c}$ have different modules that depend on the lattice spacing a, asymmetry ϵ and the angle δ of the unit cell (Saltiel

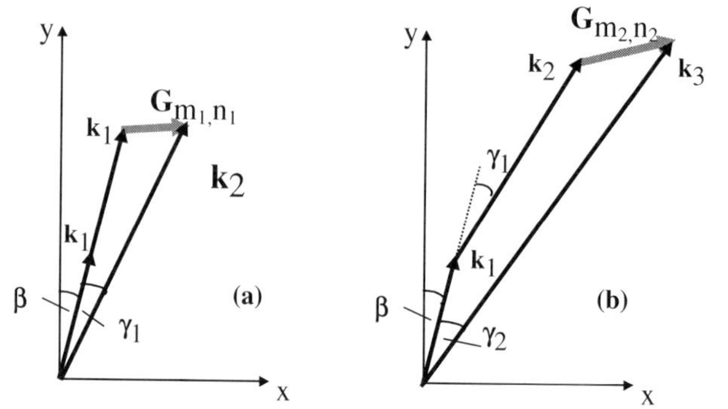

Figure 16. *Double phase-matching for single crystal third harmonic in 2D NPC: (a) the process* $\omega + \omega = 2\omega$ *phase-matched by the reciprocal vector* $\mathbf{G_{m_1,n_1}}$ *and (b) the process* $\omega + 2\omega = 3\omega$ *phase-matched by the reciprocal vector* $\mathbf{G_{m_2,n_2}}$.

and Kivshar 2000b). All reciprocal vectors of the 2D NPC crystal are formed by a simple rule, $\mathbf{G_{m,n}} = \mathbf{mG_c} + \mathbf{nG_a}$. Any two vectors of this set $\{\mathbf{G_{m,n}}\}$ can be used to compensate for the bulk mismatches $\mathbf{\Delta b_1}$ and $\mathbf{\Delta b_2}$; however in most of the cases such phase-matching conditions assume non-collinear interactions. Phase-matching conditions for single crystal cascade third harmonic generation processes are shown in Figure 16. The first step $\omega + \omega = 2\omega$ is phase-matched by the reciprocal vector $\mathbf{G_{m_1,n_1}}$ and the second step $\omega + 2\omega = 3\omega$ is phase-matched by the reciprocal vector $\mathbf{G_{m_2,n_2}}$. Phase-matching is achieved by choosing a, the lattice spacing, and β, the angle of incidence. 2D NPC can be also used to simultaneously phase-match *three nonlinear processes*, for example, second, third, and fourth-harmonic generation (Saltiel and Kivshar 2000b) or generation of pair SH waves and fourth harmonic generation (de Sterke *et al.* 2001). Experimentally, the simultaneous second-, third-, and fourth-harmonic generation was recently observed in poled $LiNbO_3$ 2D NPC (Broderick *et al.* 2000).

5 Cascaded second-order processes in 2D NPC

5.1 Collinear third and fourth-harmonic generation in 2D NPC

In the general case when a nonlinear process is realized in 2D NPC, as shown above, all the interacting waves are non-collinear. When the input beam(s) have finite width the non-collinearity of the interaction leads to reduced efficiency. However with special design of the unit cell it is possible to construct cascaded interactions in which the final generated wave is collinear to the input fundamental wave. Such interactions for fourth harmonic generation (de Sterke *et al.* 2001) and third harmonic generation (Karaulanov and Saltiel 2002) have been proposed. Both proposed interactions start with two channels of SHG that propagate at small angles at both sides of the direction of the fundamental wave. The possibility for two-channel harmonic- generation is a characteristic only for 2D NPC, when the unit cell is symmetrical (i.e. asymmetry $\epsilon = 1$). For this kind of 2D NPC in certain directions $|\mathbf{G_{m,n}}| = |\mathbf{G_{m,-n}}|$ and when these directions are used for the propagation of the

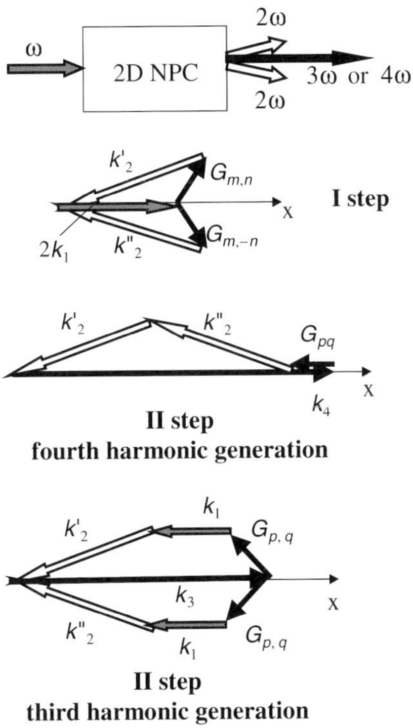

Figure 17. *Double phase-matching schemes for the generation of collinear third and fourth-harmonic in single 2D NPC.*

fundamental beam two channel SHG can be generated. The idea for collinear third and fourth harmonic generation in 2D NPC is clarified in Figure 17. The two channel SHG results in four times higher efficiency for the process of SHG (in comparison with single channel and suggesting the same nonlinearity). The higher efficiency for the SH process results in a higher efficiency for the total process of fourth-harmonic generation. As seen from Figure 17 the second step of the suggested scheme for third-harmonic generation is also "two channeled", and should therefore lead to increased efficiency. Indeed, for the process of fourth-harmonic generation on the assumption of non-depletion of the SH wave due to generated fourth harmonic wave one finds overall efficiency (de Sterke *et al.* 2001):

$$\eta_{4\omega} = \sigma_2^2 \left[A_0 L - \frac{1}{\sqrt{2}\sigma_1} \tanh\left(\sqrt{2}\sigma_1 A_0 L\right)\right]^2, \tag{43}$$

where σ_1 and σ_2 are the two coupling coefficients that control the first and the second step. For comparison, the result for the single-channel case can be obtained from Equation (43) by replacing $\sqrt{2}\sigma_1$ with σ_1.

5.2 Wavelength routing with the interchange of information

The non-collinearity as a characteristic for the interactions in 2D NPC can be useful for interactions where splitting of several information channels is required. Such an example is the reported experimental investigation of wavelength routing with the interchange of information (Chowdhury *et al.* 2000, 2001).

In the proposed scheme two information signals one at wavelengths ω_a, ω_b, modulated with information $\mathbf{I_1}$, $\mathbf{I_2}$ respectively, and pump beam enter a 2D NPC. As a result of two simultaneously phase matched second order frequency mixing processes at the output of the crystal two deflected signals will be generated at the same frequencies, but with exchanged information: the output deflected signal at wavelength ω_a is modulated with information $\mathbf{I_2}$ and the output deflected signal at wavelength ω_b carries information $\mathbf{I_1}$. The balance of the wave-vectors and the frequencies is given by :

$$k_p - k_a = k_b' + G_{ij}, \tag{44a}$$
$$k_p - k_b = k_a' + G_{mn}, \tag{44b}$$
$$\omega_p - \omega_a = \omega_b, \tag{45a}$$
$$\omega_p - \omega_b = \omega_a, \tag{45b}$$

where the prime indicates the wave-vectors of the output signal waves and $\mathbf{G_{ij}}$ and $\mathbf{G_{mn}}$ are the two reciprocal vectors that phase match the two interactions. The equations that describe these double phase-matched interaction are:

$$\frac{dA_1}{dz} = -i\sigma P B^*, \tag{46a}$$
$$\frac{dA}{dz} = -i\sigma P B_1^*, \tag{46b}$$
$$\frac{dB_1}{dz} = -i\sigma P A^*, \tag{46c}$$
$$\frac{dB}{dz} = -i\sigma P A_1^*. \tag{46d}$$

The solution of this system gives not only exchange of information, but also amplification of the two output deflected signals (A_1, B_1) and depletion (T) of the output non-deflected signals (A, B):

$$\eta_{A_1} = (A_1/B_0)^2 = \eta_{B_1} = (B_1/A_0)^2 = \sinh^2(\sigma P x), \tag{47}$$
$$T_A = (A/A_0)^2 = T_B = (B/B_0)^2 = \cosh^2(\sigma P x). \tag{48}$$

5.3 All-optical deflection by second-order cascading in 2D NPC

In our last example we will consider the possibility of the all optical pump induced deflection of signals in 2D NPC (Saltiel and Kivshar 2002). The idea is shown in Figure 18a. Input pump and probe waves are at the same wavelength and are polarized in perpendicular planes. The result of the cascaded interaction is that the signal is deflected only when the strong pump is present. In this sense this device can be considered to be a nonlinear optical polarizer that is controlled by the pump. The cascading interaction includes two steps: in the first one the pump is frequency-doubled with a process that is phase matched with a reciprocal vector collinear to the pump wave-vector. A generated SH wave collinear with the pump mixes with the input signal (also collinear with the pump) generating a deflected signal at the same frequency and polarized the same way as the input one. The

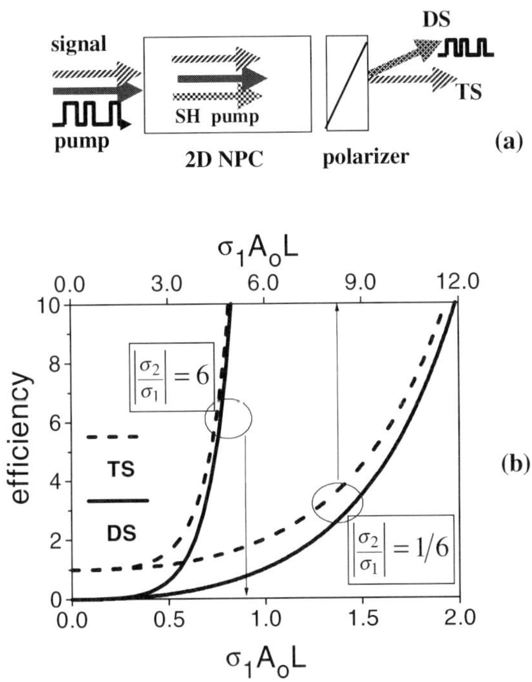

Figure 18. *(a) Schematic of the suggested cascading interaction for deflection in 2D NPC. Signal and pump are cross polarized. TS, transmitted signal; DS, deflected signal. (b) Efficiency of the deflected (DS) and non-deflected (TS) signals relative to normalized input pump amplitude for two different values of the ratio* σ_2/σ_1.

reason for the deflection is that phase matching is achieved with an arbitrarily oriented reciprocal vector. The balance of the wave-vectors and the frequencies for the two steps discussed is given by:

$$\omega_p + \omega_p = 2\omega_p = \omega_s, \tag{49a}$$

$$\omega_2 - \omega_s = \omega_s', \tag{49b}$$

$$k_p + k_p = k_2 + G_{ij}, \tag{50a}$$

$$k_2 = k_s + k_s' + G_{mn}. \tag{50b}$$

The amplitude of the deflected and transmitted signal is given by

$$\eta_s\,(\eta_{s'}) = \frac{1}{4}\left\{\,[\mathrm{sech}(\sigma_1 A_p L)]^{\,|\sigma_2/\sigma_1|} \pm [\cosh(\sigma_1 A_p L)]^{\,|\sigma_2/\sigma_1|}\right\}^2, \tag{51}$$

where $\eta_s = [A_s(L)/A_s(0)]^2$, A_p is the input pump's amplitude, $A_s(0)$ is the input signal's amplitude, L is the crystal length; σ_1 and σ_2 are the two coupling coefficients that control the first and the second step ; signs "+" and "−" correspond to transmission efficiency of the non-deflected and deflected output signals, respectively. The dependences obtained from Equation (51) are shown on Figure 18b. It is seen that both deflected and non-deflected signals at certain levels of the pump and length of the medium will be amplified.

6 Conclusion

In conclusion, in this chapter a systemisation and examples of cascaded nonlinear optical processes have been presented. One subset of cascaded interactions that are controlled by a single phase mismatch parameter, are responsible for self interaction effects like self modulation, induced polarization rotation, soliton propagation, compression etc. Respectively, such interactions can be used for construction of devices for routing, all-optical switching, mode locking and others.

The second group of cascaded nonlinear optical processes are those that are controlled by two phase-matched parameters. These types of interactions are mainly responsible for generating new waves in a single nonlinear medium. For example, recently the interest toward third-harmonic generation in a single quadratic crystal has been extremely strong due to the desire to enhance compact diode-pumped femtosecond systems with integrated single crystal frequency converters in UV range. Other applications of the double-phase matched interactions are fourth-harmonic generation and frequency conversion for optical communication, cross-polarized wave generation. For best efficiency both phase matched conditions should be satisfied. A review of the methods for achieving simultaneous phase-matching was also presented in this chapter. In several experimental works it has already been proved that phase-reversed QPM gratings, quasi-periodical optical superlattices and two-dimensional nonlinear photonic crystals are suitable structures that can be artificially designed to achieve double phase matching. Several promising cascaded interactions that can exist only in two-dimensional nonlinear photonic crystals have been discussed.

Acknowledgments

The author is greatly indebted to the following colleagues and graduate students, who made major contribution to this work: Kaloian Koynov, Georgi Petrov, Nikolay Minkovsky, Ivan Buchvarov, Yana Deyanova , Stoian Tanev; Yuri S. Kivshar, Andrey A. Sukhorukov, Tristram J. Alexander; Jean Etchepare, Olivier Albert; C Martijn de Sterke and Alan Boardman. The author acknowledges the financial support of the Bulgarian Science Fund (grant F-1201).

References

Akhmanov S A and Khokhlov R V, 1964, *Problems of Nonlinear Optics*, (Akad. Nauk USSR, Moscow) [English Edition 1973, Gordon and Breach, New York].

Akhmanov S A, Dubovik A N, Saltiel S M, Tomov I V, Tunkin V G, 1974, 'Fourth order nonlinear optical effects in LFM crystal' , *Pis'ma Zh. Eksp. Teor. Fiz.* **20** 264 [English translation in JETP Lett. **20** (1974) 117].

Akhmanov S, Martinov V, Saltiel S, Tunkin V, 1975, 'Nonresonant Six Photon Processes in Calcite Crystal', *Pisma v JETF* **22** 143 [English translation in JETP Lett. **22** 65]

Assanto G, Stegeman G, Sheik-Bahae M and Van Stryland E W, 1995, 'Coherent interactions for all optical signal processing via quadratic nonlinearities', *IEEE J.Quantum. Electron.* **31**, 673.

Azzam R.M., Bashara N.M., 1977, *Ellipsometry and polarized light* (North-Holland Publ.).

Bang O, Clausen C B, Christiansen P L, and Torner L, 1999, 'Engineering competing nonlinearities', *Opt. Lett.* **24** 1413.

Banks P S, Feit M D, and Perry M D, 1999, 'High-intensity third-harmonic generation in beta barium borate through second-order and third-order susceptibilities', *Opt. Lett.* **24** 4.

Berger V, 1998, 'Nonlinear Photonic Crystals', *Phys. Rev. Lett.* **81** 4136.

Broderick N G R, Ross G W, Offerhaus, H L, Richardson D J, and Hanna D C, 2000, 'Hexagonally Poled Lithium Niobate: A Two-Dimensional Nonlinear Photonic Crystal', *Phys. Rev. Lett.* **84** 4345.

Buchvarov I, Saltiel S, Iglev Ch, Koynov K, 1997, 'Intensity dependent change of the polarization state as a result of nonlinear phase shift in type II frequency doubling crystals', *Opt. Commun.* **141** 173.

Chou M H, Hauden J, Arbore M A and Fejer M M, 1998, '1.5μm-band wavelength conversion based on difference-frequency generation in LiNbO3 waveguides with integrated coupling structures', *Opt. Lett.* **23** 1004.

Chou M H, Parameswaran K R, Fejer M M, and Brener I, 1999, 'Multiple-channel wavelength conversion by use of engineered quasi-phase-matching structures in LiNbO3 waveguides', *Opt. Lett.* **24** 1157.

Chowdhury A, Hagness S C, and McCaughan L, 2000, 'Simultaneous optical wavelength interchange with a two-dimensional second-order nonlinear photonic crystal', *Opt. Lett.* **25** 832.

Chowdhury A, Staus Ch, Boland B, Kuech Th F and McCaughan L, 2001, 'Experimental demonstration of 1535–1555-nm simultaneous optical wavelength interchange with a nonlinear photonic crystal', *Opt. Lett.* **26** 1353.

de Sterke M, Saltiel S M and Kivshar Yu S, 2001, 'Efficient collinear fourth harmonic generation by two-channel multistep cascading in single two dimensional nonlinear photonic crystal', *Opt. Lett.* **26** 539.

Egorov O A and Sukhorukov A P, 1998, *Izvestiya Rossiiskoi Akademii Nauk. Seriya Fizicheskaya* **62** 2345. [Translated in: *Bulletin of Russian Academy of Sciences –Physics*, **62** 1884].

Fejer M M, Magel G A, Jundt D H, and Bayer R L, 1992, 'Quasi-phase matched second harmonic generation: tuning and tolerances', *IEEE J. Quant. Electron.* **28** 2631.

Fradkin-Kashi K and Arie A, 1999, 'Multiple-Wavelength Quasi-Phase-Matched Nonlinear interactions', *IEEE J. Quantum Electron.* **35** 1649.

Fradkin-Kashi K, Arie A, Urenski P, Rosenman G, 2002, 'Multiple Nonlinear Optical Interactions with Arbitrary Wave Vector Differences' , *Phys. Rev. Lett.* **88** 023903.

Gallo K and Assanto G, 1999, 'Analysis of lithium niobate all-optical wavelength shifters for the third spectral window', *J. Opt. Soc. Am.* **16** 741.

Gu B Y, Zhang Y, Dong B Z, 2000, 'Investigations of harmonic generations in aperiodic optical superlattices', *Journal of Applied Physics* **87** 7629.

Hooper B A., Gauthier D J and Madey J M J, 1994, *Appl.Opt.* **33** 6980.

Karaulanov T and Saltiel S, 2002, 'Efficient collinear third-harmonic generation in a single two-dimensional nonlinear photonic crystal', *Proceeding of IQEC-2002*, June 22-28 , Moscow, Russia.

Kildal H and Iseler G W, 1979, 'High order nonlinear processes in CdGeAs2', *Phys. Rev B* **19** 5218

Kivshar Yu S, 1998, 'Bright and dark spatial solitons in non-Kerr media', *Opt. Quant. Electr.* **30** 571.

Lefort L and Barthelemy A, 1995, 'Intensity dependent polarization rotation associated with type II phase-matched second-harmonic generation: application to self-induced transparency', *Opt. Lett.* **20** 1749.

Liu Z W, Zhu S N, Zhu Y Y, Qin YQ, He J L, Zhang C,Wang H T, Ming N B, Liang X Y and Xu Z Y, 2001, 'Quasi-CW Ultraviolet Generation in a Dual-periodic LiTaO3 Superlattice by Frequency Tripling', *Jpn. J. Appl. Phys.* **40** 6841.

Liu Z W, Du Y, Liao J, Zhu S N, Zhu Y Y., Qin Y Q, Wang H T, He J L , Zhang C, Ming N B, 2002, 'Engineering of a dual-periodic optical superlattice used in a coupled optical parametric interaction', *J. Opt. Soc. Am.* **B 19** 1676.

Louis S, Couderc V, Louradour F, Faugeras P and Barthelemy A, 2001, 'Nonlinear polarization evolution in type I and type II second-harmonic-generation crystals applied to the mode locking of a pulsed Nd:YAG laser', *J. Opt. A: Pure Appl.* Opt. **3** 139.

Minkovsky N, Saltiel S M, Petrov G I, Albert O, Etchepare J, 2002, 'Polarization Rotation Induced By Cascaded Third-Order Processes', *Opt. Lett*, **27** 2025.

Misoguti L, Backus S, Durfee C G, Bartels R, Murnane M M, and Kapteyn H C, 2001, 'Generation of Broadband VUV Light Using Third-Order Cascaded Processes', *Phys. Rev. Lett.* **87** 013601.

Ostrovskii L A, 1967, 'Self action of light in crystals', *JETP Lett.* **5**, 272.

Petrov G I, Saltiel S M, Ivanova A B, 1999, 'Measurement of $\chi^{(2)}$ components by comparing polarization resolved second-order cascade processes', *SPIE* **3733** 112.

Petrov G I, Albert O, Etchepare J and Saltiel S M, 2001, 'Cross-polarized wave generation by effective cubic nonlinear optical interaction', *Opt. Lett.* **26** 355.

Pfister O, Wells J S, Hollberg L, Zink L, Van Baak D A, Levenson M D, and Basenberg W R, 1997, 'Continuous-wave frequency tripling and quadrupoling by simultaneous three-wave mixing in periodically poled crystals: application to a two-step 1.19-10.71-μm frequency bridge', *Opt. Lett.* **22** 1211.

Saltiel S, Koynov K and Buchvarov I, 1995, 'Analytical and numerical investigation of opto-optical phase modulation based on coupled second order nonlinear processes', *Bulg. J. Phys.* **22**, 39.

Saltiel S, Koynov K and Buchvarov I, 1996a, 'Analytical formulae for optimization of the process of low power phase modulation a quadratic nonlinear medium', *Appl. Phys..* **B 63** 39.

Saltiel S, Koynov K, Buchvarov I, 1996b, 'Self-induced transparency and self-induced darkening with nonlinear frequency doubling polarization interferometer', *Appl. Phys..* **B 63** 371

Saltiel S, Tanev S, Boardman A D, 1997, 'High-order nonlinear phase shift caused by cascaded third-order processes', *Opt. Lett.* **22** 148.

Saltiel S and Deyanova Y, 1999, 'Polarization switching as a result of a cascading of two simultaneously phase-matched processes', *Opt. Lett.* **24** 1296.

Saltiel S M and Kivshar Yu S, 2000a, 'Phase-matching for nonlinear optical parametric processes with multistep-cascading', *Bulg J. Phys.* **27** 57.

Saltiel S and Kivshar Yu S, 2000b, 'Phase matching in $\chi^{(2)}$ nonlinear photonics crystals', *Opt. Lett.* **25** 1204, 1612(E).

Saltiel S and Kivshar Yu S, 2002, 'All-optical deflection and splitting by second-order cascading', *Opt. Lett.* **27** 921

Stegeman G I, Hagan D J and Torner L, 1996, 'Cascading phenomena and their applications to all-optical signal processing, mode locking, pulse compression, and solitons', *Opt. Quantum. Electron.* **28** 1691.

Torner L and Stegeman G, 2001, 'Multicolor solitons', *Opt. Photon. News*, **12** 36.

Zhang C, Wei H, Zhu Y Y, Wang H T, Zhu S N, and Ming N B, 2001, 'Third-harmonic generation in a general two-component quasi-periodic optical superlattice', *Opt. Lett.* **26** 899.

Zhu Y Y and Ming N B, 1999, 'Dielectric superlattices for nonlinear optical effects', *Opt. Quantum. Electron.* **31** 1093.

Spatial beam instabilities due to instantaneous nonlinear mechanisms

George I Stegeman

University of Central Florida, USA

1 Introduction

Plane waves are the eigenmodes of electromagnetic fields in linear homogeneous media, that is, they are solutions to the wave equation in which the polarization induced in a medium is strictly linear in the applied field. However all optical media exhibit a nonlinear response at high enough intensities so that the induced polarization is proportional to the product of electromagnetic fields (Shen 1984, Boyd 1992). The discipline dealing with this nonlinear polarization is called "nonlinear optics" and it is normally taught in textbooks as the mixing of multiple plane wave fields (Shen 1984, Boyd 1992). This nonlinear response has the potential over some propagation distance to destabilize plane wave fields with superimposed amplitude or phase noise of any non-zero magnitude so that plane waves are unstable solutions to the nonlinear wave equation (Akhmediev and Ankiewicz 1997, Enfield and Rowlands 1990, Nayfeh and Mook 1979). This was first realized in the early days of nonlinear optics when it was found that high-intensity beams broke up into higher intensity filaments of finite cross-section due to the self-focusing associated with the third order nonlinearity (reviewed in Akhmanov *et al.* 1972) It was given the name "modulational instability", MI, because it was the modulation of plane wave fields by noise that led to the beam break-up. Perhaps not as well known is that beam break-up into filaments was also predicted to occur via second order nonlinearities (Trillo and Ferro 1995a, 1995b). And in fact, all orders of nonlinearity are believed to give rise to plane wave instabilities.

In the laboratory nonlinear optics experiments are carried out with beams of finite cross-section, not plane waves. The same nonlinear field-matter interaction that results in MI for plane waves leads to MI for finite cross-section beams, but now with an intensity threshold which varies inversely with the beam's width. (Henceforth the word "beams" will refer only to waves with a finite cross-section.) When dealing with beams, diffraction, i.e. beam spreading due to the angular spectrum of wavevector components associated with a beam, occurs. Diffraction is inherent to any wave equation and is always present. The nonlinear eigenmodes in which diffraction is balanced by some nonlinear beam narrowing mechanism are called "spatial solitons" (recent work summarized in Trillo and Torruellas 2001). They are the "steady state", stable solutions to the wave equation(s) driven by a nonlinear polarization field. Note that solitons are not only the solutions to a nonlinear wave equation, but they are *also* stable against perturbations, a property called "robustness". (Many of the so-called "solitons" appearing in the literature are not solitons because they are not stable!)

This chapter deals with both the mechanisms that lead to the evolution in shape of beams that travel through nonlinear media, and the nonlinear eigenmodes of the nonlinear medium-electromagnetic field interaction, namely solitons. For historical reasons, the case of third order nonlinearities is dealt with first, and then second order nonlinearities. Only local, ultra-fast electronic nonlinearities are dealt with which excludes the popular field of photo-refractive nonlinear effects. This chapter is written from an experimental perspective, using primarily results from the author's own laboratory, and detailed theory is kept to a minimum. Finally, the references are meant to be representative and not a comprehensive review of the literature.

2 Beams in $\chi^{(3)}$ media

The simplest form of a nonlinear response occurs in Kerr media where the index change can be expressed as $\Delta n = n_2 I = n_{2,E}|E_0|^2$ where I is the local intensity. The incident optical field and nonlinear polarization field are written as

$$E = \frac{1}{2}E_0 \exp[i(\omega t - kz)] + c.c. \tag{1a}$$

$$P^{\rm NL} = \frac{1}{2}P_0^{\rm NL} \exp[i(\omega t - kz)] + c.c. \tag{1b}$$

with

$$P_0^{\rm NL} = \frac{3}{4}\epsilon_0 {\rm Re}\left\{\chi^{(3)}(-\omega;\, \omega, -\omega, \omega)\right\}|E_0|^2 E_0, \tag{2}$$

which is appropriate for intensity- and field-dependent refractive index coefficients defined by

$$n_2 = \frac{3}{4n_0^2}c\epsilon_0 {\rm Re}\left\{\chi^{(3)}(-\omega;\, \omega, -\omega, \omega)\right\}, \tag{3a}$$

$$n_{2,E} = \frac{3}{8n_0}\chi^{(3)}(-\omega;\, \omega, -\omega, \omega). \tag{3b}$$

The nonlinear polarization driven wave equation is

$$\nabla^2 E + \frac{n_0^2}{c^2}\omega^2 E = -\omega^2 \mu_0 P^{\rm NL} \tag{4}$$

When the slowly varying field and amplitude approximation is applied, this gives

$$-2ik\frac{\partial}{\partial z}E + \nabla_\perp^2 E = -3\frac{\omega^2}{c^2}\chi^{(3)}|E|^2 E. \tag{5}$$

which can be solved in both the plane wave limit ($\nabla_\perp^2 E = 0$ and with diffraction included ($\nabla_\perp^2 E \neq 0$) in slab (1D) waveguides.

2.1 Modulational instability

Consider the *plane wave* limit in 1D (slab waveguides). The solution to Equation 5 is

$$E = \frac{1}{2} E_0 \exp\left\{ i\left(\omega t - \left[k + k_{\text{vac}} n_{2,E} |E_0|^2\right] z\right)\right\} + c.c. \tag{6}$$

The net result is that the wavefront experiences an additional nonlinear phase shift given by $\Delta\phi^{\text{NL}} = k_{\text{vac}} n_{2,E} |E_0|^2 z$. However, just because there is a solution to the nonlinear wave equation does not mean that it is stable. Stability is determined by introducing a small amplitude and phase perturbation represented by a complex amplitude $\delta \ll 1$ with a spatial periodicity $2\pi/\kappa = \Lambda$.

$$E = E_0 \left[1 + \delta(\kappa) \cos(\kappa x) \exp(\gamma z)\right] \exp\left\{ i \left[\omega t - (k + k_{\text{vac}} n_{2,E} |E_0|^2) z\right]\right\}. \tag{7}$$

This can be thought of as a Fourier component of noise present on the wave front. If this solution, when inserted into the wave equation which includes diffraction along x, leads to a real value for γ, the noise grows exponentially with distance and the solution is unstable. This analysis leads to

$$\gamma^2 = \frac{\kappa^2}{2k} \left[2k_{\text{vac}} n_2 |E_0|^2 - \frac{\kappa^2}{2k} \right], \tag{8}$$

which predicts that γ is real when $|E_0|^2 > \kappa^2 / 4k_0^2 n_0 n_{2,E}$. Note that for a given value of κ, there is a threshold intensity needed for the instability to occur.

Consider an AlGaAs 1D waveguide (Figure 1) whose nonlinear response has been shown to be the classical Kerr response when operated with photon energies just below one half the semiconductor's [$Al_{0.18}Ga_{0.82}As$] bandgap (Kang *et al.* 1996). Experimental results are available for this geometry and the calculated value of γ is shown in Figure 2a for two different intensities (Malendevich *et al.* 2002). The higher the input intensity, the smaller the minimum period of the noise that can lead to filamentation, the larger the maximum gain and the smaller the periodicity at which the maximum gain occurs. Note that the gain is relatively flat near its maximum and strong gain occurs over a range of periodicities.

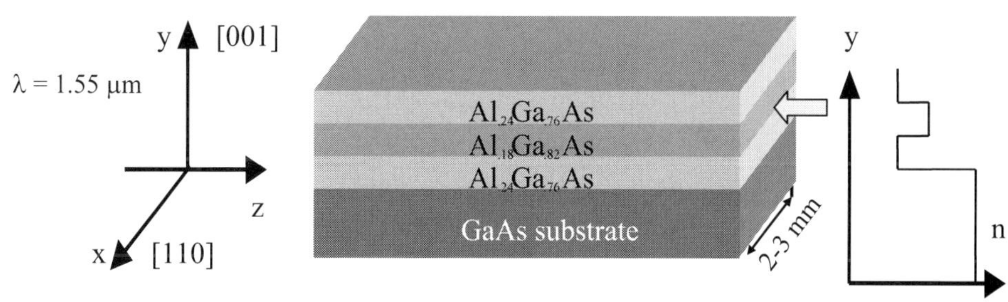

Figure 1. *AlGaAs waveguide for investigating Kerr effects in 1D (Kang et al. 1996).*

In any real experiment, the input beam has some noise superimposed on its envelope. When a noisy plane wave is injected into a Kerr slab waveguide, all the Fourier components with κ smaller than the threshold value are amplified, with the largest amplification

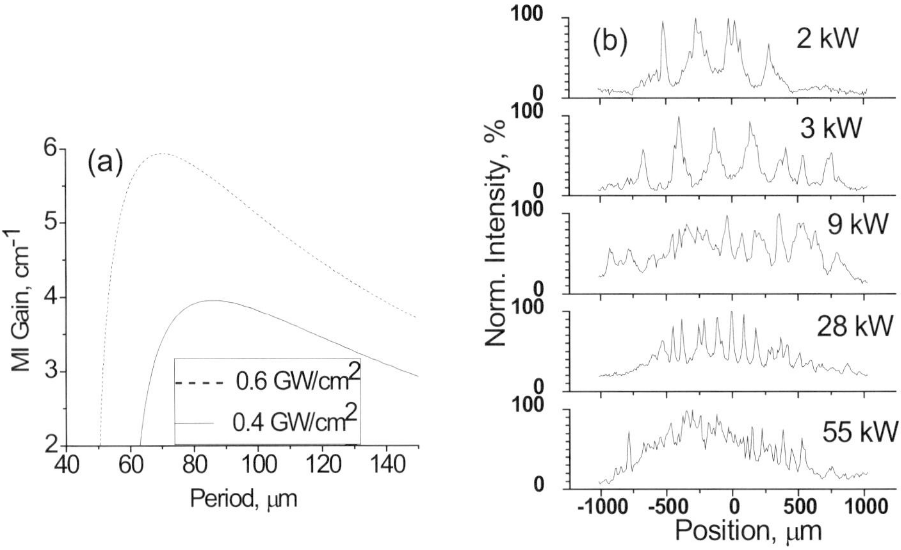

Figure 2. *(a) The calculated values of the MI gain coefficient γ versus noise periodicity for the AlGaAs waveguide in Figure 1. (b) Experimentally observed beam break-up in AlGaAs waveguide for 2.5mm wide input Gaussian beams versus total pulsed power at 1550 nm.(Malendevich et al. 2001).*

occurring at the peak of the gain curve. Assuming "white noise", the periodicities associated with the maximum in the gain will be preferentially amplified. In fact that is exactly what is observed experimentally for beams a few millimetres wide, as shown in Figure 2b. The periods measured experimentally are in excellent agreement with theory (Malendevich *et al.* 2001). Note that the gain coefficients are calculated in the small growth limit and the results in Figure 2b are already well beyond this limit. Although the discussion is in terms of a "gain" coefficient, of course there is no new source of optical power and what actually occurs is a re-distribution of the power. Because there is always "noise" on any beam, these results indicate that plane waves are *not* robustly stable solutions to the nonlinear wave equation.

2.2 Spatial solitons

There are, however, robustly stable solutions to the nonlinear wave equation for *finite* beams (Akhmediev and Ankiewicz 1997, Enfield and Rowlands 1990, Trillo and Torruellas 2001). They are the eigenmodes of the nonlinear wave equation, known as spatial solitons, and are given for $n_2>0$ by

$$E_0(x,z) = \sqrt{\frac{n_0}{n_{2,E}}}\frac{1}{n_0 k_{\text{vac}} w_0}\,\text{sech}\left(\frac{x}{w_0}\right)\exp\left[\frac{-iz}{2n_0 k_{\text{vac}} w_0^2}\right]. \tag{9}$$

Note that there is an additional nonlinear phase shift, just as for the unstable nonlinear plane waves. Nonlinear phase shifts always also occur on finite beams due to n_2 at lower (than solitonic) intensities but in those cases the phase shift at any point of an intensity distribution varies with the local intensity, that is, the phase shift varies along the beam profile. But for solitons the phase shift has the *same* value right across the intensity distribution of the soliton. Furthermore, there is a trade-off between the peak intensity ($\propto |E_0|^2$) and the beam width w_0, reducing the number of free parameters from two to just one. The product $P_0 w_0$ is a constant where P_0 is the maximum beam power. The "robustness" of these solutions stems from the fact that an increase in power can be compensated by a decrease in the beam width, or vice versa.

Spatial soliton excitation is achieved by focusing a beam of appropriate power and width onto the input facet of a Kerr nonlinear waveguide. If the power is too low, the beam diffracts as shown in Figure 3 at the output facet of the waveguide shown in Figure 1. As the input power is increased, the output beam narrows until it equals the input beam. Further increases in the power lead to narrower output beams, etc.

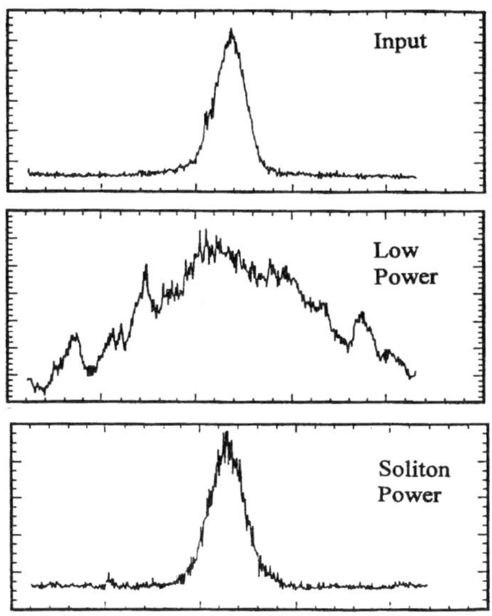

Figure 3. *The input beam (upper trace, ~70μm full width at half maximum) and the output from an AlGaAs waveguide at low input powers (middle trace, diffracted beam) and at the solitonic power (lower trace). As the input power is increased further, the peak output power increases and the beam width decreases (Kang et al. 1996).*

2.3 Spatial solitons versus modulational instability

What happens to real (some level of noise) input beams of gaussian shape whose peak intensity (power) is much larger than that required for their width to form a single spatial soliton? In fact that is the case already given in Figure 2 where break-up was shown to occur and to be described very well by plane wave modulational instability theory. The nonlinear self-focusing mechanisms are of course a property of the light-material and are identical in the two cases, beam break-up and spatial soliton formation.

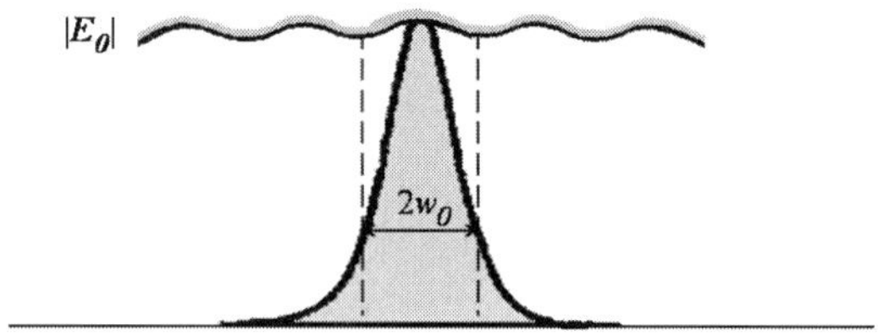

Figure 4. *A plane wave with a modulation period corresponding to the maximum gain for an intensity corresponding to a soliton of width* $2w_0$.

A simple analysis of the two limits leads to the following relationship between modulational instability and solitons. Plotted in Figure 4 is a plane wave with a periodic modulation corresponding to the peak in the gain coefficient and a spatial soliton of the same peak intensity. To within ~10%, the MI period matches the width ($2w_0$) of the soliton. This suggests that there should be a transition for real (noisy) beams from single soliton generation to multi-filament formation as the peak intensity is increased. In fact the interpretation is that a wide, noisy beam leads to multiple filaments, each of which attempts to evolve into a nonlinear eigenmode of the system. This will happen when a number of periods "fit" within the width of an input beam. What hinders this evolution is the excess radiation that is trapped between the filaments by self-focusing forces etc. For a finite beam, this excess energy should be able to "leak out" the sides of the beam leading ultimately to a train of spatial solitons. This is an experiment asking to be performed. The only problem is identifying the appropriate geometry and material system that allows enough diffraction lengths of propagation for this result to be "clean".

2.4 Beams in bulk media

2.4.1 Kerr media

There are no stable solutions to the NLSE for beams with two freely diffracting transverse dimensions, that is, there are no stable solitons (Trillo and Torruellas 2001). To understand why, it is useful to go back to a simple physical picture for spatial solitons in 1D, namely the robust balance between the characteristic lengths associated with diffraction and self-focusing. Solitons are formed when the diffraction length $L_{\rm diff} = \pi w_0^2 n_0/\lambda_{\rm vac}$ is comparable to a nonlinear length $L_{\rm NL}$ defined by $\Delta\phi^{\rm NL} = k_{\rm vac} n_2 L_{\rm NL} P_0/h w_0$ where h is an effective waveguide thickness. Based on the 1D case, $\Delta\phi^{\rm NL} = \pi/2$. Thus equating $L_{\rm NL} = L_{\rm diff}$ for a circular beam of radius w_0 gives $P_0 = \lambda_{\rm vac}^2/2n_0 n_2$. The key here is that diffraction can only be balanced by self-focusing at precisely only one value of the beam power, independent of the beam width. The response to any fluctuations (which defines stability) does not lead to a stable beam but to either diffraction or catastrophic self-focusing, depending on the fluctuation.

In fact the situation in bulk Kerr media is much more complex than discussed here. When beam diameters, for example due to self-focusing, become of the order of a few wavelengths and less, the scalar NLSE is no longer adequate and the electromagnetic field contains all six components. The discussion about the behaviour of beams with these and other complications present is still under active discussion.(see for example Ciattoni *et al.* 2002) For example, a periodic focusing-defocusing dependence on propagation distance has been predicted, accompanied by radiative losses. The one thing that is clear is that the beam dynamics is quite complex!

2.4.2 Saturating bulk media

A material with an index change that is indefinitely linear in the intensity just does not exist. At high enough intensities, material damage always occurs. But even below damage thresholds, the index change frequently exhibits at least the onset of saturation, especially in the spectral vicinity of real transitions between electronic or vibrational levels. This introduces additional parameters into the description of $\Delta n(I)$. This is sufficient to make spatial solitons with two transverse dimensions (2D), in addition to 1D, robustly stable.

This result can be easily deduced from a simple model for index saturation. For example, assume

$$\Delta n(I) = \Delta n_{\text{sat}}\left[1 - \frac{1}{1 - I/I_{\text{sat}}}\right] \cong \left(\frac{\Delta n_{\text{sat}}}{I_{\text{sat}}}\right) I - \frac{1}{2}\left(\frac{\Delta n_{\text{sat}}}{I_{\text{sat}}^2}\right) I^2 + \cdots \tag{10}$$

in which only the two leading terms in the expansion for $I/I_{\text{sat}} < 1$ are included. There are now two parameters which describe the index change, namely $\Delta n_{\text{sat}}/I_{\text{sat}}$ (equal to n_2) and I_{sat}. Again equating the nonlinear length to the diffraction length,

$$\frac{L_D}{L_{\text{NL}}} = \frac{1}{\lambda_{\text{vac}}^2}\left[2n_0 n_2 P_0 - \frac{n_3 P_0^2}{\pi w_0^2}\right] \approx 1, \tag{11a}$$

(11b)

and setting

$$n_3 = \frac{\Delta n_{\text{sat}}}{2I_{\text{sat}}^2}, \tag{11c}$$

leads to the equation

$$\left[n_2 - 2n_3\frac{P_0}{\pi w_0^2}\right] dP_0 = -2n_3\frac{P_0^2}{\pi w_0^3} dw_0. \tag{12}$$

Note that $dP_0/dw_0 < 0$ for $n_2 > 2n_3P_0/\pi w_0^2$ so that there can be a physically sensible trade-off in which an increase in peak power is compensated by a decrease in beam width, and vice-versa. Thus these solitons are robustly stable. Note also that $dP_0/dw_0 > 0$ if $n_2 < 2n_3P_0/\pi w_0^2$, apparently implying instability. However, at these power levels the next higher order term in the expansion must be included and the solitons are still stable. One might also argue that this is a consequence of this particular model for $\Delta n(I)$ but the real key is that there are two parameters describing the material system's response to the optical field.

Spatial solitons in 2D based on saturable third order nonlinear media were actually the first spatial solitons to be observed (Bjorkholm and Ashkin 1974). (For that matter, they were the first optical solitons of any kind to be observed, spatial or temporal.) The medium of choice has been metal vapors near an electronic resonance, and the simplest cases have involved sodium and rubidium (Bjorkholm and Ashkin 1974, Tikhonenko *et al.* 1995). Operating at wavelengths near the saturation of the absorption has led to an effective $n_2 > 0$ for wavelengths shorter than the transition wavelength.

3 Beams in χ^2 media

Although the concept that spatial solitons could exist due to second order nonlinearities actually dates back to the 1970s, it was not widely known until the 1990s when such solitons were first observed (Karamzin and Sukhorukov 1975, Torruellas *et al.* 1995, R. Schiek *et al.* 1996). The key question is the physical origin of the beam narrowing mechanisms involved since $\chi^{(2)}$ interactions at optical frequencies do not lead to index changes. Although this situation is in fact quite complex because multiple nonlinear equations are involved which do not have analytical solutions, the basic mechanisms themselves are relatively straightforward.

3.1 Beam narrowing mechanisms

In order to illustrate the mechanisms involved, consider the simplest possible case of Type I second harmonic generation (SHG). The nonlinear polarization induced due to the presence of an orthogonally polarized fundamental (FW) and second harmonic (SH) is written as

$$P_i^{\mathrm{NL}} = \epsilon_0 \chi_{ijk}^{(2)} \left\{ E_j(\omega) E_k(\omega) + E_j^*(\omega) E_k(2\omega) \right\}, \tag{13}$$

where $2\chi_{ijk}^{(2)} = d_{ijk}^{(2)}$ is the second order susceptibility and the two terms correspond to up-conversion ($\omega + \omega \to 2\omega$) and down-conversion ($2\omega - \omega \to \omega$) respectively. Assuming FW and SH fields of the form

$$E(\omega) = \frac{1}{2} a_1 \exp[i(\omega t - k_1 z)] + c.c. \tag{14a}$$

$$E(2\omega) = \frac{1}{2} a_2 \exp[i(2\omega t - k_2 z)] + c.c. \tag{14b}$$

in which $|a_1|^2$ and $|a_2|^2$ are normalized to the appropriate intensity (W/cm for 1D and W/cm^2 for 2D). Applying the standard slowly varying amplitude and phase approximation, including diffraction, leads to the coupled mode equations

$$i\frac{\partial}{\partial z} a_2(z) = \frac{1}{2k_2} \left\{ \frac{\partial^2}{\partial y^2} + \frac{\partial^2}{\partial x^2} \right\} a_2(z) + \Gamma a_1^2(z) e^{-i\Delta k z}, \tag{15a}$$

$$i\frac{\partial}{\partial z} a_1(z) = \frac{1}{2k_1} \left\{ \frac{\partial^2}{\partial y^2} + \frac{\partial^2}{\partial x^2} \right\} a_1(z) + \Gamma a_1^*(z) a_2(z) e^{i\Delta k z}, \tag{15b}$$

in which the $\partial^2/\partial y^2$ term is zero here for the 1D (waveguide) case, $\Delta k = 2k_1 - k_2$ is the wavevector mismatch and $\Gamma \propto d_{\mathrm{eff}}^{(2)}$ which takes into account the polarizations of the fundamental and harmonic in the usual way (Shen 1984, Boyd 1992). These equations contain all of the physics associated with nonlinear changes in the beam shapes in $\chi^{(2)}$ media, including beam instabilities, beam narrowing and spatial solitons.

3.1.1 "Cascading" phase shifts

It has been known since the mid-1960s that there is a nonlinear phase shift introduced onto interacting waves during second harmonic generation (Ostrovskii 1967, reviewed in

Stegeman *et al.* 1996). This effect is called "cascading". There is a well-known periodic energy exchange between the fundamental and harmonic with distance for $\Delta k \neq 0$, as indicated in Figure 5. For SHG where only the fundamental is input, the energy flows out of, and back into the fundamental. However, the two waves propagate at different phase velocities, i.e. $V_\omega \neq V_{2\omega}$, so that the phase fronts are no longer exactly $\pi/2$ out of phase with one another. As a result the down-converted fundamental has a net phase shift which is not equal to π with respect to the fundamental that was not up-converted.

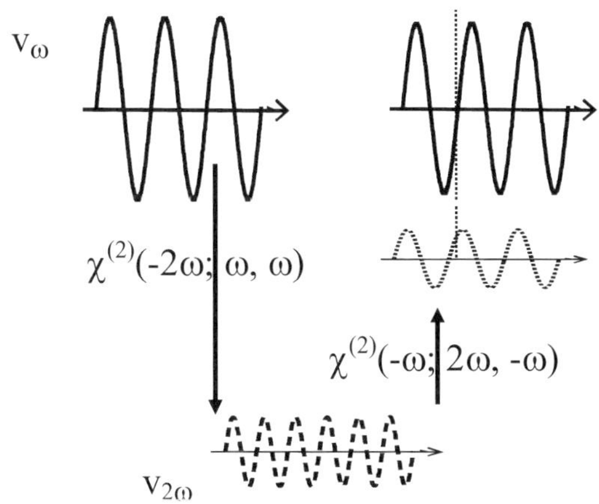

Figure 5. *Pictorial representation of the "cascading" process. The second harmonic is generated via $\chi^{(2)}(-2\omega;\,\omega,\omega)$. After some propagation of the harmonic, the SH is down-converted back to the fundamental via $\chi^{(2)}(-\omega;\,2\omega,-\omega)$.*

This is the origin of the nonlinear phase shift and leads to either a slowing down ($\Delta k > 0$) or a speeding up ($\Delta k < 0$) of the fundamental phase-front. Bearing in mind that there is no real index change here that can be measured in any of the usual ways, the effect can be written as an effective $\bar{n}_{2,\text{eff}}$, averaged over one cycle of the periodic energy exchange,

$$\bar{n}_{2,\text{eff}} = \frac{k_{\text{vac}} \left[d_{\text{eff}}^{(2)} \right]^2}{2\epsilon_0 c n^3 \Delta k}. \tag{16}$$

This is a plane wave result. The situation is more complex for finite beams for two reasons.

1. A finite beam contains an angular spectrum of optical wavevectors. As a result, cascaded phase shifts occur even on phase-match due to the off-axis wavevectors. Furthermore, these phase shifts can be spatially anisotropic. That is, near and at phase-match, different wavevector directions in the angular spectrum can experience different Δk, and even different signs for Δk. Far off phase match, only a single sign for Δk occurs.

2. The nonlinear phase shift which depends on the local intensity along a beam's intensity profile induces curvature on the phase front. Near and at the peaks, the curvature change is the strongest, leading to self-focusing for $\Delta k > 0$ and self-defocusing for $\Delta k < 0$.

3.1.2 Beam narrowing through wave mixing

The second beam narrowing mechanism is associated with all nonlinear wave mixing interactions (reviewed in Stegeman *et al.* 1996). Consider the fundamental to be a (1D for simplicity) Gaussian beam of the form $a_1 \exp[-x^2/w_0^2]$. The up-conversion nonlinear polarization is proportional to $a_1^2 \exp[-2x^2/w_0^2]$ and the resulting second harmonic (in the absence of diffraction) is $a_2 \propto a_1^2 \exp[-2x^2/w_0^2]$, i.e. the harmonic beam has a width given by $w_0/\sqrt{2}$. This results in a narrowing of the harmonic relative to the fundamental. When this second harmonic mixes with the unconverted fundamental to generate down-converted fundamental, that is, $\{a_1^2 \exp[-2x^2/w_0^2]\} \times \{a_1^* \exp[-x^2/w_0^2]\} \propto \exp[-3x^2/w_0^2]$, the fundamental is also narrowed in space. Therefore, in the absence of diffraction, this process could potentially lead to mutual beam collapse.

This wave mixing mechanism always leads to beam narrowing, independent of the sign of Δk. In principle the cascading mechanism can interfere constructively or destructively with this wave mixing mechanism, however, as will be shown later, this is an oversimplified picture of what happens when the equations are solved fully (Di Trapani *et al.* 2001).

3.2 Modulational instability

Given the fact that beam narrowing occurs through $\chi^{(2)}$ nonlinearities, then it should not be surprising that plane-wave beams in a $\chi^{(2)}$-active medium are inherently unstable. The approach used mimics that for the $\chi^{(3)}$ case, i.e. find the nonlinear plane wave (no diffraction) solutions to the nonlinear wave equations (15) and then introduce periodic perturbations. (Trillo and Ferro 1995a, 1995b) The situation is complicated by the presence of two field components at different frequencies.

The solutions to the plane wave version of Equations 15 can be nicely separated into an amplitude part and a phase part. Writing $a_1 = \rho_1(z)a_1(0)\exp[-i\varphi_1(z)]$ and $a_2 = \rho_2(z)a_1(0)\exp[-i\varphi_2(z)]$, the new variables are the amplitudes $\rho_1(z)$ and $\rho_2(z)$ and the phases $\varphi_1(z)$ and $\varphi_2(z)$ respectively. Furthermore, defining $\mu = \omega d_{\text{eff}}^{(2)} a_1(0)/nc = L_{\text{pg}}^{-1}$, $\Delta s = \Delta k/\mu$ and $\xi = \mu z$, the coupled mode equations can be re-written as

$$\frac{d}{d\xi}\rho_1(\xi) - i\rho_1(\xi)\frac{d}{d\xi}\varphi_1(\xi) = -i\rho_1(\xi)\rho_2(\xi)\exp\{i[\Delta s\xi + 2\varphi_1(\xi) - \varphi_2(\xi)]\}, \tag{17a}$$

$$\frac{d}{d\xi}\rho_2(\xi) - i\rho_2(\xi)\frac{d}{d\xi}\varphi_2(\xi) = -i\rho_1^2(\xi)\exp\{-i[\Delta s\xi + 2\varphi_1(\xi) - \varphi_2(\xi)]\}. \tag{17b}$$

Here L_{pg} is the characteristic parametric gain length, that is, the distance required for a large energy exchange between the fundamental and harmonic. Note that although the phase can change with z, the "steady state" solutions have constant amplitudes, i.e. $\rho_1 a_1(0)$ and $\rho_2 a_1(0)$ are constants. Equations (17) give four equations when separated into real and imaginary parts. There are two distinct solutions (which turn out later to be unstable). The fields and phases are related by $4\rho_2 = -\Delta s \pm [\Delta s^2 + 4\rho_1^2]^{1/2}$, $\varphi_1 = -\rho_2\xi$ and $\varphi_2 = -\rho_1^2\xi/\rho_2$. For the positive sign (in the $\pm$ option), $\varphi_1 = \varphi_2$ so that the fields are *in phase* for all Δk. As $\Delta s \equiv +\infty$, $\rho_2/\rho_1 \equiv 0$ which means that the solution is dominated by the fundamental for large $\Delta k>0$. This is the "Kerr limit" and indeed the fundamental fields resemble in shape closely those associated with Kerr solitons. (This is the nonlinear solution which will mimic closely a quadratic soliton when diffraction is included.) When the negative sign is used, $\varphi_1 - \varphi_2 = \pi$ so that the fields are π *out of* phase. And for $\Delta s \equiv -\infty$, $\rho_1/\rho_2 \equiv 0$ which means that the harmonic dominates in this limit.

However, one can now show that neither solution is stable against perturbations. Consider the + sign first. Defining

$$A_1 = a_1(0)\left[\rho_1 + F_1 \cos(\kappa x)\exp(\gamma z)\right]\exp[-i\rho_2\xi],$$
$$A_2 = a_1(0)\left[\rho_2 + F_2 \cos(\kappa x)\exp(\gamma z)\right]\exp[-i\rho_1\xi/\rho_2^2],$$

with $F_{1,2} \ll \rho_{1,2}$ and substituting back into Equations (15) with diffraction only along the x-axis (waveguide case), leads to

$$i\gamma F_1 - \kappa^2 2k_1 F_1 - \rho_1\mu(F_1 - F_1^*) + \rho_2\mu F_2 = 0, \quad (18a)$$
$$i\gamma F_2 - \kappa^2/2k_2 F_2 - \rho_2\mu F_2 + 2\rho_2\mu F_1 = 0. \quad (18b)$$

Again separating the real and imaginary parts leads to four equations. Real (physical) values for κ and γ are obtained over a range of ρ_1 (Intensity). Thus these nonlinear solutions are inherently unstable. The eigenvectors F_1 and F_2 are complex which indicates that both phase and amplitude fluctuations trigger MI. Finally, it is noted that a similar analysis leads to the same conclusions for the second solution set, namely with the − sign.

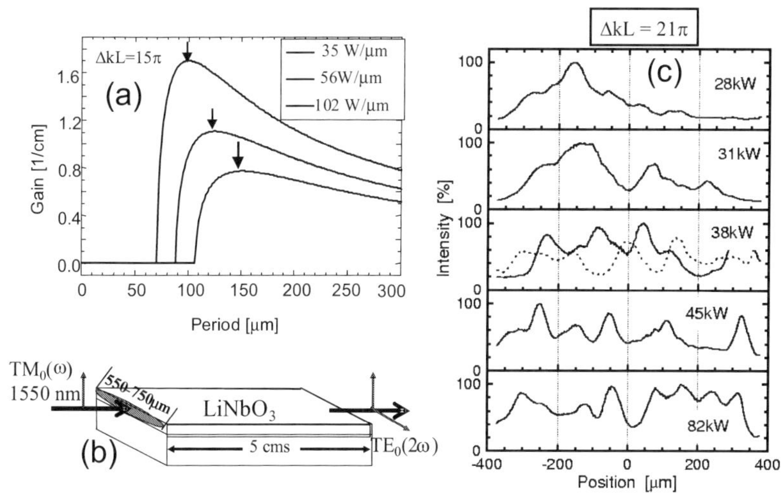

Figure 6. *1D modulational instability. (a) MI gain γ versus modulation period for different linear intensities of the input fundamental beam at 1550nm in a birefringence phase-matched $LiNbO_3$ waveguide. (b) The experimental waveguide geometry. (c) Output beams from the waveguide for different input power levels. The middle panel shows the output pattern for two successive laser shots. (Fang et al. 2000)*

It is not possible to get a simple analytical expression for the dependence of the gain coefficient γ on the modulation period $\Lambda=2\pi/\kappa$. However, numerical solutions have been obtained, an example of which is shown in Figure 6a (Fang *et al.* 2000). The curves are very similar to those for the Kerr case. As expected they depend on the phase mismatch:

The larger ΔkL, the higher the intensity required for MI, in keeping with the decrease in the effective nonlinearity with increasing ΔkL (Equation 16)).

The predicted break-up of noisy beams has been observed experimentally (Fang *et al.* 2000). The experimental geometry is shown above in Figure 6b. The sample is a birefringently phase-matched, 1D, $LiNbO_3$ waveguide investigated with 1550nm radiation as the fundamental input. Just as for the Kerr case, the beams (fundamental *and* the harmonic generated in the sample) break up, with the number of filaments increasing with the input power, reproduced in Figure 6c. Note also the patterns obtained for two successive shots at 38KW input power. The noise origin of the filamentation is clear since the output pattern changes from shot to shot. Finally the period agrees very well with the theoretical predictions.

In brief summary, nonlinear plane waves are *not* stable eigenmodes of $\chi^{(2)}$-active media near and at phase-matching conditions. The larger the phase mismatch, the higher the input intensity needed to bring on the instability.

3.3 1D spatial solitons

Equations (17) do have steady state solutions which are robustly stable, that is, stable quadratic spatial solitons exist (Pelinovsky *et al.* 1995, Torner *et al.* 1995). Unfortunately it is necessary to use numerical techniques to find them, even for the simplest Type I phase-match geometry in which only two distinct waves are present. They exhibit the usual soliton properties, a fixed set of *in phase* amplitudes (for a given total power) with a nonlinear phase shift on each spectral component that is uniform across the field profiles and accumulates linearly with distance (i.e. an effective wavevector shift). Note that these effective wavevector shifts compensate for any $\Delta k \neq 0$ so that the field components remain in phase. Although the relative peak amplitudes are fixed, the widths of the two spectral components are different. Finally, all of these properties, namely the threshold intensity, the ratio of peak amplitudes and the relative beam widths vary with the wavevector mismatch Δk. Examples of some of these properties are shown in Figure 7 (Torner 1995).

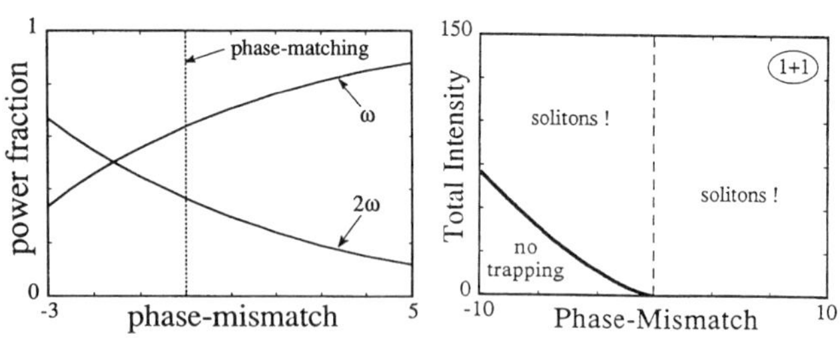

Figure 7. *(a) The fraction of power carried by the fundamental and second harmonic components of a 1D quadratic spatial soliton versus phase mismatch. (b) The threshold peak intensity versus phase mismatch (Torner 1995b).*

It is possible to excite these solitons by just injecting a fundamental field into a sample and relying on the SHG process as described by the coupled mode equations to generate

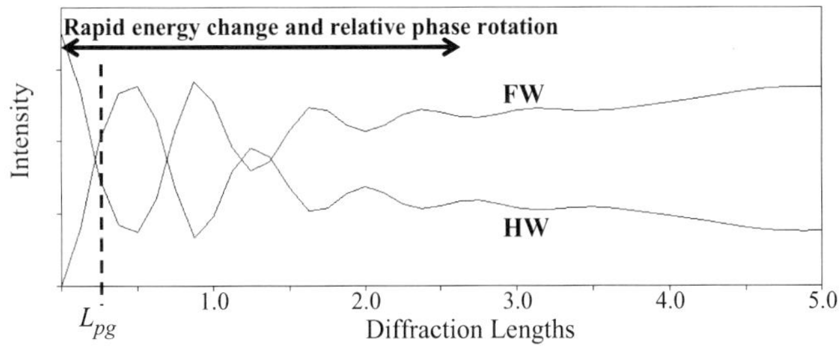

Figure 8. *Typical evolution with distance of the intensity of an input fundamental and the generated second harmonic on phase-match.*

the required SHG at the correct relative phase. Because these solitons are the nonlinear eigenmodes of the system, even though such excitation conditions are far from the stationary solutions, spatial solitons will be generated (Torner *et al.* 1994). However, the process is not adiabatic and requires propagation distance for the evolution to occur. An example of the calculated evolution for a QPM $LiNbO_3$ sample is shown in Figure 8. In QPM (quasi-phase-matching) the nonlinearity $d_{33}^{(2)}$ was periodically reversed with period Λ_{QPM} to produce the condition $2k_1 - k_2 + 2\pi/\Lambda_{QPM}=0$. Note the large oscillations in amplitude that occur and the long distances required (multiple L_{pg}) for the asymptotic limit of stationary quadratic solitons to be reached. This process is not adiabatic and only a fraction of the input fundamental power is converted into the soliton: The balance is radiated away at both the fundamental and harmonic frequencies. Although the relative phase is not plotted, it also evolves from a value of $\pi/2$ to zero with distance.

It is not possible experimentally to monitor the evolution with distance of the fields into a soliton. However, the intensity can be monitored at the output facet of the waveguide. It is shown in Figure 9 for a QPM $LiNbO_3$ waveguide (Stegeman *et al.* 2003).

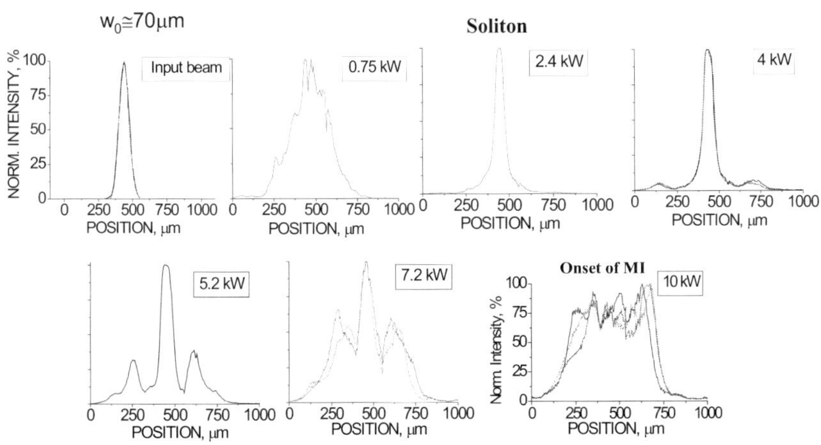

Figure 9. *The input beam and output beam profiles for increasing powers input into a* $LiNbO_3$ *waveguide.*

The various stages can be easily identified. The beam diffracts at low powers. As the input power is increased, the output beam waist collapses until a soliton is formed at the output. Note that underlying the soliton is a broad background corresponding to the weak diffracting tails of the pulses used in the experiment: They are too weak to be trapped as part of the soliton. Further increases in the input power lead to well-defined radiation lobes by which excess energy from the evolution process is shed from the soliton's vicinity. The second set of radiation lobes is sufficiently intense to form additional solitons, see for example the plot for a 7.2KW input power. Simulations (cw) show that these solitons continue to interact as they slowly separate from the central soliton, exchanging energy along the way. An example is shown in Figure 10 (Stegeman *et al.* 2003)

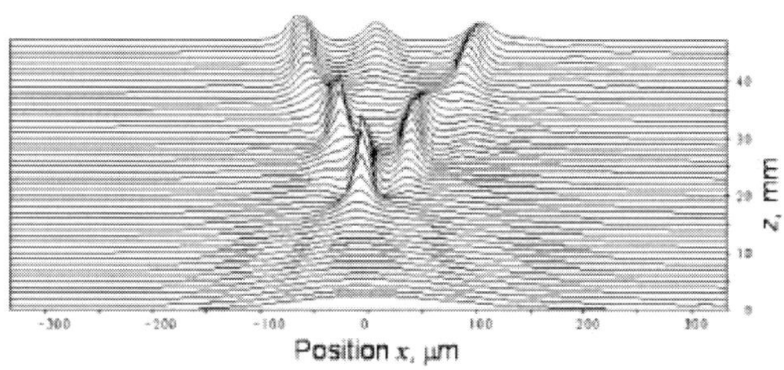

Figure 10. *CW simulations of multi-soliton generation in a $LiNbO_3$ slab waveguide. (Stegeman et al. 2003)*

The last picture in Figure 9 illustrates an interesting transition from multi-soliton generation to modulational instability. Noise is always present and leads to modulational instability when a beam is wide enough and the intensity high enough so that multiple MI periods are contained within the input beam width. The relationship was discussed previously in the section on third order effects. Note that the intensity in this case is about 4 times that required for a single soliton and multiple (3–4) peaks appear at the output. The output pattern changes from shot-to-shot, indicating that it is the non-deterministic (random) noise on the input beam profile which is responsible for the break-up. It is for these reasons that we identify this output with the onset of MI.

A problem with exciting quadratic solitons in waveguides by inputting a fundamental field only is that the excess energy is dissipated sideways in the plane of the waveguide and this process takes a long propagation distance to be efficient, as shown earlier in Figure 8. However, the QPM process can be engineered to increase the fraction of input power trapped in the soliton and decrease the distance required for steady state fields. This is achieved by varying the QPM period as a function of distance from the input facet, i.e. $\Lambda_{\mathrm{QPM}}(z)$. The evolution of the fields for the optimum case is shown in Figure 11. This design led to the generation of almost ideal quadratic solitons. For example, shown below in Figure 12 is the soliton generated near the phase-matching condition. Both the shape and fraction of power in the two components is in excellent agreement with theory. Thus this approach reduces the distance over which the soliton is formed, increases the generation efficiency and produces ideal quadratic solitons.

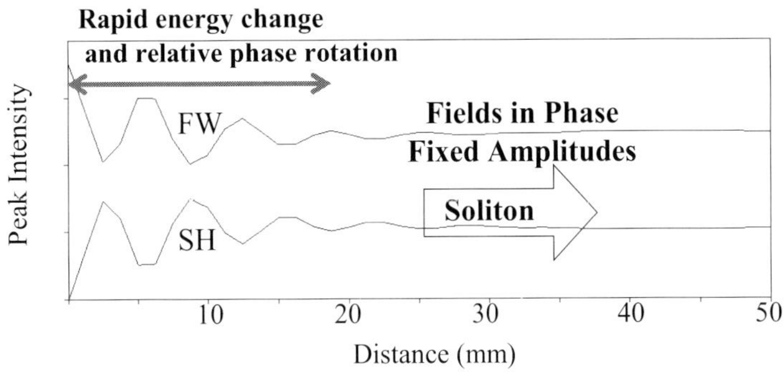

Figure 11. *Evolution of an input fundamental and generated second harmonic with optimized QPM.*

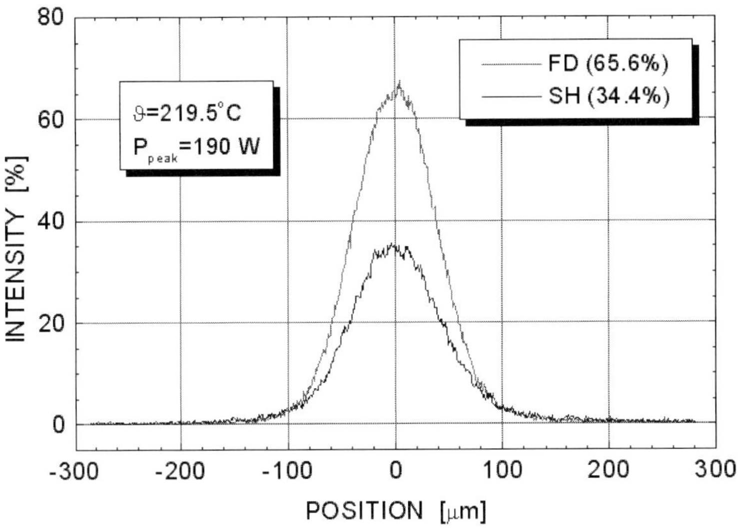

Figure 12. *The fundamental and second harmonic soliton fields measured at the output for optimized QPM.*

3.4 2D quadratically nonlinear media

Both MI and quadratic solitons also exist in bulk media. In many ways, quadratic systems behave *like* saturable nonlinearities although there is no saturation in the second order nonlinearity. Because bulk samples of frequency doubling media are commercially available, many more experiments have been performed in 2D than in 1D. However, the physics in bulk samples is the same as that just discussed for 1D slab waveguides. Thus experiments that show effects similar to those already discussed will not be reviewed here. The focus will on new or deeper insights relevant to beam distortion in quadratically nonlinear media obtained in 2D samples.

3.4.1 Modulational instability

As stated previously, the same mechanisms that drive soliton formation are also responsible for beam break-up via modulational instability. It was postulated that given the opportunity to shed the excess electromagnetic power, the filaments would evolve into spatial solitons, the nonlinear eigenmodes of the system. An experiment has been performed with very wide elliptical beams in a 2D sample of $KTiOPO_4$ using Type II phase-matching which supports this postulate (Fuerst *et al.* 1997). The incident beam, and the output beam at a large peak input intensity are both shown in Figure 13. The key point is that a quasi-periodic series of circular beams with all the characteristics of quadratic solitons, i.e. shape, beam widths, total intensities etc., exit the sample. In this case, as shown pictorially at the bottom of Figure 13, the excess radiation can dissipate (diffract) away easily leaving behind solitons as the product of the MI process.

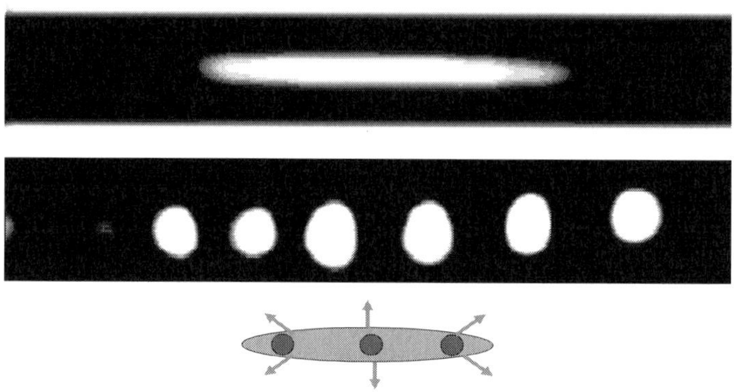

Figure 13. *Upper: Input beam into the KTP sample. Middle: Output beam for a peak input intensity of 57 GW/cm^2. Bottom: Visualization of how the excess radiation escapes in order to get solitons at the output.*

3.4.2 2D quadratic soliton formation

The conclusion could be drawn from the previous discussions of the beam narrowing mechanisms that all beams narrow monotonically down to their stationary soliton values. This is not in fact the case as has been shown recently in two different cases (DiTrapani *et al.* 2001, Kim *et al.* 2003). Shown in Figure 14 are detailed measurements of the fundamental and harmonic beams output from a PPKTP (periodically poled KTP) sample on and near phase match (Kim *et al.* 2003). These show that, although eventually both spectral components do collapse to their solitonic values, the detailed behavior is quite complex and depends on the sign of the wavevector mismatch. As the input intensity increases, the fundamental beam waist decreases monotonically for $\Delta k \geq 0$, but for $\Delta k < 0$ it first increases and then decreases. The behaviour of the harmonic is exactly the opposite. This basically shows that the beam dynamics are, in their details, complex processes.

In bulk media the multi-soliton generation process can produce results which depend

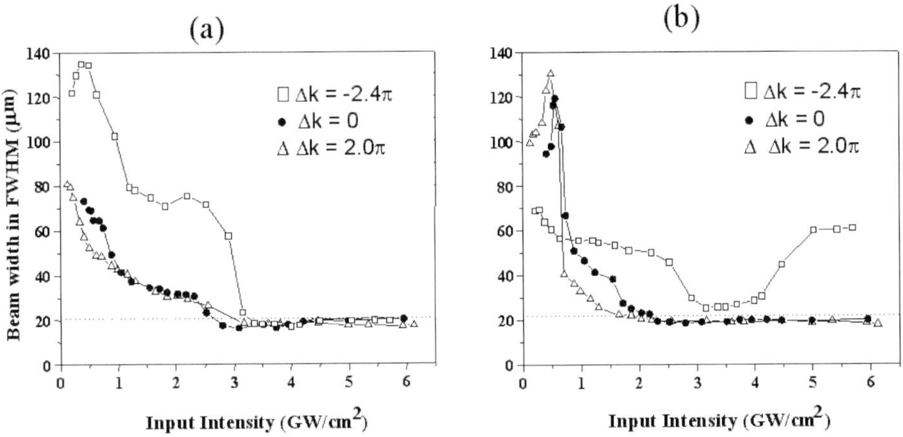

Figure 14. *The output beam widths for the fundamental (a) and second harmonic (b) as a function of input fundamental intensity for three values of phase mismatch in PPKTP on the non-critical phase-match axis.*

on a number of variables such as beam quality, the crystal symmetry associated with the propagation axis, beam polarizations etc. The results shown below in Figure 15 are for $KNbO_3$ for a 983nm fundamental beam incident along the b-axis (Malendevich *et al.* 2002). This corresponds to a non-critical phase match (NCPM) geometry in this biaxial crystal. As the input intensity increases, first a single soliton is generated, then two solitons along a line and then finally at very high intensities multiple solitons. It is believed that there are competing nonlinearities ($\chi^{(2)}$ and $\chi^{(3)}$, including both nonlinear refraction and absorption), as well as non-cylindrically symmetric beams and spatially anisotropic diffraction, all contributing to the complexity of these results.

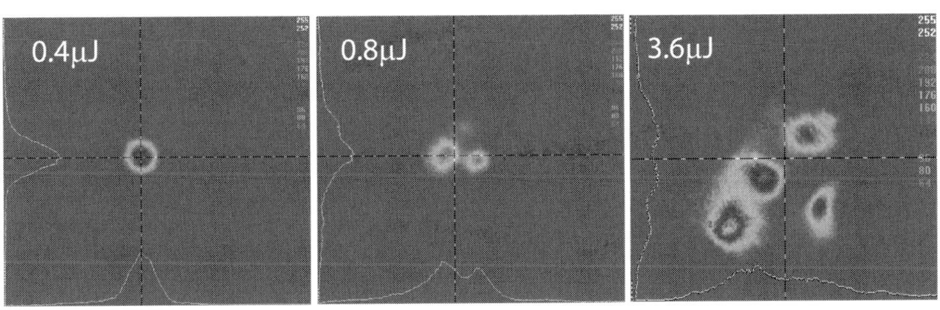

Figure 15. *The output beam profiles from $KNbO_3$ for different fundamental input energies in 20ps pulses. The input beam was a Gaussian*

4 Summary

Optical beams undergo complex beam dynamics at sufficiently high input intensities due to strong nonlinear interactions between the electromagnetic waves and matter. The most important mechanisms are those which lead to beam narrowing, and these occur with both second and third order nonlinearities. In all cases it is the interplay between diffraction and beam narrowing that controls the beam distortion.

Two basic effects are encountered, spatial soliton formation and beam filamentation (modulation instability). The first is the formation of spatial solitons which propagate without change in shape in the presence of spatial diffraction, but with an additional nonlinear phase shift which accumulates linearly with distance. These solitons are the nonlinear eigenmodes of nonlinear media and are robustly stable solutions to an appropriate set of nonlinear wave equations. The second leads to the break-up of “wide” beams into periodically spaced filaments. Under appropriate conditions, these filaments can evolve into spatial solitons.

Acknowledgment

This research was supported by a US Army Research Office MURI.

References

Akhmanov S A, Khokhlov R V and Sukhorukov A P, 1972, in *Laser Handbook*, eds. Arecchi F T and Schulz-DuBois E O, 1151 (North Holland, Amsterdam).

Akhmediev N N and Ankiewicz A, 1997, *Solitons: Nonlinear Pulses and Beams* (Chapman and Hall, London).

Bjorkholm J E and Ashkin A, 1974, ‘CW self-focusing and self-trapping of light in sodium vapor’, *Phys. Rev. Lett.* **32** 129.

Boyd R W, 1992 *Nonlinear Optics*, (Academic Press, Boston).

Ciattoni A, Conti C, DelRe E, Di Porto P, Crosignani B and Yariv A, 2002, ‘Polarization and energy dynamics in ultrafocused optical Kerr propagation’, *Opt. Lett.* **27** 734.

Di Trapani P, Bramati A., Minardi S, Chinaglia W, Conti C, Trillo S, Kilius J and Valiulis G, 2001, ‘Focusing versus defocusing nonlinearities due to parametric wave mixing’, *Phys. Rev. Lett.* **87** 183902.

Enfield E and Rowlands G, 1990, *Nonlinear Waves, Solitons and Chaos* (Cambridge University Press, Cambridge).

Fang H, Malendevich R, Schiek R and Stegeman G I, 2000, ‘Spatial modulational instability in one-dimensional $LiNbO_3$ slab waveguides’, *Opt. Lett.* **25** 786.

Fuerst R A, Baboiu D-M, Lawrence B, Torruellas W E, Stegeman G I and Trillo S, 1997, ‘Spatial modulational instability and multisoliton-like generation in a quadratically nonlinear optical medium’, *Phys. Rev. Lett.* **78** 2760.

Kang J U, Stegeman G I, Villeneuve A and Aichison J S, 1996, ‘AlGaAs below half bandgap: A laboratory for spatial soliton physics’, *J. European Optical Society , Part A Pure and Applied Optics* **5** 583.

Karamzin Yu N and Sukhorukov A P, 1975, 'Mutual focusing of high-power light beams in media with quadratic nonlinearity', *Zh. Eksp. Teor. Phys* **68**, 834 (1976 *Soviet Physics-JETP* **41** 414).

Kim H, Jankovic L, Stegeman G I, Carrasco S, Torner L, Eger D and Katz M, 2003, 'Quadratic spatial solitons in periodically poled $KTiOPO_4$' *Opt. Lett.* **280** 240.

Malendevich R, Jankovic L, Stegeman G I and Aitchison J S, 2001, 'Spatial modulation instability in a Kerr slab waveguide', *Opt.Lett.* **26** 1879.

Malendevich R, Jankovic L, Polyakov S, Fuerst R, Stegeman G, Bosshard Ch and Gunter P, 2002, 'Two-dimensional type I quadratic spatial solitons in $KNbO_3$ near non-critical phase-matching', *Opt. Lett.* **27** 631.

Nayfeh A H and Mook D T, 1979, *Nonlinear Oscillations* (J. Wiley and Sons, New York).

Ostrovskii L A, 1967, 'Self-action of light in crystals', *JETP Letters* **5**, 272.

Pelinovsky D E, Buryak A V and Kivshar Y S, 1995, 'Instability of solitons governed by quadratic nonlinearities', *Phys. Rev. Lett.* **75** 591.

Schiek R, Baek Y and Stegeman G I, 1996, 'One-dimensional spatial solitons due to cascaded second-order nonlinearities in planar waveguides', *Phys. Rev.* **E 53** 1138.

Shen, Y. R., 1984, *The Principles of Nonlinear Optics*, (J. Wiley and Sons, New York).

Stegeman G I, Hagan D J and Torner L, 1996, '$\chi^{(2)}$ cascading phenomena and their applications to all-optical signal processing, mode-locking, pulse compression and solitons', *Optical and Quantum Electronics* **28** 1691.

Stegeman G I, Roland Schiek, Fang H, Malendevich R, Jankovic L, Torner L, Sohler W and Schreiber G, 2003, 'Beam evolution in quadratically nonlinear 1-dimensional media: $LiNbO_3$ slab waveguides', *Laser Physics*, **13** 137.

Tikhonenko V, Christou J, Luther-Davies B, 1995, 'Spiraling bright spatial solitons formed by the breakup of an optical vortex in a saturable self-focusing medium', *J. Opt. Soc. Amer.* **B 12**, 2046.

Torner L, Menyuk C R and G.I. Stegeman G I, 1994, 'Excitation of soliton-like waves with cascaded nonlinearities', *Opt. Lett.* **19** 1615.

Torner L, Mihalache D, Mazilu D and Akhmediev N N, 1995, 'Stability of spatial solitary waves in quadratic media', *Opt. Lett.* **20** 2183.

Torner L, 1995, 'Stationary solitary waves with second-order nonlinearities', *Optics Communications* **114** 136.

Torruellas W E, Wang Z, Hagan D J, VanStryland E W, Stegeman G I, Torner L and Menyuk C R, 1995, 'Observation of two-dimensional spatial solitary waves in a quadratic medium', *Phys. Rev. Lett.* **74** 5036.

Trillo S and Ferro P, 'Modulational instability in second harmonic generation' 1995a, *Opt. Lett.* **20** 438.

Trillo S and Ferro P, 1995b, 'Periodical waves, domain walls, and modulational instability in dispersive quadratic nonlinear media', *Phys. Rev.* **51** 4994.

Trillo S and Torruellas W, 2001, *Spatial Solitons*, eds. Trillo S and Torruellas W (Springer-Verlag, Berlin).

Planar photonic crystal microcircuit elements

Thomas F Krauss

University of St Andrews, UK.

1 Introduction

Photonic crystals provide a fascinating platform for a new generation of integrated optical devices and components. Circuits of similar integration density as hitherto only known from electronic VLSI can be envisaged, finally bringing the dream of true photonic integration to fruition. What sets photonic crystals apart from conventional integrated optical circuits is their ability to interact with light on a wavelength scale, thus allowing the creation of devices, components and circuits that are several orders of magnitude smaller than currently possible. Apart from miniaturisation, a key property of photonic crystal components is their "designer dispersion", that is, their ability to implement desired dispersion characteristics into the circuit. Functionalities include optical delay lines ("slow light"), optical pulse compression/dilation and wavelength splitters. These dispersive aspects will increase in importance, since no other approach can deliver the degrees of freedom available with photonic crystals in terms of spectral (wavelength selection) and temporal (pulse shaping) properties on a miniature scale. Studying photonic crystals in the short pulse regime is therefore clearly a fruitful pursuit and the present paper describes the groundwork and conceptual understanding required for many of the forthcoming short pulse experiments.

In order to realise these functions, we need to understand the fundamental properties of photonic crystals and be able to fully master the technology of making them. Another key aspect is the understanding of the origin of propagation losses and their reduction to acceptable levels. While competitive values in terms of "loss per device" have now been demonstrated, the full potential of photonic crystal circuits will only be realised once many components can be cascaded together in complex circuits.

In the following, the underlying concepts, technology and device realisations of planar photonic crystals will be discussed. Another theme of the paper is the identification of areas where photonic crystals have particularly favourable properties. The discussion is restricted to semiconductor-based structures, mainly GaAs/AlGaAs and silicon on insulator (SOI), which provide the required high refractive index contrast (Δn>2) and have a mature fabrication technology. Among the many possible realisations of photonic crystals, the triangular lattice of holes is chosen because of its high degree of symmetry that leads to large and robust bandgaps. All experimental work is conducted in the "holes in dielectric" instead of the reverse "pillars in air" geometry because of its connected nature and the fact that the dielectric provides the required vertical confinement.

2 Fabrication

2.1 Lithography

Photonic crystals consist of features well below 1μm in size, and need to be fabricated with a precision of tens of nanometres. In photonic crystal research, the most commonly used pattern definition technique is e-beam lithography, where a focussed beam of electrons is scanned across the surface to generate the pattern. The resolution of this process tends to be limited by the resists and substrates used. Although the beam can be focussed down to less than a 2nm spot, minimum feature sizes are typically of order 10–20 nm, because of electron scattering in the resist and electron backscattering from the substrate. Despite this limitation, the achievable resolution is very impressive, as illustrated in Figure 1.

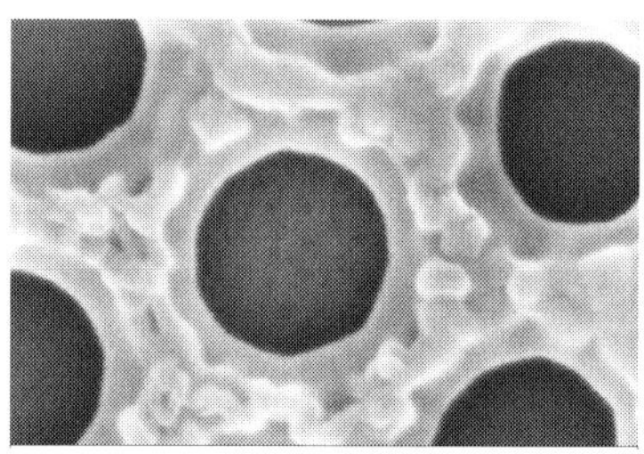

Figure 1. *Photonic crystal lattice exposed by electron-beam lithography after transfer into an SiO_2 mask layer. The central hole (200nm diameter) highlights the high resolution achievable with the process, as it was generated as a dodekanon (12-sided polygon) and most of the 12 corners are still visible. The hole at the top right is distorted, possibly due to some resist-related fluctuations. The uneven layer on the ridges between the holes is partially eroded resist and the smooth layer underneath is patterned SiO_2.*

State-of-the-art e-beam pattern generators use 16 bit D-A converters, which means that patterns with a resolution of 65k by 65k pixels can be written; in other words, for a resolution of 10nm, the maximum write-field that can be exposed is a 650μm square, although this size may also be limited by write-field distortions that are machine-dependent. For larger writing areas, write-fields are "stitched" together, using high-resolution laser interferometry stages. These stages operate with high precision, and have remaining stitching errors of typically 30–50nm, which lead to unacceptable phase-jumps in the photonic crystal pattern. The critical parts of a photonic crystal circuits are therefore always placed in a single write-field.

E-beam lithography, being a serial process, presents inherent limitations to high-volume mass-manufacture, so it is worthwhile to explore other routes to generating photonic crystal patterns. Classical photolithography, which is used for the current generation of planar lightwave circuits, lacks the resolution to fabricate these submicron structures with sufficient precision. Extending photolithography into the deep ultra-violet (DUV), however, where excimer lasers are used as light sources, changes the picture. With illumination wavelengths of 248nm, moving on to 193nm and eventually 157nm, DUV lithography offers both the resolution and speed required for the high volume mass-manufacture of photonic circuits. Recent results obtained using 8in SOI wafers are very promising indeed and highlight the viability of the process (Bogaerts *et al.* 2002). See Figure 2

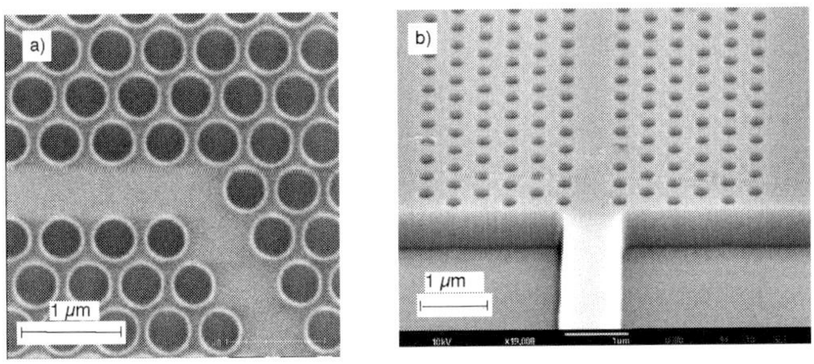

Figure 2. *SEM micrographs of photonic crystal waveguide patterns generated by DUV lithography. The period in both cases is 500nm, the hole diameter is (a) 380nm and (b) 230nm. Courtesy of W. Bogaerts, University of Gent, Belgium.*

2.2 Pattern transfer

Once written, the pattern is transferred into the semiconductor by dry etching techniques. Silicon-on-insulator based photonic crystals can be etched using the resist directly as a mask, because the required etch-depth is relatively low (200–300nm). The etch chemistry is typically fluorine-based, (e.g. SF_6), but chlorine-based chemistries have also been used (Bogaerts *et al.* 2002a). III–V based waveguides require a two-stage etching process because of the greater etching depth required (Krauss *et al.* 1996). A 200nm-thick SiO_2 layer is typically used as an intermediate mask and is etched with fluorine chemistry (e.g. CHF_3). The second, deep etching step is carried out with chlorine chemistry, which forms volatile compounds with the respective constituent atoms, e.g. $GaCl_3$ (GaAs) and $InCl_3$ (InP). Since $InCl_3$ is not volatile until about 130–150°C (Youtsey and Adesida, 1995), etching of InP compounds requires elevated temperatures.

Dry etching is a combination of physical and chemical components. The physical component arises from the bombardment of the reactive ions whereas the chemical component describes the chemical reaction between these ions and the etched material. Etching processes where the plasma is generated in a separate part of the etching chamber and then accelerated towards the specimen, for example, in an inductively coupled plasma (ICP), an electron-cyclotron resonance plasma (ECR) or an ion gun (Ion beam etching), allow separate control of the physical and chemical parameters. More conventional reactive ion etching, where the plasma is generated in a parallel-plate configuration directly in the etching chamber, gives less independent control, as most of the parameters are interlinked, but is inherently stable and robust. An illustration of the effects of the two etching

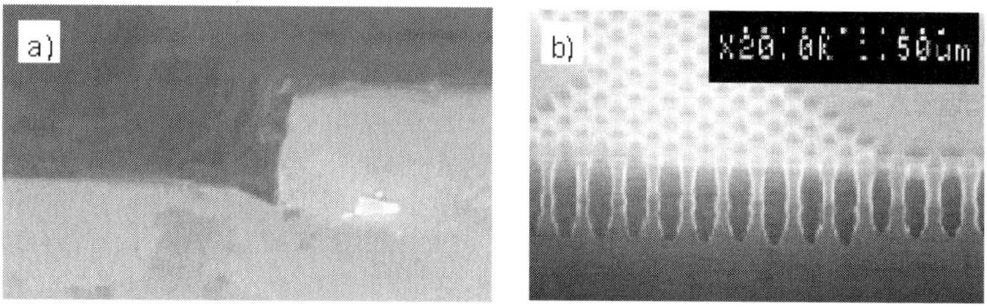

Figure 3. *Dry etching. (a) Excessive physical component. The ion bombardment dominates the etching process which leads to an overcut profile and a notch in the etch floor ("Ricochet") effect caused by ions deflecting off the sidewall). (b) Excessive chemical component leading to undercutting and layer selectivity in a GaAs/AlGaAs heterostructure. The chemical etching has preferentially removed layers of high Al-content that form the waveguide cladding, whereas the waveguide core half-way down the hole bulges out due to a lower lateral etch rate.*

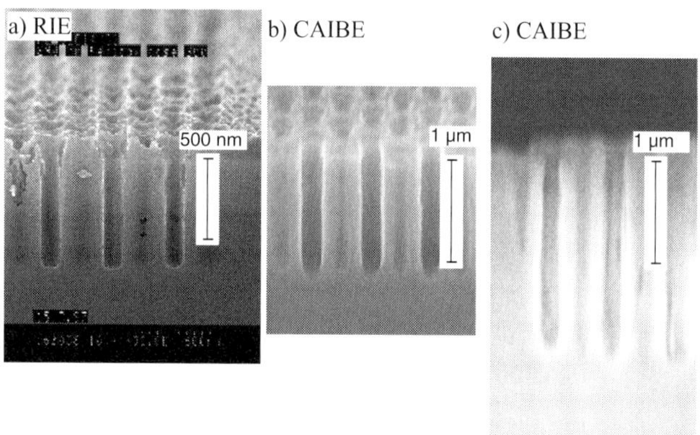

Figure 4. *Cross-sections of photonic crystal lattices etched using different techniques. The respective lattices are all of different size (260–340nm range) and have been scaled to allow a direct comparison of the aspect ratio. (a) Reactive Ion Etching (RIE). Having optimised all parameters, an aspect ratio of 1:6 (hole diameter: etch depth) is the best we were able to achieve while maintaining verticality of the holes. (b) and (c) Chemically assisted ion beam etching (CAIBE). By allowing independent control of the relevant parameters, aspect ratios of up to 10 have been achieved, with further improvement possible. Parameters are beam voltage 300V, beam current 17mA, substrate temperature 105° C and chlorine flow 10sccm. Etching time in (b) is 15min, whereas it is 30min in (c), enabled by the use of a more durable mask.*

components is shown in Figure 3. Balancing the physical and chemical components allows one to etch vertical holes as shown in Figure 4. Once a coarse balance has been achieved, the parameters such as gas flow, pressure, acceleration voltage and substrate temperature need to be optimised individually to optimise the process. Some results arising from such an optimisation are shown in Figures 4b,c.

3 Waveguide propagation losses

While two-dimensional photonic crystals have the advantage of being amenable to fabrication via the sophisticated planar technology developed by the silicon industry, their disadvantage is the lack of periodicity in the third dimension, so some other form of confinement is required. The most obvious approach is also the most commonly used, whereby the photonic crystal is formed in a slab waveguide structure, so photonic crystal effects in the plane are combined with total internal reflection (TIR) guiding in the vertical dimension. This seemingly simple"hybrid" combination of photonic crystals and total internal reflection has two major problems. Firstly, it assumes that the periodic structure extends infinitely deep into the waveguide and secondly, it suffers from the fact that there is no waveguiding in the low-index region, that is, the holes. Both of these effects lead to out-of-plane leakage and diffraction, which provide the major source of propagation loss. Additionally, there is roughness scattering arising from fabrication imperfections, lattice positioning and hole size variations. Finally, the finite size of waveguide components leads to losses, as every change in waveguide geometry constitutes a discontinuity that acts like a point scatterer. Which of these is dominant depends on the particular case and is difficult to assess in general, but some principal trends can be observed.

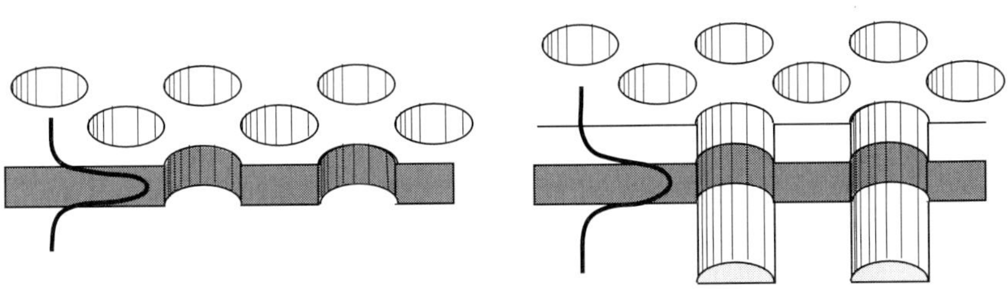

Figure 5. *The two principal types of planar photonic crystal waveguides. (left) Membrane-type, where the waveguide is a semiconductor membrane suspended in air or silica. (right) Heterostructure-type, where the light is guided by a "laser-like" heterostructure and sits on a high index substrate.*

Before discussing the different loss mechanisms, we need to introduce the two types of waveguides commonly used, that is, membrane and hetero-structure. (Figure 5). Membranes are typically made by patterning the waveguide core and completely removing the cladding, for example, by selective wet etching, thus producing a suspended structure with air on both sides. Silicon on insulator (SOI) waveguides can also be classified as membranes, because the silicon waveguide core sits on a silica cladding that offers equally high vertical confinement.

In contrast, the "laser-like" z semiconductor heterostructures only offer moderate vertical confinement (typically between n=3.4 (core) and n=3.2 (cladding)). This leads to leakage losses into the high-index substrate, but offers all the possibilities inherent to semiconductor opto-electronics, such as gain, carrier effects and electro-optic tuning as well as better heat-sinking. Channel waveguides are formed in these geometries by removing a single or multiple line of holes from the otherwise regular lattice.

3.1 Out-of-plane leakage

Out-of-plane leakage is the dominant loss mechanism for heterostructure waveguides. As all waveguide modes are situated above the light line, there is always the possibility of coupling to radiation modes. In simple terms, the origin of these losses is (a) the lack of waveguiding in the holes, with resulting diffraction loss and (b) the fact that the holes are not etched sufficiently deeply, so the tail of the mode radiates into the substrate. Both of these effects have been used to describe the losses in recent numerical treatments. A model by Hadley based on solving the 3D Helmholtz equation numerically (Hadley 2002) observes good agreement between the frequency ranges of high waveguide propagation losses and those where coupling to dominant radiation modes takes place. Correspondingly, there are also regions where coupling to radiation modes is weak and propagation loss is low. An alternative semi-analytical model proposed by Benisty *et al.*, (1999) describes the losses of the structure by placing a material with complex refractive index into the holes, which combines the effects of diffractive scattering at the holes and of radiation into the substrate. The model achieves reasonable agreement with experimental data and concludes that losses increase parabolically with vertical index contrast. In other words, it suggests that the losses are lower for heterostructure waveguides than for membranes, which appears surprising – we would expect that higher vertical index contrast leads to better confinement because it offers operation below the light line (Figure 6b). This apparent contradiction is addressed by Bogaerts *et al.* (2002b) who have studied losses

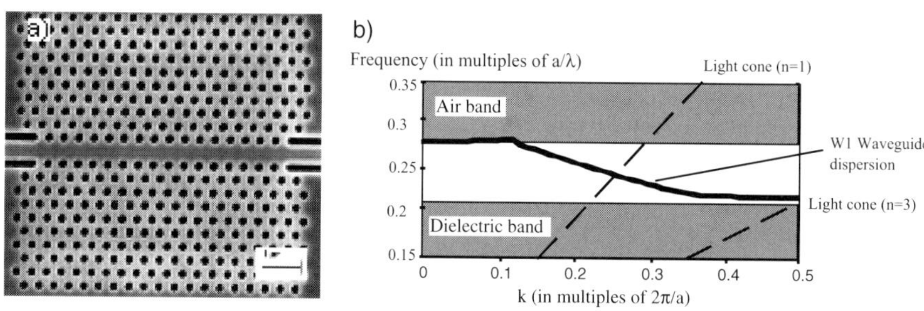

Figure 6. *SEM micrograph of a W1 waveguide. The term "W1" refers to a waveguide consisting of a single line of missing holes. Here, the light enters from the left and exits to the right via ridge waveguides that have been tapered down to the size of the W1. The dark lines are deeply-etched trenches that laterally confine the ridge waveguides. (b) Dispersion and the light cone for a W1 waveguide. For membrane-type waveguides, there is a region below the light cone (Here: frequency range between* $0.21 < a/\lambda < 0.23$*) where propagation is principally loss-less. Heterostructure waveguides tend to operate inside (or above) the light cone, i.e. there is the possibility of coupling to radiation modes at all operating frequencies.*

with an eigenmode expansion code and concluded that the losses, for small index contrast, indeed have the parabolic dependence proposed by Benisty, but then drop again once the index contrast is high enough for the waveguide to operate below the light line. Therefore, the model by Benisty appears correct for small index contrast, but the inherent assumptions do not hold for high index contrast and do not account for loss-less Bloch modes. (Figure 7)

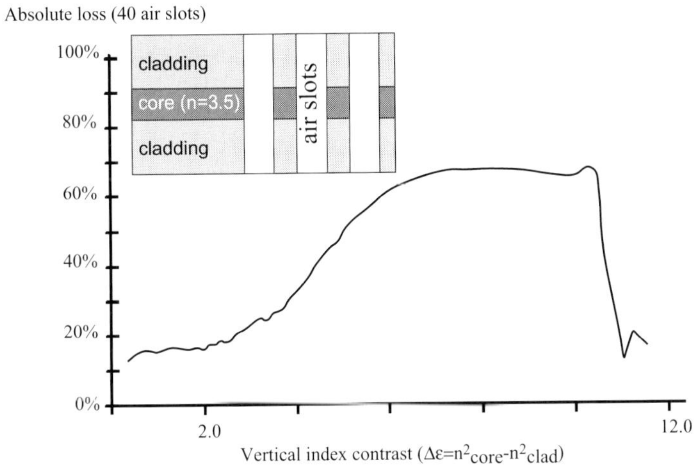

Figure 7. *Propagation loss of a waveguide mode after traversing 40 air slots as a function of vertical index contrast calculated numerically using an eigenmode expansion code. Please note that this is a 2D model, i.e. it assumes that the air features are slots rather than holes, so only gives a qualitative indication of losses in real photonic crystal waveguides. The losses increase quadratically from an initial low value, in agreement with Benisty et al. (1999), then saturate and eventually drop with the onset of Bloch modes around* $\Delta\epsilon \approx 11$. *Graph courtesy of W.Bogaerts, University of Gent, Belgium.*

All of the models conclude that low loss propagation above the light line is indeed possible and that current waveguide designs are not yet optimised. The key parameters to improve are diffractive loss at the holes (reduced by smaller hole size), scattering into the substrate (reduced by deeper holes) and improvement of the waveguide geometry, away from the current asymmetric surface waveguides. As can be seen from the origins of these losses, most of the possible improvements are technology-dependent, which explains the large amount of effort currently devoted to dry etch technology with the aim of etching holes with higher aspect ratio. Another method involves embedding the waveguide in a 3D photonic crystal material, as recently proposed by Toader and John, (2002), which would eliminate the possibility of substrate leakage altogether. Tentative projections suggest that values around 10dB/cm or lower will indeed be possible, initially in W3 waveguides (Lalanne, 2002), but should eventually also be achieved in W1 waveguides with these improvements in both design and technology, as well as sidewall roughness, which is considered next.

3.2 Roughness scattering

Waveguides that operate below the light line and that are in principle loss-less, such as air-clad membranes and silicon on insulator (SOI) types mainly suffer from roughness scattering. Roughness scattering is a well-known challenge in high-contrast waveguides, since the loss scales with $(\Delta n)^3$. Typical values of roughness are of order 10nm RMS, which is well sub-wavelength, but still significant because of the $(\Delta n)^3$ dependence. In silicon-based waveguides, the surface roughness can be reduced significantly by wet oxidation and subsequent wet etching, as demonstrated recently on narrow ridge SOI waveguides (Lee *et al.* 2001) where a loss-figure of 0.8dB/cm was achieved. While this result provides a useful benchmark, it can not, however, be compared directly with photonic-crystal based waveguides. The 0.8dB/cm result was obtained with ridges that had been oxidised down a 50nm Si-core. The effective index of such a guide is essentially that of the cladding, and it behaves mainly like a low-index waveguide, so the sharp bends and other attractive features of high-contrast waveguides are not possible. Nevertheless, this figure is clearly encouraging and points to the fact that low losses are ultimately achievable, even with high-index materials.

In photonic crystal waveguides, oxidation-smoothing has also shown promising results, as exemplified by W1 waveguides fabricated in SOI (Arentoft *et al.* 2002). The losses achieved in this system are 4dB/mm or lower, whereas the best results achieved with similar waveguides without oxidation are 6dB/mm (Notomi *et al.* 2002). While it is difficult to compare results from different groups that use different measurement techniques, the lower loss obtained with oxidation-smoothed guides supports the evidence that oxidation indeed reduces the sidewall roughness and leads to lower propagation losses.

3.3 Finite structures

The assumption of loss-less guiding breaks down as soon as discontinuities are introduced into the structure (Figure 8). A bend, for example, can be understood as the superposition of an infinite guide and a point defect. Propagation is then only possible where there is an allowed state in both the waveguide and the point defect, that is, where the two dispersion

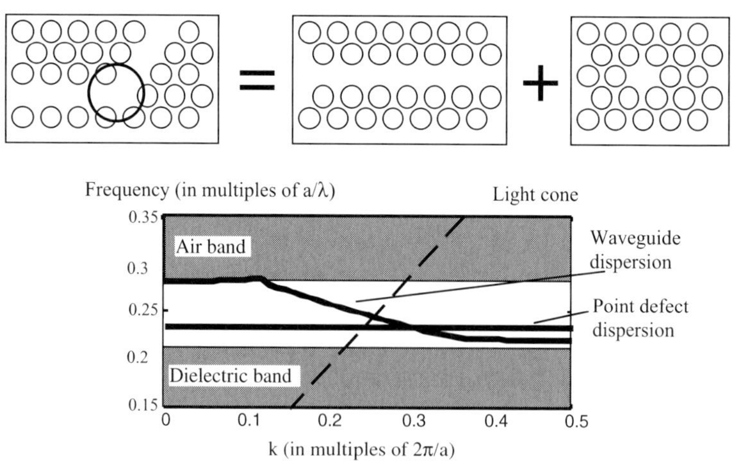

Figure 8. *Real structures are finite. A bend can be understood as the superposition of a line defect and a point defect. The point defect leads to a reduction of the bandwidth and a projection of the waveguide dispersion into the light cone causing radiation losses.*

curves overlap. This limits the bandwidth of the bend (here: to the Q of the defect, which is indicated by the width of the horizontal dispersion line) and opens an exit route for radiation modes, because the dispersion of the point defect extends into the light cone.

It is difficult to assess the magnitude of this effect, but according to (Bogaerts *et al.* 2002a), the out-of-plane scattering at discontinuities is larger for higher vertical index contrast structures, so such losses are expected to be higher in membranes than in heterostructures. Currently, there is little experimental evidence or 3D modelling available to draw a more quantitative conclusion, and we can also expect that these effects will be minimised by improved design.

4 Y-junctions

While straight waveguides and bends have now been studied extensively, the very important problem of bends and junctions that is essential for the operation of more complex circuits has only recently received attention (Lin *et al.* 2002; Boscolo *et al.* 2002a; Wilson *et al.* 2003).

Photonic crystals offer a real advantage in this case, as they allow, in principle, to construct 60° or even 90° bends at junctions, unlike in conventional integrated optics, where splitting angles are restricted to values around 2°. Let us consider the problem in more detail. If we join three W1 waveguides together, the 60° bend (or 120° split) is the most natural configuration in a triangular lattice. As light enters the junction, it experiences an optical volume that is slightly larger than that of the constituent waveguides and expands laterally (Figure 9a). This mode expansion is analogous to that in a multimode interference (MMI) coupler and leads to the excitation of higher-order modes at the output port. This issue can be addressed by placing a smaller hole at the centre of the junction. The optical volume is then reduced, the mode cannot expand and the excitation of higher order modes is suppressed (Figure 9b). As a result, efficient transmission over a wide bandwidth is observed (Boscolo *et al.* 2002a).

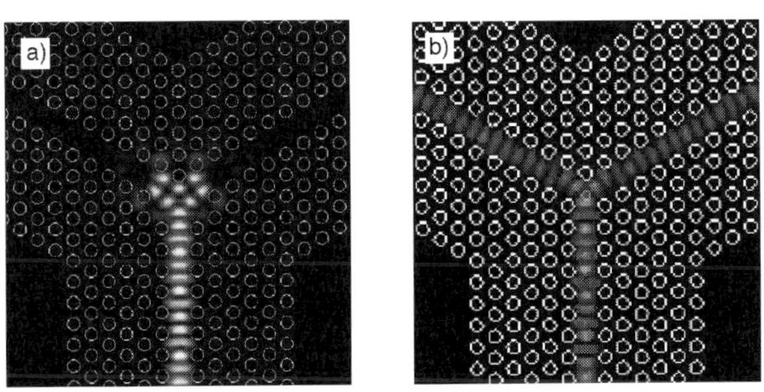

Figure 9. *Mode expansion at a Y-junction. Two additional holes have been reduced in order to highlight the mode expansion effect. Mode expansion still occurs at a simple junction, but not as obviously. (b) By adding a single, smaller hole at the centre of the junction, the mode expansion is suppressed, resulting in clean and efficient splitting. Figure courtesy of M. Midrio, University of Udine, Italy.*

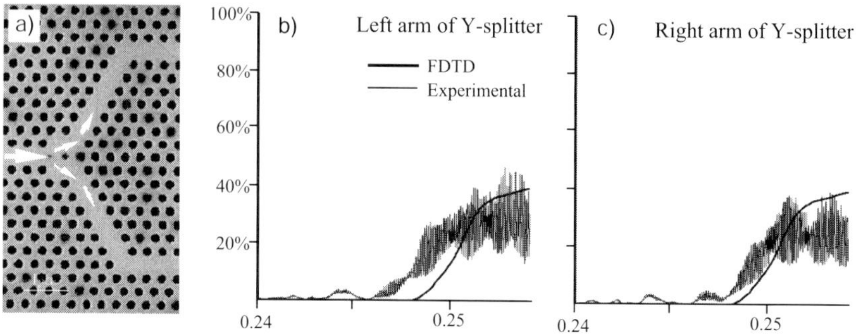

Figure 10. *SEM Micrograph and performance of a Y-junction with additional holes at the centre to avoid mode expansion. The vertical scale in (b), (c) represents transmission relative to a W1 waveguide. The oscillations in the experimental curves are due to Fabry-Perot resonances between the cleaved facets with superimposed resonances arising from reflections at the junction. The fact that the 2D FDTD agrees well with the 3D experiment indicates low out-of-plane losses. The total transmission is between 70–80% over a relatively broad spectral range (30–40 nm @ 1300 nm). The bandwidth is limited by the bend rather than the junction.*

This effect has also been observed experimentally. We used a configuration similar to Figure 9b and placed two smaller holes of increasing size at the centre of the junction in order to achieve a graded transition (Figure 10a). The results are very encouraging, with a good balance between the two ports and a total transmission around 80% being observed (Figure 10b,c).

Two of these Y-junctions mounted "back-to-back" constitute a Mach-Zehnder interferometer, which has also been operated successfully (passive, no tuning) with 28% transmission observed relative to a W1 waveguide. The design of the interferometer is extremely compact, with overall dimensions of the order of $10\times10\mu m^2$.

4.1 Impedance matching

Each change in waveguide geometry presents a discontinuity. Even the symmetric, adjusted Y-junction of Figure 9b has a transmission that is theoretically limited to 44.4% per branch, the remainder of the light being reflected back. In order to improve the transmission further, one can take measures to match the junction to the incoming mode. The general idea is based on the concept of impedance matching that is very familiar to microwave engineers: if the incoming mode is presented with the same impedance in every part of the circuit, it will propagate without reflections. If the impedances of the different circuit elements are different, matching sections have to be incorporated in order to transform the impedances. One such impedance transformer that can be realised in the photonic crystal platform is a "double-stub tuner". A single stub is analogous to a $\lambda/4$ impedance transformer, also known as an anti-reflection coating in optics. A double stub offers more flexibility and generally features a broader bandwidth. Before discussing the design of such a double stub tuner, we need to understand the concept of impedances in photonic crystal waveguides. A mode of a given polarisation propagating in a waveguide is described by its electric and magnetic fields. The impedance of the waveguide is then simply determined by the ratio of the two types of field, that is,

$Z = E(x, y)/H(x, y)$, assuming propagation in the z-direction (Boscolo *et al.* 2002b). By calculating a weighted average over the x, y dependence, we can obtain a single value for the impedance of the waveguide. If we now change the geometry of the waveguide, for example, by adding holes, and recalculate the impedance in the same way, we can obtain the characteristic impedance of the additional hole. Using this method, it turns out that the impedance of an additional hole is mainly inductive, with a small capacitive component, that is, it behaves like an L-C element. We now have a handle to tuning the circuit by designing holes with the appropriate L-C values to match the incoming waveguide an the junction. An example for a Y-junction design incorporating a double-stub tuner is shown in Figure 11.

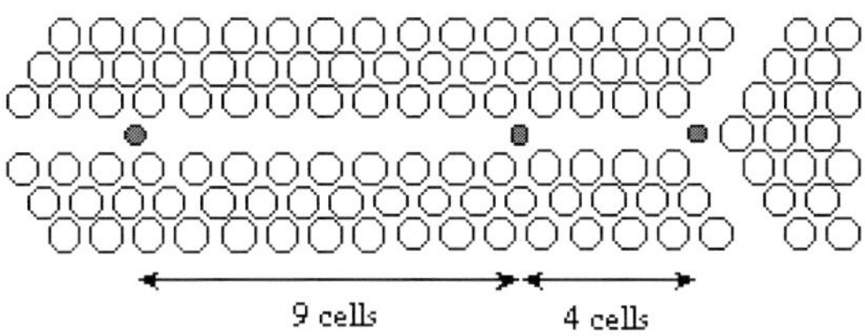

Figure 11. *Y-junction incorporating a hole at the junction for mode matching and two tuning holes in the waveguide for impedance matching. Simulations using the multiple scattering technique (Boscolo et al. 2002a) indicate that the tuning holes increase transmission of the system from 88.8% to 98.6%. Because the L-C components of the tuning hole impedance cannot be adjusted independently, 100% transmission is difficult to achieve.*

5 Dispersion

Amongst the most exciting aspects of photonic crystals are their dispersive properties that manifest themselves as strong temporal and spatial effects. Spatially dispersive structures are essential for wavelength-selective functions, such as WDM filters and multiplexers, and are discussed in the next section. Temporal dispersion is essential for functions such as optical delay lines and for the compression and dilation of optical pulses "pulse shaping". This high degree of dispersion occurs because of the strong interaction between the propagating mode and the lattice. It is a general feature of periodic structures, such as Bragg gratings, that dispersion is high in close proximity to the band-edge. The range of this interaction grows as the index increases, which is the reason why photonic crystals interact strongly with the guided mode over a large frequency range.

The strong dispersion is also one of the key distinctions between photonic crystal waveguides and high index contrast ridge waveguides "photonic wires". While photonic wires can confine light equally strongly as photonic crystal waveguides and also guide light sharply around bends, they do not offer the many possibilities offered by the dispersive nature of the photonic lattice.

The reason for the strong dispersion, as mentioned above, is that the propagation characteristics change dramatically at points where the mode strongly interacts with the

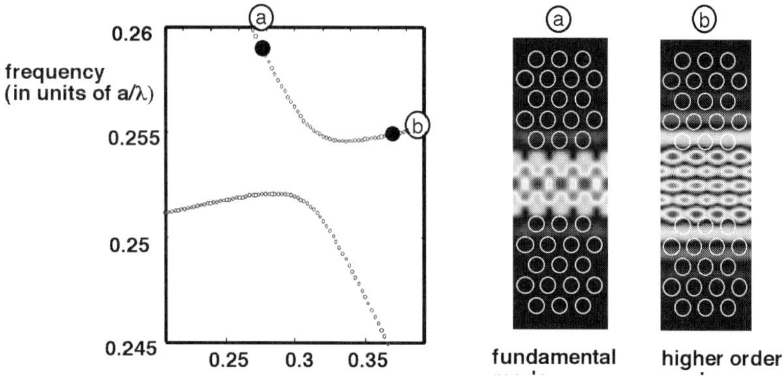

Figure 12. *Detail of the anti-crossing phenomenon observed in W3 photonic crystal waveguides. The mode profile and thereby the propagation constant changes dramatically near the mini-stopband, which highlights the dispersive nature of the waveguide.*

lattice, for example, at band-edges, where the mode changes from a propagating to a standing and then an evanescent wave. The same effect also happens at mini-band edges, due to the anti-crossing between modes of the similar symmetry, where the mode changes from one type to another, with a large resulting change in propagation constant (Olivier *et al.* 2002). See Figure 12.

5.1 Dispersion engineering

Coupled cavity waveguides (CCWs) provide a good example of the possibilities offered by photonic crystal waveguides for dispersion engineering. CCWs consist of chains of defects in an otherwise perfect lattice (Stefanou & Modinos 1998; Bayinder *et al.* 2000). They allow independent tuning of the spectral position, bandwidth and group velocity dispersion of the waveguide via the shape and size of the individual cavities and via the distance and lattice orientation between cavities. We have studied the propagation of light through a variety of CCWs and have already found good transmission for closely coupled systems (Karle *et al.* 2002). As an example, Figure 13 illustrates the example of a CCW for optical pulse-compression.

5.2 Slow wave structures

Another exciting concept that can be realised with photonic crystal waveguides is that of slow-wave structures, that is, propagation of light at greatly reduced speed. Optical delay lines or even optical waveguide memory can be envisaged, as well as electro-optic slow wave modulators that are more or efficient because of the longer interaction time. A recent realisation of slow wave propagation in photonic crystal waveguides has already demonstrated a more than ten-fold reduction in propagation speed (Notomi *et al.* 2001). A major issue that needs to be addressed for these devices to become practical, however, is that of impedance or mode matching between the conventional and the slow wave structure, as mismatch inevitably leads to reflection losses.

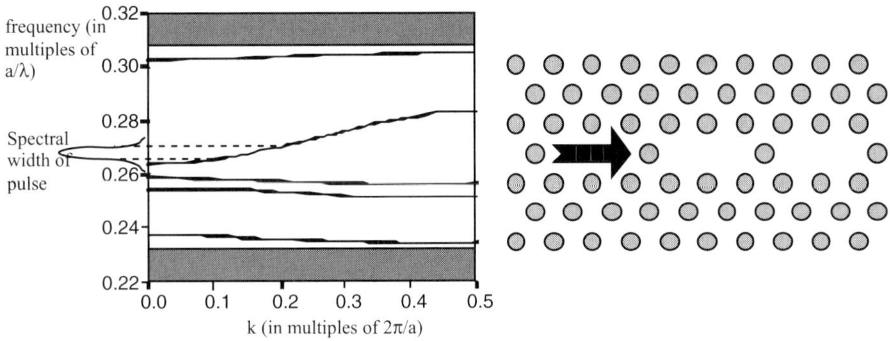

Figure 13. *Example of pulse compression in a coupled cavity waveguide. Assume a 100fs long pulse at a wavelength of 1.55μm enters the structure. The spectral width of the pulse corresponds to a frequency of 25nm, or 0.004 a/λ for a lattice of a=420nm. The pulse has been dilated, e.g. by travelling through a fibre, to a length of 1ps with the low-frequency "red" part of the pulse leading. Since the lower frequency experiences a lower group velocity (slope of the dispersion curve, dk/dω) in the CCW, it is now being delayed relative to the high frequency "blue" end of the pulse. Assuming that the ratio in group velocity across the pulse is approximately 2, red travels effectively twice as far as blue. In a material with index n=3, 100μm correspond to a time of 1ps, so the pulse can be re-compressed to its original length over a distance of 100μm.*

6 Spectral dispersion

One of the remarkable properties of photonic crystals is their high spectral dispersion at selected operating points, which is exploited in devices such as "superprisms" (Kosaka *et al.* 1999). The operating point of these devices typically lie near a band edge. Figure 14 shows such an operating point schematically, calculated for a triangular lattice of holes with a=320nm and 40% fill-factor, that was realised in a GaAs/AlGaAs heterostructure. The stopband for Γ-M direction opens up at $a/\lambda = 0.248$ as indicated. The same bandstructure is shown in Figure 14b as a wavevector diagram, whereby the thick curves represent the dispersion surfaces at $a/\lambda = 0.238$ and 0.254, respectively. Whereas the bandstructure represents the crystal by only showing the major symmetry axes, the wavevector diagram represents the different crystallographic axes as they occur in real space, so repeat every 60°. In other words, the wavevector diagram is obtained by "slicing" the bandstructure horizontally, at a given frequency.

What is then the origin of the superprism phenomenon is the fact that the dispersion surface , that is, the shape of the wavevector diagram, changes dramatically with a change in frequency. In the example, the dispersion surface changes from a near circular shape at $a/\lambda = 0.238$ to a set of triangles located at each K-point because, at $a/\lambda = 0.254$, a stopband has opened up in Γ-M direction and the only remaining solutions are clustered around the K-point. This change of shape of the dispersion surface yields a large change in output direction as illustrated in Figure 14b.

Following the demonstration of the superprism effect in the autocloned 3D system

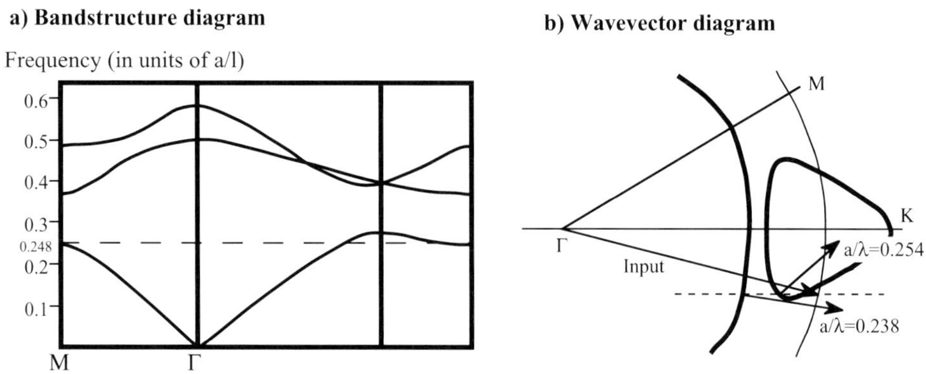

Figure 14. *Bandstructure and wavevector diagram for a triangular lattice of holes with 40% air fill-factor. The wavevector diagram highlights the change of the dispersion surface around a/λ=0.24−0.25. The angular dispersion is obtained as follows. The input wave travels in the un-patterned semiconductor waveguide, which is isotropic and therefore represented by a circular dispersion surface. (dotted line). At the interface, the parallel wavevector is conserved, so all possible solutions lie on the dashed line. The intersection with the dispersion surface then determines the direction of the output wavevector. For a/λ=0.238, the output is almost collinear with the input wavevector and there is little change in the propagation direction. For a/λ=0.254, the shape of the dispersion surface has changed dramatically, with a corresponding large change in the output direction.*

(Kosaka *et al.* 1999), we have now also been able to observe it in a planar geometry (Wu *et al.* 2002; Wu *et al.* 2003). Figure 15 shows the experimental layout and resulting output as a function of wavelength. At 1290nm, the light is mainly coupled into guide A.

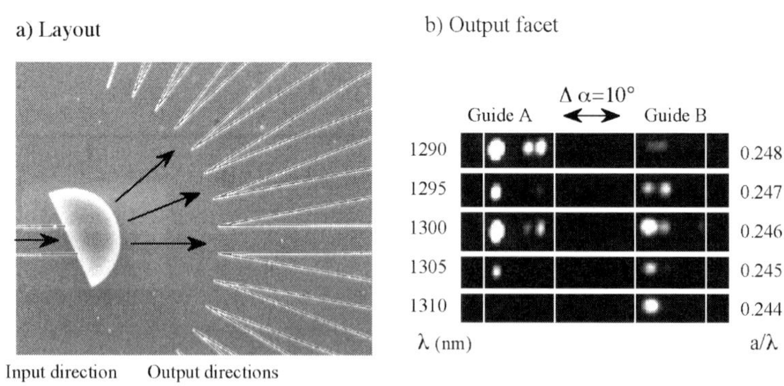

Figure 15. *Experimental geometry used to observe the superprism effect. The light is incident from the left, impinges on the photonic lattice and is dispersed into different output waveguides. The guides then carry the light to the output facet and are imaged onto a vidicon camera. (b) is a montage of the signal on two subsequent output waveguides as a function of wavelength.*

It then swings across to guide B as the wavelength is changed to 1310nm. Considering the angular difference between the waveguides of 10°, the wavelength swing corresponds to a dispersion of 10°/20nm. For comparison, the wavelength swing predicted from Figure 14b is approximately 50°/80nm, which is of similar magnitude.

6.1 The mechanism of high spectral resolution

In order to understand the mechanism of this high dispersion, it is instructive to compare the mode of operation with that of a regular grating. For simplicity, let us assume a 1-D grating of the same periodicity, that is, a=320nm in Figure 16.

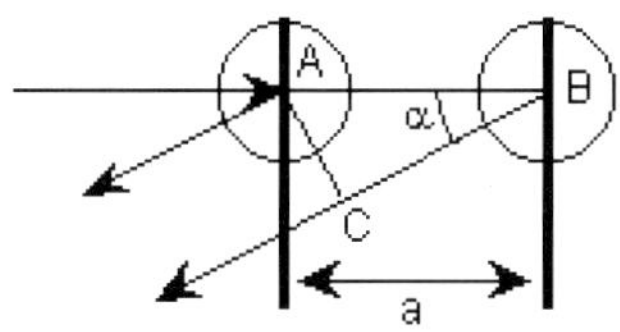

Figure 16. *Resolution of a grating of similar period as the superprism. The two circles at A and B represent the positions of the holes in an equivalent photonic lattice and are shown for illustrative purposes only.*

In order for constructive interference to occur, the path-length difference between the beam reflecting directly from point A and that reflecting from point B should be equal to a wavelength (or integer multiples), so that $\lambda = AB + BC = a + a\cos\alpha$. Inserting the same parameters as those used to calculate the superprism resolution (λ_1=1290nm, λ_2=1310nm, a=320nm and using n=2.5 for the average refractive index of the lattice), we find that α_1 =52.23° and, α_2 = 50.39°, so the angular resolution is approximately 2°/20nm or 5 times lower than that of the superprism. How is it possible that the superprism photonic lattice, which is effectively a 2-D grating, has a significantly higher resolution than a regular 1-D grating? The answer is related to the group velocity, and is best understood via time-of-flight considerations.

Fundamental quantum mechanics postulates that $\Delta p \Delta x \geq \hbar/2$ and $\Delta E \Delta t \geq \hbar/2$ (Heisenberg uncertainty principle). In terms of a normal grating, this tells us that if the grating is large (large Δx), we can resolve a small Δp, which is equivalent to a small Δk or a large $\Delta\lambda$, so large gratings have a high wavelength resolution (because the light can sample many periods). In the case of a cavity (Figure 17a), the Δp Δx explanation is more difficult, as the cavity is small (small Δx) yet we can achieve high resolution (small Δk), which is an apparent contradiction. This contradiction can be resolved by considering the ΔE Δt pair, that is, high resolution (small ΔE) is obtained via a long time-of-flight (large Δt). Light goes around the cavity many times, so has a long pathlength and therefore high resolution. The superprism works in a similar way. Light inside the lattice is reflected many times and is thereby slowed down, so the high resolution (small ΔE) is again achieved via a long interaction time (large Δt). It is no surprise, therefore, that the best operating point of the superprism is at the band-edge, where group velocity is lowest (Figure 14a). so long interaction times are achieved.

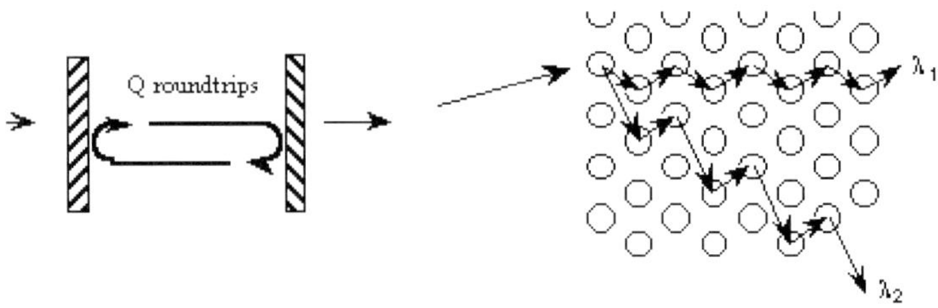

Figure 17. *Resolution of a cavity (left) and a superprism (right). High resolution in a cavity is achieved via a high number of round-trips that effectively slow the light down. Similarly, the light in the superprism experiences many reflections and is also slowed down, leading to a low group velocity.*

6.2 Limitations

While the fundamental superprism phenomenon is fascinating and has now been demonstrated in both 3-D and 2-D geometries, there are several design aspects that need to be addressed before superprism multiplexers can find practical applications. Firstly, the picture shown in Figure 14b is a simplification because it only shows a single propagating mode being excited in the lattice. In reality, multiple modes exist due to the fact that the construction line intersects multiple dispersion surfaces (not shown in Figure 14b).Maximising the coupling efficiency to the desired mode is therefore a necessary, yet difficult design challenge. Secondly, any practical spectrometer combines collimating and dispersive functions, for example, a lens at the input and output sides with a grating at the centre. The present superprism design only incorporates the dispersive function, so collimation needs to be added. Our recent work in this area has shown (Wu *et al.* 2003) that both collimation and superprism effects can be achieved in planar photonic crystals, and that the parameters of both effects lie relatively closely together. A combined collimator-superprism photonic crystal should therefore be possible, ideally with a gradual transition in-between.

7 Discussion and conclusion

Several examples of miniaturised photonic crystal circuit elements have now been demonstrated, such as sharp bends, Y-junctions and cavity waveguides. While this makes a clear argument in favour of photonic crystal circuits, miniaturisation alone may not be sufficient to ensure their major impact. Take WDM components as an example; only if miniaturisation can be achieved with comparable or better performance than already offered by existing devices will photonic crystal components succeed. State-of-the-art Arrayed-Waveguide Gratings (AWGs), for example, offer 50 GHz channel spacing with a channel accuracy of 2.5GHz and a cross-talk in excess of 30dB. These are benchmarks that are presently well out of reach of photonic crystal circuits based either on superprisms or cavities. Rather than simply copying existing functionalties, we need to look for applications that come natural or offer solutions that cannot otherwise be achieved. On the other hand, miniaturisation offers redundancy, that is, many circuits can be manufactured on the same piece with only the ones required for a given application being utilised, thus offering simplicity and flexibility at the same time.

If photonic crystal circuits are to succeed, we need to make more use of the inherent properties of the materials out of which we fabricate them. These materials are typically semiconductors, and we use them because of their high index contrast. The fact that the material properties can be adjusted with current injection and electric field effects, however, is mainly ignored in photonic crystal research. These effects offer tunability, but also gain and absorption (for amplifiers and detectors, respectively). A generic application would therefore be a modulator with integrated dispersion control, or a detector with dispersion compensation at the input side. Alternatively, the effective light-matter interaction in a modulator could be increased via slow wave structures, that is, photonic crystal waveguides that operate in a low group velocity regime. This should allow modulators to be realised with effective electro-optic coefficients that scale with the reduction in group velocity and be much shorter. Another aspect is that present-day integrated optical devices rely almost exclusively on thermal tuning, which is rather slow and scales badly due to thermal cross-talk and overall power budget.

Dispersion of photonic crystals is another fascinating property which can be engineered to suit a given application. Photonic crystal fibres have been made, for example, with either very low dispersion at unusual wavelengths or very high dispersion for "Supercontinuum generation". This high dispersion can be exploited in photonic crystal waveguides, for example in integrated dispersion compensation functions such as the compression and dilation of optical pulses or for slow wave propagation in optical delay lines.

Nonlinearities present further opportunities. The strong confinement achieved in photonic crystal waveguides and cavities offers itself for applications that require high peak fields for nonlinear phenomena, such as all-optical switching, but also wavelength conversion with the phase-matching provided by the periodic structure. While these prospects are very exciting, they are currently held back by the fact that typical nonlinearities are weak on the length scale of photonic crystals, and losses are too high. The obvious solution is to reduce propagation losses, a point already made repeatedly. Alternatively, one could investigate new materials; the search for compounds with more suitable electro-optic and nonlinear coefficients has only just started in a field that is almost exclusively concerned with III–V's and silicon-based materials. Whereas photonic crystals offer miniaturisation of passive components, it is the material coefficients of the hosts that need to provide the strong light-matter interaction that will allow us to create both passive and active functionalities on a small length-scale. Much work remains to be done at this frontier.

Finally, the high Q cavities that may lead to integrated quantum information processing devices rely on another unique property of photonic crystals, their ability to localise light. Although very high Qs in excess of 10,000 have already been predicted for planar photonic crystals (Vuckovic & Scherer 2002), higher values may require 3-dimensional lattices. Such lattices would have the additional benefit of full three-dimensional control over spontaneous emission and finally enable the dream of no-threshold, no-noise lasing that motivated the pioneers in this field.

In conclusion, we have shown some examples of the favourable properties of photonic crystals that can be usefully exploited in communications, namely miniaturisation and dispersion. Once the losses have been reduced to more acceptable levels, which appears realistic, and the electro-optic properties of the constituent materials are being utilised, we can indeed expect to see high density integrated photonic crystal circuits.

Acknowledgments

I would like to acknowledge the fruitful discussions and experimental results provided by T.J. Karle, L.J. Wu, M. Mazilu, M.V. Kotlyar and R. Wilson of St. Andrews University. We are indebted to the Nanoelectronics Research Centre at Glasgow University for access to the e-beam writer and dry etching facility, as well as their general technical support. The work is funded by the EU-IST "PICCO" programme, the UK-EPSRC "Ultrafast Photonics Collaboration" and the Royal Society via a University Research Fellowship.

References

Arentoft J, Sondergaard T, Kristensen M, Boltasseva A, Thorhauge M and Frandsen L, 2002, *Electron. Lett.* **38**, 27.

Bayindir M, Temelkuran B and Ozbay E, 2000, *Phys. Rev.* **B 61**, 11855.

Benisty H, Weisbuch C, Labilloy D, Rattier M, Smith C J M, Krauss T F, De la Rue R M, Houdre R, Oesterle U, Jouanin C and Cassagne D, 1999, *J. Lightwave Technol.* **17**, 2063.

Bogaerts W, Bienstman P, Taillaert D, Baets R and De Zutter D, 2002b, *Opt. Quantum Electron.* **34**, 195.

Bogaerts W, Wiaux V, Taillaert D, Beckx S, Luyssaert B, Bienstman P and Baets R, 2002a, *IEEE J. Select. Topic. Quantum Electron.* **8**, 928.

Boscolo S, Conti C, Midrio M and Someda C G, 2002a, *J. Lightwave Technol.* **20**, 304.

Boscolo S, Midrio M and Krauss T F, 2002b, *Opt. Lett.* **27**, 1001.

Hadley G R, 2002, *IEEE Photonics Technol. Lett.* **14**, 642.

Karle T J, Brown D H, Wilson R, Steer M and Krauss T F, 2002, *IEEE J. Select. Topic. Quantum Electron.* **8**, 909.

Kosaka H, Kawashima T, Tomita A, Notomi M, Tamamura T, Sato T and Kawakami S J, 1999, *Lightwave Technol.* **17**, 2032.

Krauss T F, De la Rue R M and Brand S, 1996, *Nature* **383**, 699.

Lalanne P, 2002, *IEEE J. Quantum Electron.* **38**, 800.

Lee K K, Lim D R, Kimerling L C, Shin J and Cerrina F, 2001, *Opt. Lett.* **26**, 1888.

Lin S Y, Chow E, Bur J, Johnson S G and Joannopoulos J D, 2002, *Opt. Lett.* **27**, 1400.

Notomi M, Shinya A, Yamada K, Takahashi J, Takahashi C and Yokohama I, 2002, *IEEE J. Quantum Electron.* **38**, 736.

Notomi M, Yamada K, Shinya A, Takahashi J, Takahashi C and Yokohama I, 2001, *Phys. Rev. Lett.* **87**, art. no. 253902.

Olivier S, Benisty H, Smith C J M, Rattier M, Weisbuch C and Krauss T F, 2002, *Opt. Quantum Electron.* **34**, 171.

Stefanou N and Modinos A, 1998, *Phys. Rev.* **B 57**, 12127.

Toader O and John S, 2002, *Phys. Rev.* **E 66**, art. No. 016610 Part 2.

Vuckovic, J and Scherer, A. 2002, *IEEE J. Quantum Electron.* **38**, 850.

Wilson R, Karle T and Krauss T F, 2003, *J. Phys. A* July 2003.

Wu L J, Mazilu M, Karle T and Krauss T F, 2002, *IEEE J. Quantum Electron.* **38**, 915.

Wu L J, Mazilu M and Krauss T F, 2002, accepted for *IEEE J. Lightwave Technol.*

Youtsey C and Adesida I, 1995, *J. Vac. Sci. Technol* **B 13**, 2360.

Propagation of short light pulses in photonic band-gap structures

A M Zheltikov

M V Lomonosov Moscow State University, Russia

1 Introduction

Recent years have witnessed revolutionary achievements in the generation and application of ultrashort pulses of electromagnetic radiation. High-order harmonics have been shown to be ideally suited to the synthesis of sub-femtosecond pulses (Paul *et al.* 2001, Hentschel *et al.* 2001). This breakthrough into the attosecond range of pulse durations opens exciting new opportunities for the investigation of ultrafast processes in matter. However, convenient and compact components for the transportation and manipulation of ultrashort pulses still have to be developed. Problems related to the dispersive spreading of waveforms with a few-cycle field waveforms need to be solved on both the conceptual and technical levels in order to make these pulses an efficient tool for practical applications. The light-localization abilities and dispersive properties of photonic band-gap (PBG) structures (for an overview, see, e.g., Yablonovitch 1993; Joannopoulos *et al.* 1995) are of considerable interest in this context, offering elegant and practical solutions to the problems of short-pulse control and transportation, together with spectral transformation and frequency conversion of such pulses.

In this chapter, we highlight examples of how the remarkable properties of PBG structures and waveguides can be used to transport short pulses and to enhance nonlinear-optical interactions with such pulses. We will focus our attention on the possibilities offered by photonic-crystal and microstructure (MS) fibres. The development of such fibres in recent years (Knight *et al.* 1996, 1998, Bennett *et al.* 1999, Zheltikov 2000) has resulted in several important breakthroughs in nonlinear optics. Due to the high degree of light confinement attainable in the core of such fibres, Fedotov *et al.* 2001a), nonlinear-optical effects can now be observed with sub-nanojoule femtosecond pulses and fibre samples of several centimeters length (Ranka *et al.* 2000a, Fedotov *et al.* 2002a). Supercontinuum

(SC) generation using low-energy light pulses (Ranka *et al.* 2000a, Wadsworth *et al.* 2002) is one of the most impressive achievements of nonlinear optics in microstructure fibres with important implications in optical metrology (Diddams *et al.* 2000, Holzwarth *et al.* 2000, and optical coherence tomography (Hartl *et al.* 2001). This is being studied intensely in the context of pulse compression and numerous spectroscopic applications (Fedotov *et al.* 2002b). Microstructure fibres have been shown also to produce enhancement of self- and cross-phase modulation, Fedotov *et al.* 2001a, as well as third-harmonic generation (Ranka *et al.* 2000b, Naumov *et al.* 2002). The possibilities of using hollow-core microstructure fibres for high-order harmonic generation are now being explored (Naumov and Zheltikov 2002a). The capability of enhancing the whole catalogue of basic nonlinear-optical phenomena is receiving wide recognition in various areas of modern science since it opens a new chapter in low-energy nonlinear optics. Due to progress in fibre technologies, this direction of nonlinear optics is developing in parallel with high-field nonlinear optics, allowing very efficient spectral transformations of light pulses, opening the ways to control the phase and the spectrum of short pulses, and competing with high-field physics in the generation of ultrashort pulses and synthesis of few-cycle field waveforms.

This chapter will demonstrate that PBG structures and materials and especially photonic-crystal and microstructured fibres open a new chapter in nonlinear optics, making the whole catalogue of nonlinear-optical interactions accessible to unamplified femtosecond pulses and low-power few-cycle field waveforms. We will demonstrate, in particular, that extremely broad spectral emission sometimes spanning several octaves, can be generated with nano- and even sub-nano-joule femtosecond laser pulses as a result of nonlinear-optical interactions of laser pulses localized in guided modes of microstructured and photonic-crystal fibres (PCFs). This supercontinuum generation process involves several nonlinear-optical effects and is highly sensitive to the dispersion properties of guided modes of electromagnetic radiation in microstructure and photonic-crystal fibres. Providing a deeper physical insight into supercontinuum generation and, generally, spectrum transformation processes in photonic-crystal materials and waveguides will, therefore, require at least a brief introduction into the basic nonlinear-optical processes, as well as into the theory of light-pulse propagation in dispersive media. Such an introduction is provided in the next section.

2 Physical introduction into the ultrafast optics of photonic crystals

The basic physical processes behind the propagation, spectral transformation, and frequency conversion of ultrashort light pulses in photonic-crystal materials and waveguides include effects related to material dispersion and dispersion of waveguide modes, self- and cross-phase modulation processes, sum- and difference-frequency generation due to multiwave mixing, optical harmonic generation, stimulated Raman scattering, and sometimes ionization processes. In the regime of strong spectral broadening, modulation instabilities and shock-wave formation should be taken into consideration also. These effects distort the spectrum and the waveform of a short light pulse propagating through a fibre. However, in first-order approximation, such effects can often be ignored. Each of the above-mentioned processes is examined in a more detailed way below.

2.1 Dispersion effects

Dispersion effects play the key role in the propagation of ultrashort pulses through PBG structures and photonic crystals. Generally, the dispersion frequency dependence of the refractive index of materials leads to the distortion and spreading of short light pulses. Dispersion effects also limit the efficiency of nonlinear-optical interactions, as they give rise to phase and group-velocity mismatches of light pulses.

Dispersion effects can be qualitatively understood in terms of the expansion of the wave number k as a power series in frequency around the central frequency of the pulse ω_0 (obviously, such an approach may encounter difficulties in the case of light pulses as short as several field cycles):

$$k(\omega) = k_0 + \frac{1}{u}(\omega - \omega_0) + \frac{1}{2}k_2(\omega - \omega_0)^2 + \cdots \tag{1}$$

where we have $k_0 = k(\omega_0)$, and $u = (\partial\omega)/(\partial k)|_{\omega_0}$ is the group velocity of the pulse, and $k_2 = (\partial^2 k/\partial\omega^2)|_{\omega_0}$ is the group-velocity dispersion. The first term in Equation 1 describes effects related to the frequency dependence of the phase velocity of radiation in a medium, including phase mismatches in nonlinear-optical interactions. The second term in Equation 1 results in group-delay effects in linear and nonlinear-optical processes. The third term is responsible for group-velocity dispersion, which is especially important for ultrashort light pulses. This term describes the spreading and compression of short pulses. For media where we can restrict our analysis to the terms up to the second order in the expansion of the dispersion relation (Equation 1), the complex amplitude of the pulse at the output of the medium $E(t)$ is related to the complex amplitude of the input pulse $E_0(t)$ by the following expression (Akhmanov and Nikitin 1997):

$$E(t) = \frac{1}{\sqrt{2\pi i k_2 z}} \exp(-ik_0 z) \exp\left[\frac{i(t - z/u)^2}{2k_2 z}\right] \int_{-\infty}^{\infty} E_0(t) \exp\left(\frac{i\xi^2}{2k_2 z} - \frac{i\eta\xi}{k_2 z}\right) d\xi\,. \tag{2}$$

Higher order dispersion effects distort the waveforms of light pulses, which often becomes a serious limiting factor in short-pulse formation and in ultrafast optics in general.

One of the main advantages of PBG structures and photonic crystals, which allows the transportation of ultrashort pulses and the enhancement of nonlinear-optical interactions of such pulses, is associated with the possibility of using the dispersion of a periodic structure, as well as the dispersion of guided modes, in the case of waveguide structures, to compensate for the material dispersion (Fedotov *et al.* 2002c). This circumstance can be employed to reduce the dispersion spreading of ultrashort pulses, phase-match nonlinear-optical interactions, and to control spectral transformation processes.

The idea of using the dispersion of structures with a spatially periodic modulation of the refractive index to phase match nonlinear-optical interactions has been discussed since the pioneering work of Bloembergen and Sievers (1970). Physically, this approach to phase matching rests on the generalized momentum conservation for periodic structures (Yariv and Yeh 1977), the period of the structure should be on the order of the optical wavelength to make such phase matching possible, which implies that photonic band gaps, arising due to the strong coupling of forward and backward waves, become essential for the propagation and phase matching of light pulses. This relation between the spatial period of the structure and the optical wavelength is also important for understanding the difference between phase-matched nonlinear-optical processes in PBG structures and quasi-phase-matched wave mixing in periodically poled materials (Armstrong *et al.* 1962, Fejer *et al.* 1992). Quasi-phase matching implies that the phase mismatch is compensated by a jump-wise phase shift, introduced through the inversion of the sign of the nonlinear-optical susceptibility, while PBG structures allow the phase mismatch to be compensated on the geometrical dispersion introduced on a spatial scale of the order of optical wavelength.

A detailed analysis of band-edge phase-matching in one-dimensional PBG structures (Scalora *et al.* 1997, Zheltikov *et al.* 2000, shows that such structures offer several remarkable opportunities. In particular, due to local-field enhancement effects occurring within a limited spectral range, the efficiency of second-harmonic generation in such structures grows as a function of the nonlinear interaction length even faster than in the case of perfect phase matching (Scalora *et al.* 1997). Previous experimental results on multi-layer samples are very encouraging (Golovan *et al.* 1999, Dumeige *et al.* 2001), showing that, in many important situations, PBG structures allow the phase-matching problem to be solved in an efficient and simple fashion.

Waveguide regimes of nonlinear-optical interactions in PBG structures add new physical aspects to the phase-matching problem and conversion efficiency, as the dispersion of waveguide modes contributes to the nonlinear-optical process (Fedotov *et al.* 2002c. Although the opportunity of using waveguides with a periodically modulated refractive index to phase-match nonlinear-optical processes has been understood a long time ago (van der Ziel *et al.* 1976), recent achievements in material-processing technology and, recent advances in the theory of nonlinear-optical interactions in waveguides with a periodically modulated refractive index (Pezzetta *et al.* 2001), make them very promising candidates for the construction of compact and efficient elements for the frequency conversion of low-energy laser pulses.

2.2 Self-phase modulation

Self-phase modulation is a phenomenon that results from the dependence of the refractive index of a medium on the intensity of a light pulse. In the case of light fields of moderate intensities, the refractive index of a medium can be written as

$$n = n_0 + n_2 I(t)\,. \tag{3}$$

In this equation n_0 is the refractive index of the medium in the absence of a light field, $n_2 = (2\pi/n_0)^2\chi^3(\omega;\omega,\omega,-\omega)$ is the nonlinear refractive index, $\chi^{(3)}(\omega;\omega,\omega,-\omega)$ is the third-order nonlinear-optical susceptibility of the medium, and $I(t)$ is the intensity of laser radiation. Then, the nonlinear (intensity-dependent) phase shift of a pulse at a distance L is given by

$$\Phi(t) = \frac{\omega}{c} n_2 I(t) L\,. \tag{4}$$

Due to the time dependence of radiation intensity within the light pulse, the nonlinear phase shift is also time-dependent, giving rise to a generally time-dependent frequency deviation:

$$\Delta\omega(t) = \frac{\omega}{c} n_2 L \frac{\partial I}{\partial t}\,. \tag{5}$$

The resulting spectral broadening of the pulse can be estimated in the following way:

$$\Delta\omega = \frac{\omega}{c} n_2 L \frac{I_0}{\tau}\,, \qquad (6)$$

where I_0 is the peak intensity of the light pulse and τ is the pulse duration.

It is physically instructive to consider the compression of chirped light pulses in the time domain. Since the frequency of a chirped pulse changes from its leading edge to its trailing edge, dispersion of our compressor should be designed in such a way as to slow down the leading edge of the pulse with respect to the trailing edge of the pulse. In other words, the group velocities for the frequencies propagating with the leading edge of the pulse should be lower than the group velocities for the frequencies propagating with the trailing edge of the pulse. This can be achieved by designing a dispersive element with the required sign of dispersion and appropriate dispersion relation. Systems of diffraction gratings and, more recently, multilayer chirped mirrors (see the review by Brabec and Krausz 2000) are now widely used for the purposes of pulse compression. In certain regimes of pulse propagation, self-phase modulation and pulse compression may take place in the same medium.

Self-phase modulation results in the spectral broadening of a light pulse propagating through a hollow fibre. This effect allows compression of light pulses through the compensation of the phase shift acquired by the pulse in a hollow fibre. Compensation of a linear chirp, corresponding to a linear time dependence of the instantaneous frequency, is straightforward from the technical point of view. Such a chirp arises around the maximum of a Gaussian light pulse, where the temporal pulse envelope can be approximated by a quadratic function of time.

2.3 Cross-phase modulation

Cross-phase modulation is a result of nonlinear-optical interaction of at least two physically distinguishable light pulses (*i.e.* pulses with different frequencies, polarizations, mode structures, etc.) related to the phase modulation of one of the pulses (a probe pulse) due to the change in the refractive index of the medium induced by another pulse (a pump pulse).

The cross-action of a pump pulse with a frequency ω_1 on a probe pulse of frequency ω_2 gives rise to a phase shift of the probe pulse, which can be written as (Agrawal 1989)

$$\Phi_{XPM}(\eta, z) = \frac{3\pi\omega_2^2}{c^2 k_2}\,\chi^{(3)}(\omega_s;\omega_s,\omega_p,-\omega_p)\int_0^z \left|A_p\left(\eta - \frac{\zeta}{\sigma}, 0\right)\right|^2 d\zeta\,, \qquad (7)$$

where $\chi^{(3)}(\omega_s;\omega_s,\omega_p,-\omega_p)$ is the third-order nonlinear-optical susceptibility of the medium; $1/\sigma = 1/u_1 - 1/u_2$; u_1 and u_2 are the group velocities of the pump and probe pulses, respectively; and k_2 is the wave number of the pump pulse. Taking the time derivative of the nonlinear phase shift, we arrive at the following expression for the frequency deviation of the probe pulse:

$$\delta\omega_{XPM}(\eta, z) = -\frac{3\pi\omega_2^2}{c^2 k_2}\chi^{(3)}(\omega_s;\omega_s,\omega_p,-\omega_p)\,\sigma\left[\left|A_p(\eta,0)\right|^2 - \left|A_p\left(\eta - \frac{z}{\sigma}, 0\right)\right|^2\right]. \qquad (8)$$

As in self-phase modulation, cross-phase modulation can be employed for pulse compression. The dependence of the chirp of the probe pulse on the pump pulse intensity can be used to control parameters of ultrashort pulses (Zheltikov 2002). Cross-phase modulation also opens the way to study the dynamics of ultrafast nonlinear processes, including multi-photon ionization, and to characterize ultrashort light pulses through phase measurements on a short probe pulse (Le Blanc and Sauerbrey 1996, Fedotov *et al.* 1997).

2.4 Multiwave mixing and high-order harmonic generation

Coherent multiwave mixing and high-order harmonic generation play an important role in the nonlinear optics of short pulses in PBG structures and fibres, allowing high efficiencies of frequency conversion and high sensitivities of methods of coherent nonlinear spectroscopy to be achieved. In the case when pump pulses with frequencies ω_1 and ω_2 are coupled into a nonlinear medium, coherent multiwave mixing leads to the generation of new spectral components with frequencies $\omega_{mn} = m\omega_1 \pm n\omega_2$, where m and n are integers. The cases of $m = 0$ or $n = 0$ then correspond to the generation of nth- and mth-order harmonics of radiation with frequencies ω_2 and ω_1.

In the case of centrosymmetric media, for example, gases filling the core of hollow photonic-crystal fibres, even-order nonlinear processes are usually characterized by much lower efficiencies as compared to odd-order nonlinear processes. However, even-order nonlinear-optical processes may take place as a result of inhomogeneities in a medium or in the light field violating the central symmetry of the problem. This may be the case, for example, in the regime of gas ionization.

Specific features of multiwave mixing related to the guided-wave propagation of ultrashort pulses in a gas filling a hollow fibre include large interaction lengths, the presence of the waveguide dispersion component, and multimode regimes of wave mixing. All these physical factors have to be taken into consideration for the optimization of nonlinear-optical frequency conversion and short-pulse formation in hollow fibres. In particular, the increase in the wave-mixing length may improve the efficiency of frequency conversion only when the nonlinear process is phase-matched on the spatial scale on the order of the fibre length (Fedotov *et al.* 2001b).

2.5 Stimulated Raman scattering

Stimulated Raman scattering (SRS) is one of the most extensively used and intensely studied phenomena in nonlinear optics. The SRS effect arises due to the modulation of laser radiation by molecular vibrations in a medium. In contrast to spontaneous light scattering, stimulated scattering involves molecular modes excited and phased by light in a large volume of a medium. This results in the generation of highly intense and well-directed radiation.

Multiple Stokes and anti-Stokes components produced by SRS are separated by the frequency Ω of molecular vibrations. A simple model of the SRS process interprets this phenomenon as a result of harmonic phase modulation of a light pulse with the frequency Ω of molecular vibrations. The spectrum of a laser pulse propagating in a Raman-active medium under these conditions becomes dressed with Stokes and anti-Stokes components.

The equidistant frequency components arising in the spectrum of such a pulse can be employed, as shown by Harris and Sokolov (1998) and Nazarkin *et al.* (1999), for the synthesis of ultrashort light pulses. As recently shown by Russell's group (2002), stimulated Raman scattering can be substantially enhanced by the use of hollow-core photonic-crystal fibres.

2.6 Effects related to gas ionization

The goals of many nonlinear-optical studies performed with short laser pulses propagating in photonic band gap structures, especially in photonic-crystal fibres with gas-filled hollow cores, usually involve efficient frequency conversion or formation of ultrashort pulses. Sometimes, high-order optical nonlinearities are employed for this purpose, and laser pulses with sufficiently high power densities can be coupled into a fibre. Under these conditions, ionization effects may have a considerable influence on both dispersion and nonlinear-optical properties of the gas filling the fibre. If the collision frequency in the plasma produced as a result of this ionization process is much lower than the frequency of optical radiation, then the nonlinear additive to the refractive index of the gas associated with plasma electrons is given by

$$\Delta n = -\frac{\omega_p}{2\omega^2}, \tag{9}$$

where $\omega^2 = 4\pi e^2 N_e/m$ is the plasma frequency, e is the electron charge, N_e is the electron plasma density, and m is the electron mass.

Dynamic variations in optical properties of the gas accompanying the propagation of a short pulse through a hollow fibre, gives rise to the nonlinear change in the refractive index of the gas, changed phase and group-velocity matching and may result in the temporal and spatial self- and cross-action of short light pulses. These phenomena, as shown by Tempea and Brabec (1998), can also be employed to produce ultrashort pulses. Such effects may also have a considerable influence on phase- and group-velocity matching in nonlinear optical interactions of ultrashort pulses (Naumov *et al.* 2001a).

2.7 Supercontinuum generation

Supercontinuum generation is one of the most exciting topics in modern laser physics; this is the area where high-field science meets the physics of low-energy unamplified ultrashort pulses in the most amazing way. Although the strong-field chapter of supercontinuum-generation story was opened back in the 1970s (a comprehensive and illuminating overview of the early phases of research on supercontinuum generation is provided in the book *The Supercontinuum Laser Source* (Alfano, 1989) many interesting and physically important aspects have been added to this part of the supercontinuum story in the past few years, as laser pulses were becoming shorter and shorter, while computers and computer codes, were becoming more and more powerful and sophisticated. The rapid technical and conceptual progress in microstructure fibres has recently resulted in the appearance of a whole family of fibres, such as photonic-crystal, holey, and tapered fibres, capable of generating supercontinuum emission with unamplified, nano- and even subnanojoule femtosecond pulses, leading to revolutionary changes in frequency metrology and opening new horizons in ultrafast science, short-pulse control, as well as spectroscopic and biomedical applications.

As a physical phenomenon, supercontinuum generation involves the whole catalogue of classical nonlinear-optical effects, such as self- and cross-phase modulation, four-wave mixing, stimulated Raman scattering, and many others, which add up together to produce emission with extremely broad spectra, sometimes spanning over a couple of octaves. This is, therefore, a challenging task to develop an adequate physical model for this phenomenon based on the understanding of all its nonlinear-optical ingredients. Such a model may, perhaps, show the ways of generating few-cycle and subcycle light pulses in a simple and compact fashion, making these pulses a routine tool for every-day experiments not only in physics, but also in other areas of modern science. A detailed understanding of the properties of supercontinuum is crucial for assessing the possibilities of using this broadband emission for spectroscopic, metrological, biomedical, and pulse-compression purposes. This is why many equally accurate and elegant measurements have been carried out recently on supercontinuum produced in both strong and weak fields and much work is in progress now.

With basic ingredients of supercontinuum generation, such as self- and cross-phase modulation, four-wave mixing, and stimulated Raman scattering considered above, we will provide now a brief overview of the theoretical framework that can be employed to qualitatively describe supercontinuum generation, sometimes allowing important aspects of this process to be understood in a physically clear and illustrative way.

The generic one-dimensional equation for the normalised envelope $u(z,t) = A(z,t)/A_0$ of a short light pulse, say, with an initially Gaussian shape, $A(z{=}0,t) = A_0 \exp(-t^2/2\tau^2)$ propagating in a nonlinear medium (e.g., in a fibre) can be represented as (Gaeta 2002)

$$\frac{\partial u}{\partial \xi} + i \sum_{k\geq 2} \frac{|L_d^{(2)}|}{k! L_d^{(k)}} \frac{\partial^k u}{\partial \eta^k} = i \left(1 + \frac{i}{\omega_0 \tau} \frac{\partial}{\partial \eta}\right) P_{nl}\,, \tag{10}$$

where $L_d^{(k)} {=} \tau^k/\beta_k$ is the kth-order dispersion length, β_k are the coefficients in the dispersion expansion series (1) at ω_0 the central frequency, $\xi = z/|L_d^{(2)}|$, $\eta = (t - z/u)/\tau$, and P_{nl} is the nonlinear polarization, which includes all the relevant nonlinear-optical processes, such as self- and cross-phase modulation, four-wave mixing, stimulated Raman scattering, etc. One, of course, has to specify the nonlinear polarization to numerically simulate supercontinuum generation, and the choice of the model for P_{nl} is of crucial importance for an adequate description of this very complicated nonlinear-optical process.

Supercontinuum-generating materials and structures now constitute an important class of broadband radiation sources, opening new horizons in various areas of optical spectroscopy, including Raman and high-precision spectroscopy. It has been shown recently (Knight *et al.* 1996, Knight *et al.* 1998, Ranka *et al.* 2000a, Zheltikov 2000) that microstructure (MS) fibres—a novel class of fibres where a solid cladding, typical of standard optical fibres, is replaced by a silica/air or glass/air microstructure cladding (Figure 1)—offer new exciting possibilities in supercontinuum (SC) generation (Ranka *et al.* 2000a, Wadsworth *et al.* 2002), suggesting several practical and elegant solutions to the problems of optical metrology and high-precision measurements, optical coherence tomography, and short-pulse generation. Supercontinuum-emitting MS fibres advantageously supplement the existing SC sources, holding much promise for broadband spectroscopy, as well as for time-resolved pump–SC probe measurements. Due to the high degree of light-field confinement in the core of MS fibres and the possibility to tailor the dispersion of guided

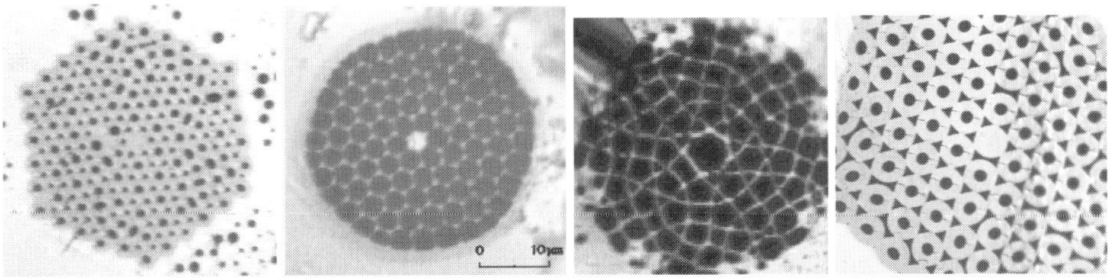

(a) (b) (c) (d)

Figure 1. *Cross sectional images of microstructure fibres: (a, b) photonic-crystal fibres with a pitch of the cladding equal to 3μm and the air-filling fraction of (a) 16% and (b) 65%; (c) a cobweb microstructure fibre; (d) a fibre with a single missing hole at the centre serving as a fibre core and additional air gaps between closely packed glass capillaries.*

modes, SC generation can be observed with low-energy, unamplified femtosecond laser pulses, suggesting the way to create a compact and convenient broadband source for numerous applications. Several important physical aspects have been recently raised in the context of SC generation in MS fibres, with the main emphasis being placed on the mechanisms behind SC in femto- (Herrmann *et al.* 2002) and pico-second (Coen *et al.* 2001) regimes, as well as on the coherence properties of SC and its compressibility and controllability.

3 Frequency-tunable supercontinuum generation in photonic-crystal fibres by femtosecond pulses

3.1 MS fibres and new solutions in supercontinuum generation

SC generation has long been used as a convenient experimental technique to produce broadband radiation for various spectroscopic applications, as well as pump–probe measurements. Light intensities required to generate a supercontinuum can be substantially lowered by using photonic-crystal fibres (PCFs) (Ranka *et al.* 2000a, Wadsworth *et al.* 2002) (also called more generally microstructure or holey fibres) and tapered fibres (Birks *et al.* 2000, Akimov *et al.* 2001). Physically, the ability of photonic-crystal and tapered fibres to enhance nonlinear-optical processes is due to the high degree of light localization in the core of such fibres, permitting high intensities of laser radiation to be achieved with relatively low energies of laser pulses. Few-nanojoule and even subnanojoule femtosecond pulses lead to supercontinuum generation under these conditions, giving rise to emission spectra spanning sometimes over two octaves. This already resulted in a major breakthrough in optical metrology (Diddams *et al.* 2000, Holzwarth *et al.* 2000), allowing the creation of compact frequency chains phase-coherently linking radio-frequency reference sources to the optical region through frequency combs generated by femtosecond mode-locked lasers. The function of PCFs in these metrological systems is to spectrally broaden femtosecond frequency combs, making them span more than an octave and allowing the offset frequency related to intracavity dispersion to be measured and controlled.

Supercontinuum generation in holey and tapered fibres also offers a method of creating new broadband sources for spectroscopic applications and suggests a new approach to pulse compression. In general, the possibility to tune and tailor dispersion of waveguide modes, offered by PCFs, makes these fibres extremely useful for ultrafast optics, including the transmission of short pulses, frequency conversion (e.g., through efficient harmonic generation), compression, and spectral control of ultrashort laser pulses, even opening the way for subfemtosecond fibre optics. Applications of such fibres in telecommunication technologies is another very important issue, which is being actively explored at the moment.

We will show below how a frequency-tunable laser source—an optical parametric amplifier (OPA), generating radiation within the range of wavelengths from 1.1 to 1.5μm—can be used to observe spectral broadening and supercontinuum generation with low-energy femtosecond pulses. Comparison of the results of these experiments with the properties of spectral broadening of femtosecond Ti:Sapphire laser pulses reveals important aspects related to the role of dispersion effects. The influence of dispersion in supercontinuum generation, as shown by the results of our measurements with OPA femtosecond pulses, can be reduced by decreasing the core radius of the photonic crystal fibre.

3.2 The physics behind SC generation in photonic-crystal fibres

3.2.1 Light confinement

SC generation generally involves a rather complicated combination of nonlinear-optical processes. The efficiency of SC generation can, therefore, be improved by enhancing all the nonlinear-optical interactions contributing to this process. The general idea of using PCFs for this purpose is based on the possibility of increasing the degree of light localization in the fibre core by changing the geometry of the fibre cladding. This would allow nonlinear-optical interactions to be enhanced without increasing the energy of light pulses.

The effective size of a waveguide mode in an optical fibre depends on the difference between the refractive indices of the fibre and the cladding and can be roughly described with the use of the formula (Snyder and Love 1983) $a = w + \lambda/(2\pi\sqrt{n_c^2 \cos\phi - n_{\text{eff}}^2})$, where w is the fibre core radius, λ is the radiation wavelength, n_c is the refractive index of the fibre core, φ is the incidence angle characterizing the mode in the core of the fibre, and n_{eff} is the effective refractive index of the cladding. Rigorously speaking this formula is applicable for standard, step-index fibres. However, it also provides a qualitative understanding of the increase in the light localization degree in microstructure fibres when the effective refractive index is introduced as $n_{\text{eff}} = \beta_{\text{cl}}/k$, where β_{cl} is the propagation constant of the fundamental space-filling mode, that is, the fundamental mode of an infinite structure obtained by periodically translating a unit cell of the PCF cladding.

Following the method of analysis proposed by Birks *et al.* (1997), we found an estimate of β_{cl} by solving the scalar wave equation for the field distribution ψ in a circular unit cell with symmetric boundary conditions $\partial\psi/\partial s = 0$, where s is the coordinate along the axis oriented in the direction perpendicular to the boundary of the unit cell. This field distribution can be expressed in terms of zeroth-order Bessel functions. The above-specified

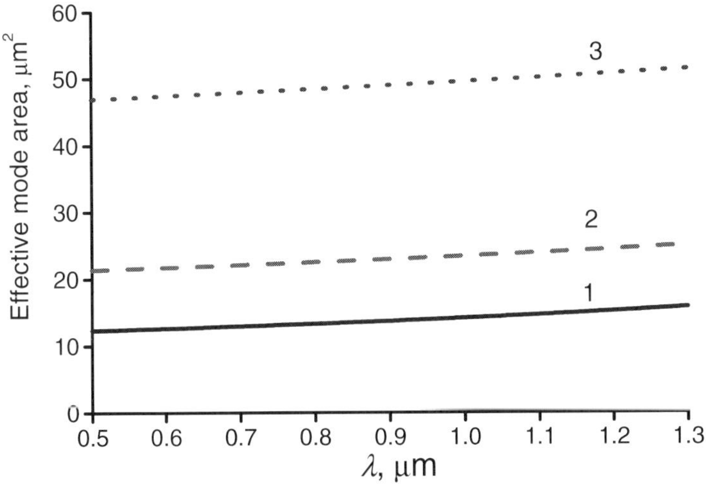

Figure 2. *The effective area of the fundamental mode of a photonic-crystal fibre with an air-filling fraction of 16% and the core radius of (1) 1.5, (2) 2, and (3) 3μm as a function of radiation wavelength. The pitch of the cladding is equal to the core radius.*

boundary condition supplemented with the continuity conditions for ψ and $\partial\psi/\partial s$ on the glass–air interface gives the characteristic equation for $\beta_{\rm cl}$.

Figure 2 displays the effective area of the fundamental mode in a PCF with an air-filling fraction of 16% as a function of radiation wavelength for different values of the fibre core radius. Dependences presented in Figure 2, provide a qualitative understanding of how nonlinear-optical processes can be enhanced by increasing the degree of light localization in a photonic-crystal fibre and show the ways of optimizing parameters of microstructure fibres for nonlinear-optical experiments. The possibility of improving the efficiency of spectral broadening of femtosecond pulses in PCFs by increasing the air-filling fraction of the cladding has been experimentally demonstrated by Fedotov *et al.* (2001a).

3.2.2 Dispersion of photonic-crystal fibres

Supercontinuum generation, as mentioned above, is often a result of the joint action of many nonlinear-optical processes, involving the generation of new spectral components and spectral broadening of these components. The efficiencies of these processes are highly sensitive to the dispersion of the nonlinear medium, which gives rise to the phase mismatch of different spectral components and leads to group-delay and pulse-spreading effects. The possibility of tailoring the dispersion of PCFs by changing their structure is, therefore, another important and very useful property of these fibres, which offers much promise for the enhancement of nonlinear-optical interactions.

To assess dispersion properties of PCFs used in our experiments, we employed standard expressions for the dispersion of a step-index fibre, replacing the propagation constant for the fibre cladding in these expressions by the propagation constant $\beta_{\rm cl}$ found by solving the scalar wave equation for the field distribution ψ in a circular unit cell with symmetric

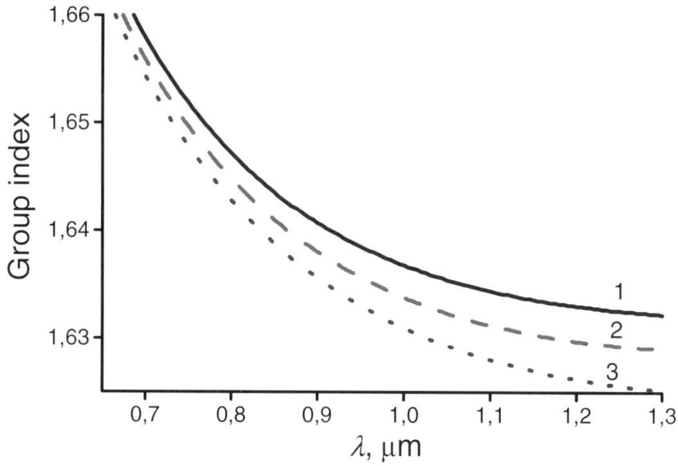

Figure 3. *The group index as a function of radiation wavelength for the fundamental mode of a photonic-crystal fibre with an air-filling fraction of 16% and the core radius of (1) 1.5, (2) 2, and (3) 3μm. The pitch of the cladding is equal to the core radius.*

boundary conditions. Figure 3 shows the group index calculated with the use of this procedure as a function of the wavelength for PCFs with fibre core radii equal to 1.5, 2, and 3μm (fibres with these core radii were employed in our experiments, see Section 3.3). The decrease in the fibre core radius, as can be seen from Figures 3 and 4, increases the influence of waveguide dispersion. In particular, zero-group-velocity dispersion can be achieved in the area of 1.3–1.5μm, depending on the fibre core diameter. It can be also seen from Figure 3 that the group-velocity dispersion for Ti:Sapphire laser pulses is much higher in our case than for radiation within the wavelength range of 1.1–1.5μm, which is typical of a femtosecond optical parametric amplifier (see Section 3.3). The dispersion length l_d estimated for 100fs pulses of 1.25μm radiation propagating in a photonic-crystal fibre with a core radius of 1.5μm ($l_d \approx 50$cm) is approximately five times larger than the dispersion length for 100fs pulses of 0.8μm radiation, with zero group-velocity dispersion achieved around the wavelength of 1.5μm for such a fibre. This observation may be crucial for understanding why femtosecond pulses of 1.1–1.5μm radiation, produced in our experiments by an optical parametric amplifier, resulted in a much more efficient spectral broadening than Ti:Sapphire laser pulses (see Section 3.4).

Since the superbroadening of femtosecond pulses involves the generation of new frequency components and nonlinear-optical interactions between these components, group-delay effects for these spectral components may become an important factor, influencing the efficiency of supercontinuum generation. Figure 4 shows the spectral dependence of the walk-off length (i.e., the length where the temporal separation of light pulses becomes approximately equal to the pulse duration). The walk-off length can be increased quite substantially, as can be seen from this figure, by reducing the fibre core radius from 3 to 1.5μm. We should expect the improvement in the efficiency of supercontinuum generation under these conditions. This expectation is verified by our experiments (see Section 3.4), which demonstrate that the efficiency of spectral superbroadening of femtosecond pulses in PCFs increases as the core radius of the fibre decreases from 3.0 down to 1.5μm.

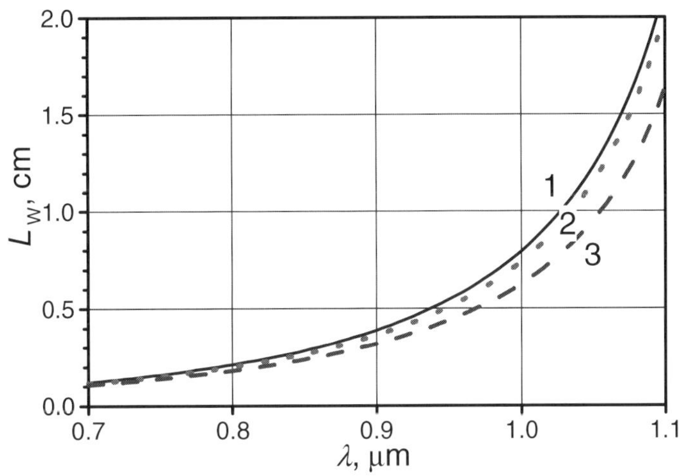

Figure 4. *The spectral dependence of the walk-off length L_w of 100fs light pulses with the wavelength λ in a photonic-crystal fibre with an air-filling fraction of 16% and the core radius of (1) 1.5, (2) 2, and (3) 3μm. The pitch of the cladding is equal to the core radius.*

3.3 Experimental

Spectral broadening and supercontinuum generation in photonic-crystal fibres were studied in our experiments with the use of two laser systems: a Ti:Sapphire laser and an optical parametric amplifier (OPA). The Ti:Sapphire laser system was employed to investigate the influence of the structure of a photonic-crystal fibre on the efficiency of self-phase modulation in the fibre core. The OPA system included its own, independent Ti:Sapphire pump laser and allowed us to generate frequency-tunable supercontinuum radiation by propagating femtosecond OPA pulses through a PCF. While many properties of spectral broadening of femtosecond Ti:Sapphire laser pulses have been explored in detail in earlier work (e.g., Ranka *et al.* 2000a, Fedotov *et al.* 2001a), the use of an OPA allowed us to add frequency-tunability aspects to numerous advantages of supercontinuum generation in PCFs.

A Ti:Sapphire laser consisting of an oscillator and a multipass amplifier pumped by a pulsed Nd:YAG laser is shown in Figure 5. Laser pulses of 800nm radiation coming out of this amplifier were used to pump an optical parametric amplifier based on a BBO crystal.

Part of the pump Ti:Sapphire laser radiation generates white light in a sapphire plate, while the remaining part is used to seed the parametric amplification process. An adjustable optical delay line is used to match the optical paths of the white-light-generating and seeding pulses. The wavelength of the signal generated as a result of this parametric process is tuned by rotating the BBO crystal. A double-pass scheme of optical parametric amplification is used to improve the efficiency of parametric frequency conversion, with an additional delay line employed to compensate for group-delay effects. The signal radiation is selected with the use of a dichroic mirror at the output of the OPA system. We were able to tune the wavelength of radiation produced by the OPA within the range

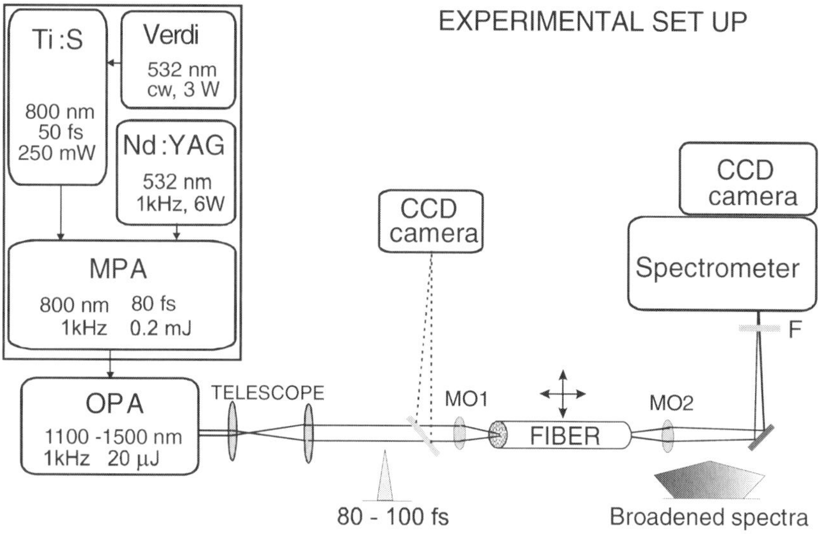

Figure 5. *Frequency-tunable femtosecond system based on an optical parametric amplifier, a Ti:Sapphire femtosecond master oscillator and a Nd:YAG pump laser. MPA is a multipass amplifier and MO1, MO2 are micro-objectives.*

of 1.1–1.5μm. The best performance of the OPA system was achieved at a wavelength of 1.25μm, where light pulses with a duration of approximately 80fs and an energy up to 20μJ were produced (Fedotov *et al.* 2002d).

The energy of light pulses coupled into PCFs was varied with the use of calibrated neutral-density filters. The laser beam was expanded with a telescope and then focused on the entrance face of a fibre sample, placed on a three-dimensional translation stage, with a Zeiss planachromat 20× micro-objective with a numerical aperture of 0.4 (MO1). Due to the used filters and optics the pulse width of the OPA may be broadened up to 100fs. The efficiency of waveguide mode excitation in a PCF was monitored by imaging the light field distribution at the output end of the fibre onto a CCD camera.

The technology employed to fabricate HFs used in our experiments, which was similar to the process developed by Knight *et al.* (1996), involved drawing identical glass capillaries stacked into a periodic preform at an elevated temperature, cutting the resulting structure into segments, and repeating the cycle again. This procedure allowed the fabrication of HFs with a cladding pitch ranging from 400nm up to 32μm (Alfimov *et al.* 2000).

To analyze the influence of the structure of a PCF on the properties of spectral broadening of femtosecond laser pulses, we employed several PCF samples with different pitches and air-filling fractions of the cladding and with different core diameters (Figures 1a, 1b). While PCFs shown in Figure 1a have small air holes in the claddings and are characterized by the air-filling fraction f equal to approximately 16%, the holes in the PCF cladding in Figure 1b are much larger, corresponding to an air-filling fraction f of about 65%. The pitch of the cladding and the core diameter of our PCFs were related to each other due to the fibre fabrication process. Photonic-crystal fibres with different pitches and core diameters allowed us to explore the influence of dispersion on the spectral broadening of femtosecond pulses. Dispersion effects, as one might expect based on Figures 3 and 4, played an increasingly important role in spectral broadening and supercontinuum generation as the core radius of the fibre was increased from 1.5 up to 3μm.

3.4 Spectral broadening of Ti:Sapphire laser pulses

To explore the ways of enhancing nonlinear-optical processes in a PCF due to light confinement in the fibre core, we employed short pieces of PCF samples, typically with a length l = 3cm, in order to reduce pulse spreading due to group-velocity dispersion (the typical dispersion length for 100fs pulses in a PCF with a core radius of 1.5μm was estimated as l_d = 10cm). Since the nonlinear lengths $L_{\rm nl}$ characterizing the efficiency of self-phase modulation were less than 0.3cm for all the experiments presented in this paper, the inequalities $L_{\rm nl} \ll l < l_d$ were satisfied with such a choice of the fibre-sample length. Figure 6 compares the spectra of input ande output pulses from a HF. The dependence

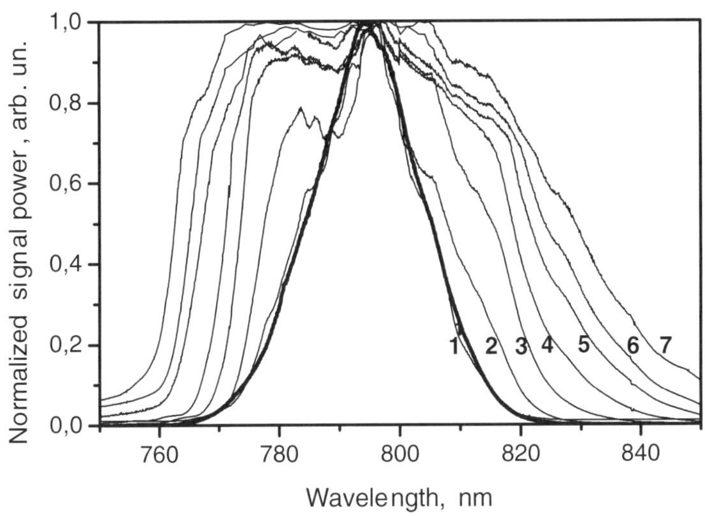

Figure 6. *The spectra of 70fs Ti:Sapphire laser pulses (bold line) at the input and (1–7) at the output of a holey fibre with a length of 3cm, the pitch of the cladding equal to 3μm, and the air-filling fraction f = 16% for pulse energies of (1) 0.5nJ, (2) 5nJ, (3) 10nJ, (4) 20nJ, (5) 30nJ, (6) 40nJ, and (7) 50nJ.*

of the spectral broadening $\Delta\omega$ of the pulse can be satisfactorily approximated under these conditions, by a linear function as can be seen from Figure 7. This regime is especially convenient for studying the enhancement of nonlinear-optical processes due to light confinement in the core of the fibre. This type of enhancement was investigated by comparing the spectral broadening of femtosecond laser pulses propagated through PCFs with different air-filling fractions of the cladding.

The enhancement in the spectral broadening efficiency can be deduced from the ratio of the slopes of cases 1 and 2 shown in Figure 7. Such an estimate yields an enhancement factor of about 1.46 in the case of 2μm-pitch PCFs.

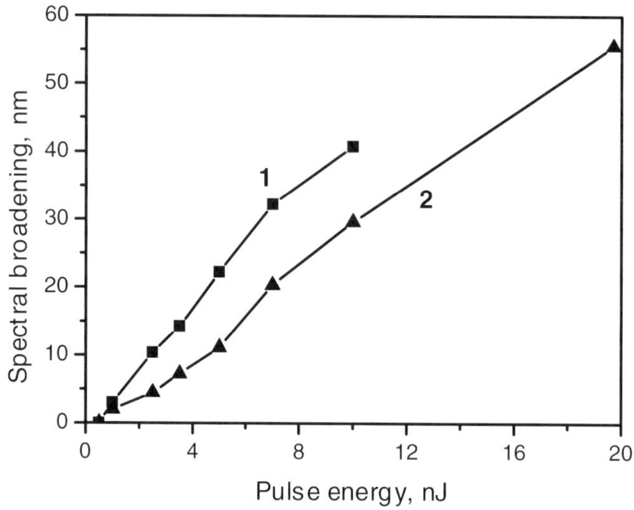

Figure 7. *The spectral width $\Delta\omega$ of 70fs pulses of 800nm Ti:Sapphire laser radiation transmitted through a 2μm-pitch photonic-crystal fibre samples with a length of 3cm and the air-filling fraction equal to (1) 65% and (2) 16% as a function of radiation energy coupled into the fibre.*

The spectral broadening of femtosecond pulses observed in the above-described experiments performed with Ti:Sapphire laser radiation even with long fibres was much less efficient than the spectral broadening achieved in other similar experiments (e.g., Ranka *et al.* 2000a), where subnanojoule femtosecond pulses were shown to allow supercontinuum generation under certain experimental conditions. One of the essential factors lowering the efficiency of spectral broadening in our experiments was related to dispersion effects, leading to the temporal walk-off of the pump pulse and the spectral components produced through various nonlinear-optical processes. Our experiments with Ti:Sapphire-laser radiation were performed in the regime of normal group-velocity dispersion, with a considerable dispersion characteristic of our fibre samples in the spectral range around 800nm (see Figures 3 and 4) preventing us from using longer fibre samples and achieving more efficient spectral broadening. Radically higher efficiencies of spectral broadening and supercontinuum generation were achieved with our PCFs for radiation wavelengths lying within the range of 1.1–1.5μm, covered by our optical parametric amplifier. The results of these experiments are discussed in the next section.

3.5 Supercontinuum generation with femtosecond OPA pulses

Propagation of 80–100fs pulses of 1.1–1.5μm radiation from an OPA results in an octave spectral broadening starting with pulse energies from several up to tens of nanojoules, depending on the pitch size and the core radius of the fibre (Fedotov *et al.* 2002d). Figure 8 shows the spectra of output radiation which become as broad as one octave starting with pulse energies of about 3, 8, and 20nJ for fibres with core radii of 1.5, 2, and 3μm, respectively.

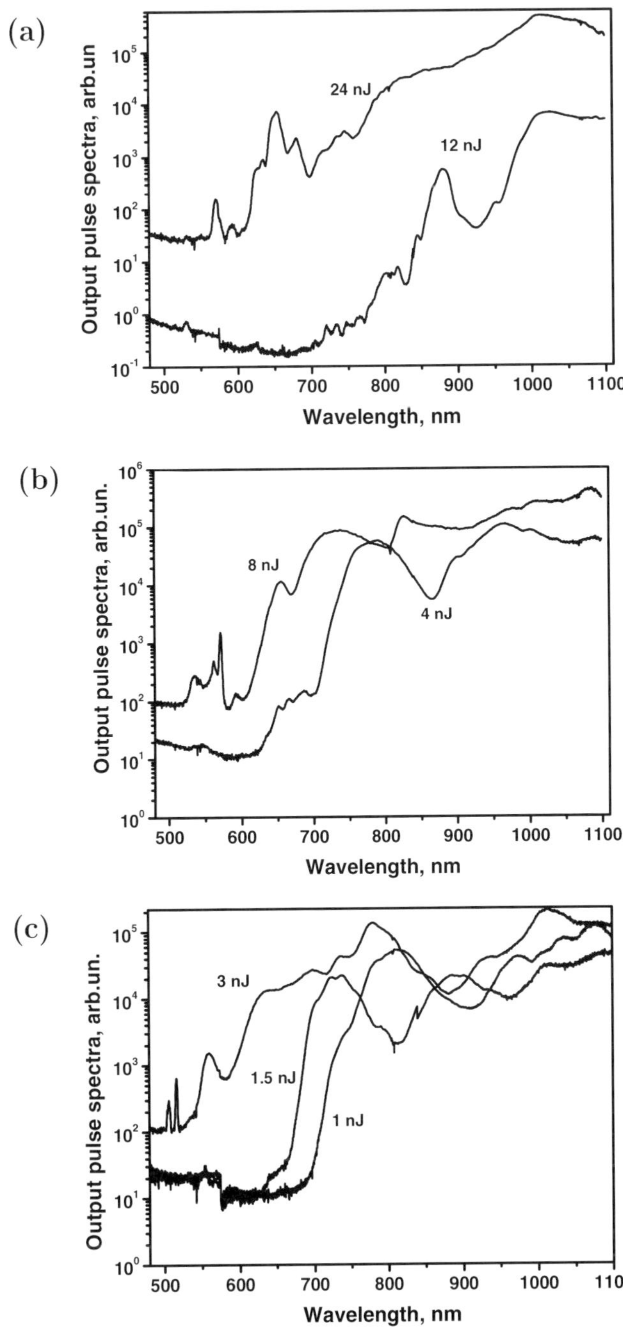

Figure 8. *The spectra of laser pulses coming out of photonic-crystal fibres with an air-filling fraction of 16%, the length of 4cm, and the core radii equal to (a) 3, (b) 2, and (c) 1.5μm. OPA pulses with a duration of 80fs and the wavelength of 1.25μm are coupled into the fibre. The energies of pulses coupled into the fibre are specified near the curves.*

Since zero group velocity dispersion was achieved within the wavelength range of 1.3–1.5μm for our PCFs with core radii of 1.5–3μm, the use of OPA radiation allowed us to generate supercontinuum in the regimes of both normal and anomalous group-velocity dispersion. This wavelength range is, therefore, very interesting for understanding the possibilities of controlling the properties of supercontinuum by tuning the wavelength of the input pulse with respect to the wavelength corresponding to zero group-velocity dispersion as shown below.

In both normal and anomalous group-velocity dispersion regimes, it was possible to precompensate the dispersion spreading of the pump pulse in the fibre with an appropriate choice of the initial chirp of the pulse. However, variation of the initial chirp also resulted in changes in the initial pulse duration, which prevented us from measuring supercontinuum parameters as functions of the initial pulse chirp in a systematic way. Even at the edges of the spectral range covered by our OPA source, the dispersion length l_d for 80–100fs pulses propagating through PCFs (approximately 50cm for a PCF with a core radius of 1.5μm) remained much longer than the fibre-sample length l, which, in its turn, was much longer than the nonlinear length L_{nl} characteristic of self-phase modulation (less than 0.3cm), leading to the following hierarchy of spatial scales: $L_{\mathrm{nl}} \ll l < l_d$. The characteristic length of shock-wave formation (estimated as $L_s \approx 50L_{\mathrm{nl}}$ for 0.8μm laser pulses and $L_s \approx 30L_{\mathrm{nl}}$ for 1.25μm laser pulses), on the other hand, was comparable with the fibre length, leading to a noticeable contribution of shock waves to the asymmetry of spectral broadening.

Spectra of supercontinuum emission produced with 80–100fs pulses of OPA radiation with different wavelengths are compared in Figures 9a and 9b. These experiments demonstrate how the spectral range covered by supercontinuum emission can be tuned by varying the wavelength of OPA radiation.

Although the processes that contribute to supercontinuum generation (self- and cross-phase modulation, four-wave mixing, self-steepening, stimulated Raman scattering, and modulation instabilities) are well known and have been extensively studied both experimentally and theoretically, it is very difficult to quantify the contributions of each of these processes to supercontinuum generation not only in experiments, but even in numerical simulations performed for physically realistic situations. In our experiments, we were able to identify quite reliably self-phase modulation as a process responsible for the spectral broadening of femtosecond pulses at very low energies of input pulses. At higher input pulse energies, several $\chi^{(3)}$ processes came into play virtually simultaneously, resulting in the generation of supercontinuum emission with very broad and asymmetric spectra. It would be important to emphasize, however, an important difference between supercontinuum generation under our conditions and supercontinuum generation with high-intensity laser pulses in gases. While supercontinuum generation in gases is often observed around the critical power of self-focusing, which makes the contribution of spatial self-action processes quite noticeable in that case (Aközbek *et al.* 2001), high degrees of light localization and long propagation lengths attainable with photonic-crystal fibres even without self-channelling and multiple refocusing phenomena, typical of high-field supercontinuum-generation experiments (Petit *et al.* 2000), allow supercontinuum emission to be generated in such fibres with laser pulse powers much lower than those leading to noticeable spatial self-action effects. In particular, as can be seen from the results of experiments presented in this paper, supercontinuum emission with spectra spanning over

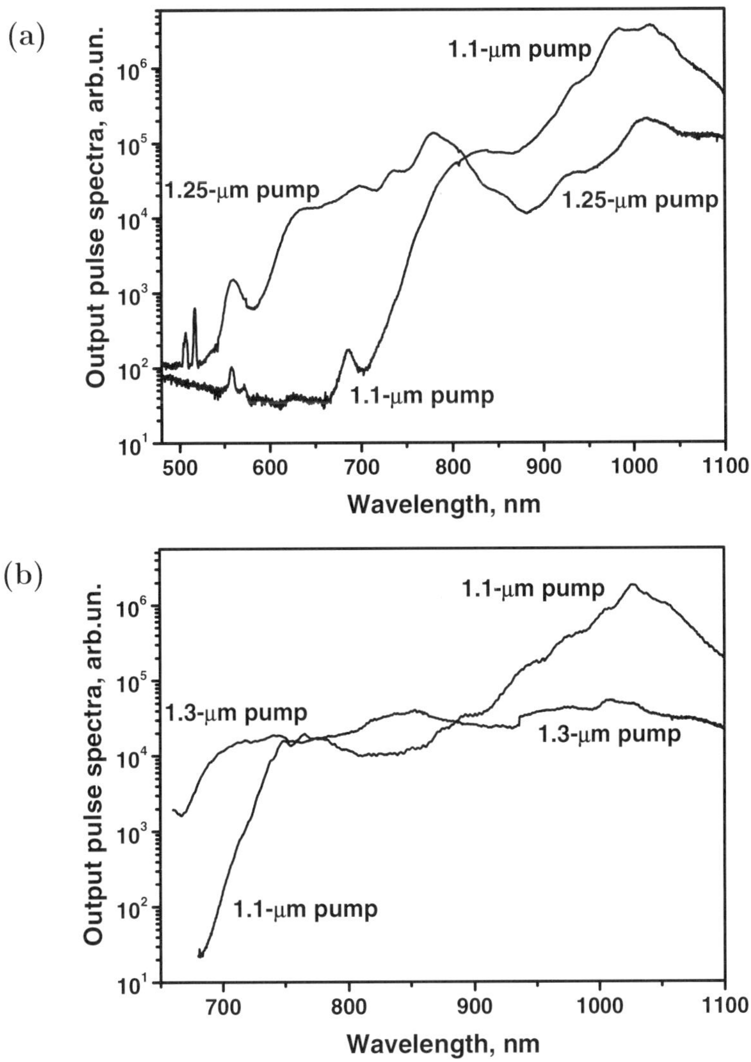

Figure 9. *The spectra of laser pulses coming out of photonic-crystal fibres with an air-filling fraction of 16%, the length of 4cm, and the core radii equal to (a) 1.5 and (b) 2μm. OPA radiation pulses with (a) the wavelengths of 1.10 and 1.25 μm and the energies of 5nJ and (b) the wavelengths of 1.10 and 1.30 μm and the energies of 8nJ are coupled into the fibre. The initial duration of light pulses is 80–100fs.*

nearly an octave can be produced with 100fs pulses whose peak power is less than 10kW. With a typical estimate for the critical power of self-focusing of 1MW, we can conclude that self-action effects (which may, of course, become quite important for higher input pulse powers) are negligible in this regime.

The results of our measurements demonstrate that the efficiency of supercontinuum generation is quite sensitive to the parameters of the fibre. In particular, the intensity of light pulses required to achieve an octave spectral broadening in the case of fibres with a core radius of 1.5μm is substantially lower than the intensity required to achieve a similar broadening in fibres with a core radius of 3.0μm. The correlation with dispersion properties of the fibres is easily observed here. Dispersion of photonic crystal fibres with a core radius of 1.5μm around the wavelength of 1.25μm is noticeably weaker than the dispersion of fibres with core radii of 2 and 3μm (see Figures 3 and 4). Group-delay and pulse-spreading effects are, therefore, less pronounced in PCFs with smaller core diameters. This allows much higher efficiencies of spectral superbroadening to be achieved with photonic crystal fibres with smaller core diameters.

The results of measurements performed with OPA pulses of different wavelengths, as can be seen from Figures 9a and 9b, generally confirm the above conclusions. The efficiency of supercontinuum generation with 1.25–1.30μm light was noticeably higher than the efficiency of supercontinuum generation with 1.10μm OPA pulses of the same initial power. This finding may also be explained in terms of dispersion properties of PCFs, as the group velocity dispersion of the fibres used in these experiments around 1.10μm is higher than the group velocity dispersion of these fibres within the range of 1.25–1.30μm (Figure 3).

An even more striking demonstration of the influence of dispersion effects on spectral broadening and supercontinuum generation in PCFs is provided by the comparison of the results of experiments performed with 800nm and 1.25μm light pulses of approximately the same pulse duration (cf. Figures 6–9). In particular, coupling 70fs Ti:Sapphire laser with an energy of 8nJ into a PCF with a pitch of the cladding equal to 2μm, we ended up with the spectral broadening of approximately 20nm at the output of the fibre (Figure 7). Much higher efficiencies of spectral broadening were achieved with 100fs, 8nJ OPA pulses, which allowed an octave-spectral broadening to be achieved with a 2μm-pitch fibre. Comparison of dispersion characteristics of photonic crystal fibres for 800nm and 1.25μm radiation suggests a natural explanation to these findings. The dispersion spreading of the pump pulse and the group delay of spectral components arising in the process of spectral broadening are much weaker around the wavelength of 1.25μm, characteristic of OPA radiation. This leads to much more efficient spectral broadening of femtosecond pulses and allows supercontinuum to be generated starting with much lower pulse energies. These results demonstrate that the efficiency of supercontinuum generation in a photonic-crystal fibre can be improved quite substantially by optimizing fibre dispersion and suggest several practical recipes for such dispersion optimization.

4 The mode structure and spectral properties of SC emission from microstructure fibres

4.1 Supercontinuum generation in the multimode regime

It is very important now, in view of numerous spectroscopic, metrological, and tomographic applications of supercontinua generated in MS fibres, to gain a deeper insight into spatial mode properties of SC emitted from MS fibres. This will provide us with a clearer idea of how to improve the spatial quality of spectral components filtered from broadband SC emission produced in MS fibres, simultaneously giving us a clue for a better understanding of the physics behind this very complicated nonlinear-optical phenomenon. Experimental studies show (Fedotov *et al.* 2002b) that SC spatial modes tend to change from a bell-shaped to a doughnut-like pattern with the increase in the detuning from the pump frequency, indicating changes in the multimode phase matching of four-wave mixing processes contributing to SC generation. We will demonstrate that, with an appropriate spectral filtering, bell-shaped modes of the long-wavelength section of the supercontinuum generated in an MS fibre can be separated from either doughnut-like or bipartite modes of the visible part of this supercontinuum. This effect can be employed for the spectral slicing of single modes of supercontinuum emission. Single-mode slices of SC emission can be then used for spectroscopic and time-resolved pump–probe measurements. To demonstrate this possibility, we will present the results of our experiments on frequency conversion of spectrally sliced spatial modes of SC emission produced in this way by wave mixing in a nonlinear crystal.

Our idea of using microstructure fibres with small fibre cores for the generation of supercontinuum emission that could be spectrally sliced in separated spatial modes is based on the fact that the difference between propagation constants for adjacent fibre modes supported by a fibre at a certain frequency ω increases with a decrease in the fibre core radius a. A physically instructive way to understand this result is to use the geometric-optic approach to the analysis of guided modes in a fibre. Fibre modes are then formed as a result of multiple reflection of radiation from the core-cladding interface, with the difference of propagation constants characterizing two adjacent guided modes with the same frequency ω being inversely proportional to the square of the fibre core radius, $\Delta\beta \approx (\pi^2 c f_n)/4a^2\omega n_c)$, where f_n is a function of the mode index and n_c is the refractive index of the fibre core. In the case of MS fibres, this simple expression, which immediately follows from the basic theory of optical fibres (Snyder and Love 1983), should be modified, of course, to include the complicated core–cladding geometry. One of the ways of doing this is to introduce the effective refractive index of the cladding through the propagation constants for cladding modes (Birks *et al.* 1997, Fedotov *et al.* 2002d), which can be found by solving the relevant wave equations. However, the physics behind the increase of $\Delta\beta$ with decreasing fibre core is clearly seen from this elementary formula.

Now, if we are interested in the regime of SC generation where multimode phase matching of four-wave mixing processes gives rise to a multimode SC signal, a large difference of propagation constants for adjacent modes of the same frequency may allow phase matching to be achieved only for one guided mode of SC emission at each frequency within some spectral range. The SC signal integrated over the spectrum should then remain multimode in this regime, but it should be possible to separate single modes for different spectral regions of SC emission. This is exactly what we observed in our experiments presented in the following sections of this paper.

4.2 Experimental

To fabricate MS fibres used in our experiments (Fedotov *et al.* 2002b), we started with a preform consisting of fused silica fibres with different diameters with one or several rings (hexagonal arrays) of identical fused silica capillaries at the centre of the structure stacked inside a fused silica capillary of a larger diameter, serving as a shell for the entire preform

(a) (b)

Figure 10. *Microscope cross-sectional images of microstructure fibres: (a) a fibre with two rings of air holes around the fibre core with a diameter of 2μm and (b) a fibre with a single ring of air holes around the fibre core with a diameter of 2μm.*

(Figure 10). This preform was then drawn at an elevated temperature on a drawing tower. This procedure allowed the fabrication of MS fibres with one or several rings of air holes around the fused silica fibre core (Figures 10a, 10b). An array of air holes of a smaller diameter around the main ring of air holes served to reduce optical losses of these MS fibres. The diameter of the fibre core in the fabricated series of fibres ranged from 1 to 5μm. The magnitude of optical losses in these MS fibres depended on the number of rings surrounding the fibre core and was estimated at 532nm as 0.4–0.5dB/m for an MS fibre with two rings of air holes (Figure 10a) and 2–3dB/m for an MS fibre with a single ring of air holes around the fibre core (Figure 10b).

Spectral broadening and supercontinuum generation in these MS fibres was studied in our experiments with the use of femtosecond pulses produced by a Ti:Sapphire laser system (Figure 11). This laser system included a Ti:Sapphire master oscillator and a regenerative amplifier and was capable of generating 40fs pulses of 800nm radiation with the energy up to 0.2mJ per pulse and a repetition rate of 1kHz. The laser beam was focused on the entrance face of a fibre sample, placed on a three-dimensional translation stage, with a microobjective. Radiation coming out of the fibre was collimated with an identical micro-objective and was split into two beams. One of these beams was delivered to a spectrograph, while the other one was used to visualize the transverse intensity distribution in the emission coming out of the MS fibre by imaging the output end of the fibre on to a CCD camera.

4.3 Mode-separable supercontinuum emission

With only a few nanojoules of Ti:Sapphire laser radiation coupled into an MS fibre sample with a length of several centimetres, we observed the generation of SC emission with a spectral bandwidth exceeding an octave. Figure 12 shows typical spectra of a SC generated by 40fs pulses with an energy of 2nJ (dotted curve) and 3nJ (solid curve) in a 1.5m MS

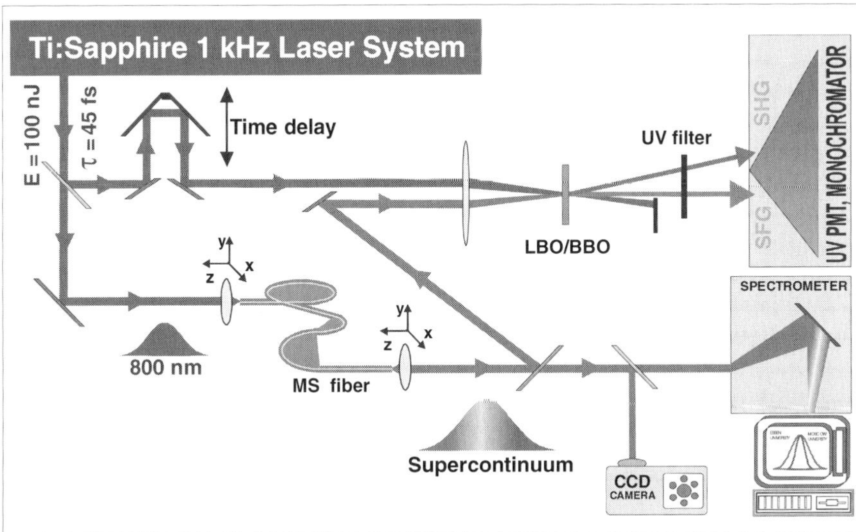

Figure 11. *Diagram of the experimental setup for studying supercontinuum generation in microstructure fibres: SFG, the signal resulting from sum-frequency in the nonlinear crystal; SHG, the signal of second-harmonic generation; PMT, photoelectric multiplier.*

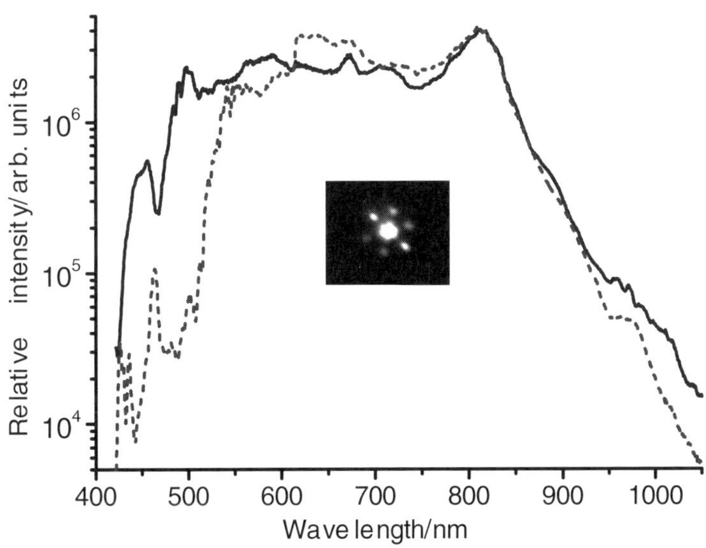

Figure 12. *Typical spectra of SC emission generated by 40fs pulses with an energy of 2nJ (dotted curve) and 3nJ (solid curve) in an 1.5m MS fibre with a single ring of air holes around the fibre core (Figure 10) with a core diameter of 3μm. The inset shows the transverse intensity distribution of pump radiation at the output end of the fibre in the low-power-pump no-SC regime.*

fibre with a single ring of air holes around the fibre core (Figure 10) with a core diameter of 3 μm. The inset in Figure 12 displays the transverse intensity distribution of pump radiation at the output end of the fibre in the regime of low-power pump, when no SC is generated. The transverse structure of the guided mode of pump radiation, as can be seen from this image, features a sixfold rotational symmetry, which is also characteristic of the fibre structure.

Using colour-glass filters, we were able to analyze the mode structure of SC emission coming out of the MS fibre within different spectral ranges. This analysis has revealed a qualitative difference between the transverse intensity distributions of SC emission measured for different wavelength regions. Figure 13 presents typical results of such measurements performed for a SC generated in a 1.5m long MS fibre with a single ring of air holes around the fibre core (Figure 10) with a core diameter of 3μm. The transverse intensity distribution of SC emission measured with a filter blocking radiation with wavelengths shorter than ~720nm (Figure 13a) has a bell-like shape, displaying a single maximum on the beam axis. The short-wavelength part of SC emission, on the other hand, has a doughnut-like spatial mode structure (Figure 13b), as indicated by measurements performed with a filter possessing maximum transmission within the range from approximately 400 to 600nm.

With a slight variation in the initial conditions of mode excitation at the input end of the MS fibre, this doughnut mode was transformed into a more complicated, bipartite pattern shown in Figure 13c. Both the doughnut-like mode of Figure 13b and the bipartite mode of Figure 13c remained stable and were observed for MS fibres with lengths ranging from several centimeters up to 2m. Apparently, because of the poorer spatial overlapping between the pump beam and the bipartite mode, the short-wavelength part of SC emission in the case of the bipartite mode was much less intense than in the case of the doughnut mode.

The changes in the mode structure of SC emission observed in our experiments for SC spectral regions far off the pump wavelength are most likely related to multimode phase matching. Phase matching of this type begins to play a predominant role in four-wave mixing contributing to SC generation when the waveguide dispersion component may compensate for the phase mismatch related to the material dispersion for frequency-conversion processes involving different waveguide modes. Supercontinuum emission is no longer single-mode under these conditions. However, with appropriate spectral filtering, we were able to separate, for example, the doughnut-like or bipartite guided modes of 400–600nm emission from the bell-shaped mode of the long-wavelength wing of the SC. The generation efficiency for the visible part of the SC spectrum and its spectral width can be varied by switching between MS fibre modes. The regime of mode-separated SC generation demonstrated in our experiments seems to be promising for numerous spectroscopic applications of SC-generating MS fibres, as it allows SC emission to be spectrally sliced in a simple and convenient way, resulting in the formation of isolated single modes. Moreover, different spectral slices of SC thus separated in space can be then separately time-delayed and polarization-prepared to be brought in a spatial coincidence again for time-resolved and pump-probe measurements. A high spatial quality of SC slices separated in that way in our experiments was proved by a high efficiency of sum-frequency generation in a 100μm thick LBO crystal, where different wavelength regions of SC emission produced in an MS fibre were mixed with fundamental radiation of our Ti:Sapphire laser.

(a)

(b)

(c)

Figure 13. *Transverse intensity distribution of SC emission generated in a 1.5m long MS fibre with a single ring of air holes around the fibre core (Figure 10) with a core diameter of 3μm: (a) with a filter blocking radiation with wavelengths shorter than ∼720nm, (b) a doughnut mode imaged with a filter possessing maximum transmission within the range of ∼400–600nm, (c) a bipartite mode imaged with a filter possessing maximum transmission within the range of ∼400–600nm.*

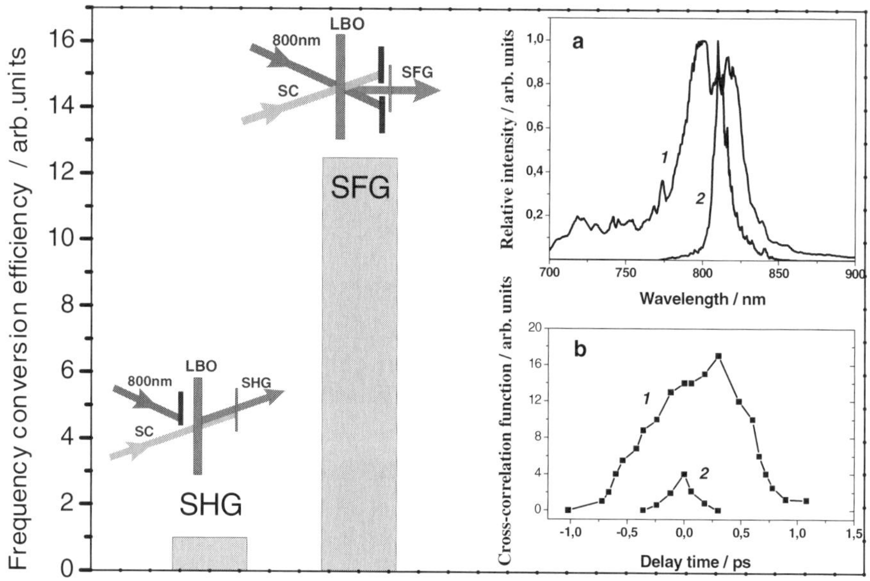

Figure 14. *Nonlinear-optical frequency conversion of spectrally sliced supercontinuum and cross-correlation measurements. (a) The spectrum of supercontinuum emission produced within the range of wavelengths from 720 up to 900nm in a microstructure fibre with a single ring of holes around the fibre core (Figure 10), a core diameter of 3μm, and the length of 1.5m (trace 1) and the spectrum of low-intensity Ti:Sapphire laser pulses transmitted through the same microstructure fibre with virtually no or very weak spectral broadening. The spectra shown by traces 1 and 2 are normalized to their maximum intensity values. (b) Cross-correlation traces for a pulse of the 720–900nm spectral slice of supercontinuum (trace 1) and a weakly broadened low-intensity Ti:Sapphire laser pulses transmitted through the same microstructure fibre with virtually no or very weak spectral broadening (trace 2). Beam diagrams of noncollinear sum-frequency generation (SFG) and collinear second-harmonic generation (SHG) are also shown. The heights of the vertical bars represent the relative efficiencies of the SFG and SHG processes.*

4.4 Frequency conversion of spectrally sliced supercontinuum

The results of experiments presented in the previous section demonstrate that SC emission can be generated very efficiently by propagating femtosecond pulses through MS fibres, and the beam profile of SC emission can be controlled by varying the initial conditions at the input end of the fibre. An important issue now is whether the beam-profile quality of these spectrally selected spatial modes of SC emission is sufficient for nonlinear-optical and spectroscopic measurements with the use of this radiation. We have studied nonlinear-optical frequency conversion of spectrally sliced spatial modes of SC emission in a nonlinear crystal to answer this question. The sum-frequency signal was produced in our experiments by mixing different parts of SC emission with the fundamental radiation of the above-described Ti:Sapphire laser in a 10μm thick LBO crystal (Figure 11). Figure 14 presents the results of these measurements performed with the use of the long-wavelength part (~720–900nm) of SC emission (the spectrum of this radiation is shown by case 1 in Figure 14a) mixed with the fundamental radiation of the Ti:Sapphire laser in the LBO crystal in the noncollinear geometry of sum-frequency generation (SFG). The broadband sum-frequency signal was produced within the spectral range from 370 up to 430nm in the direction determined by phase-matching conditions (see the diagrams in Figure 14a). This geometry of sum-frequency generation allowed an efficiency of frequency conversion of about 0.1% to be achieved. We observed also collinear second-harmonic generation using the long-wavelength part of SC emission as a pump (see the SHG diagram in Figure 14). The efficiency of this second-harmonic generation process under our experimental conditions was more than an order of magnitude lower than the efficiency of noncollinear sum-frequency generation. The results of these experiments show that the efficiency of nonlinear-optical frequency conversion of spectrally sliced SC emission can be made high enough to allow nonlinear-optical spectroscopic and time-resolved measurements and to use SC generated in MS fibres as seed radiation for optical parametric amplification.

Sum-frequency and second-harmonic generation experiments performed with spectrally sliced SC emission allow also the characteristic pulse duration to be estimated for different parts of SC emission. Figure 14b presents the results of such cross-correlation experiments, where the intensity of the sum-frequency signal was measured as a function of the delay time between the fundamental radiation pulse of the Ti:Sapphire laser and the broadband emission pulse originating from the fibre and passing through a set of optical filters. Cross-correlation traces measured by this approach were compared with the results of cross-correlation measurements performed in the same geometry for low-intensity Ti:Sapphire laser pulses transmitted through the MS fibre with virtually no or very weak spectral broadening (the spectrum of this signal is shown by trace 2 in Figure 14a). This comparison shows that different spectral components emitted as a part of supercontinuum are characterized by different delay times. This effect is associated with the dispersion properties of MS-fibre-guided modes and can be employed to temporally and spatially resolve different frequency components in pump–supercontinuum probe experiments.

4.5 Controlling the spectrum of supercontinuum emission

The spectral quality is the key property of supercontinuum for numerous spectroscopic and metrological applications, as well as for ultrashort-pulse synthesis and for using supercontinuum emission to seed optical parametric amplification. In this section, we will show that MS fibres not only allow the generation of supercontinuum with a high spectral quality, but also provide several channels to control the spectrum of supercontinuum emission. We will demonstrate, in particular, that the spectral properties of supercontinuum can be controlled by varying the initial amplitude and chirp of the input pump pulse and by changing the parameters of the MS fibre.

The results of experimental studies presented in Figures 12 and 15 demonstrate that the parameters of input laser pulses can be matched with the characteristics of an MS fibre in such a way as to allow the generation of SC emission with flat spectra spanning over nearly an octave. The possibility of achieving highly efficient conversion of pump pulse energy into the visible range is illustrated by Figure 16, which displays the spectrum of SC emission generated by 60fs pulses of Ti:Sapphire laser radiation with an energy of 8nJ (dashed line) and 12nJ (solid line) coupled into a 1m MS fibre. The pump field is

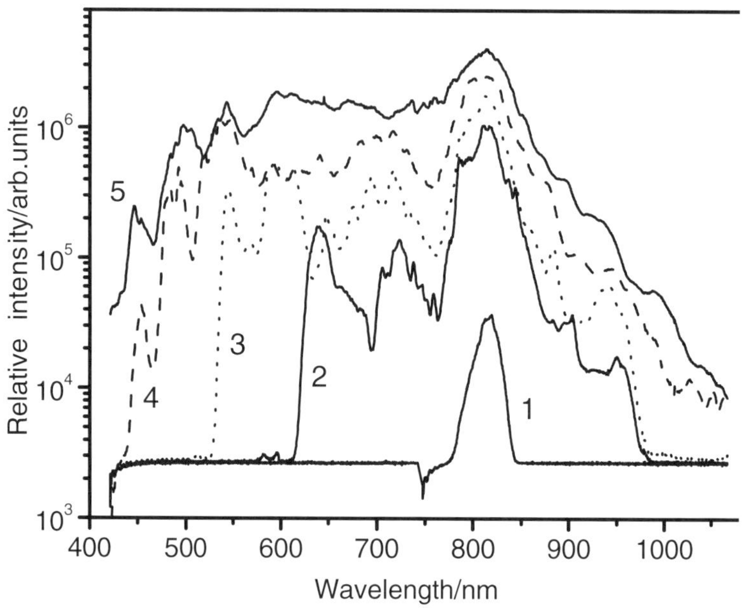

Figure 15. *Spectra of SC emission generated by a 50fs Ti:Sapphire laser pulse with an energy of (1) 1nJ, (2) 4nJ, (3) 8nJ, (4) 12nJ, (5) 16nJ in a 60cm microstructure fibre with a single ring of holes around the fibre core (Figure 10) and a core diameter of 4μm.*

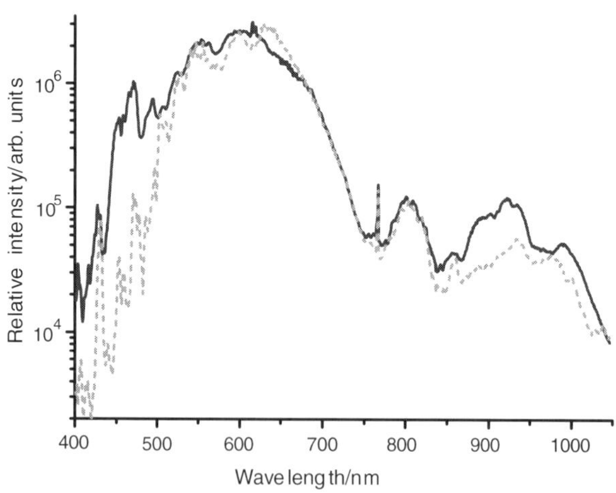

Figure 16. *Pump depletion in supercontinuum generation: spectra of supercontinuum emission generated by a 60fs Ti:Sapphire laser pulse with an energy of 8nJ (the dashed line) and 12nJ (the solid line) in a 1m microstructure fibre with a single ring of holes around the fibre core (Figure 10) and a core diameter of 2.5μm. The wavelength of input radiation is 800nm.*

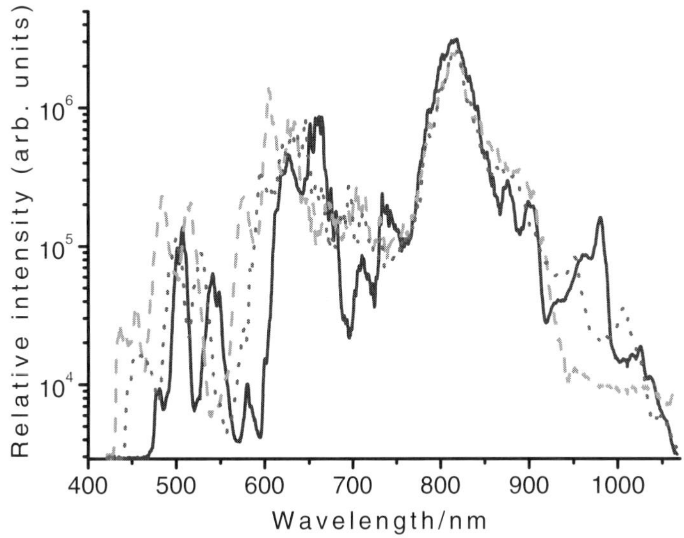

Figure 17. *The influence of the initial chirp of pump pulses on the spectrum of supercontinuum emission from a microstructure fibre with a single ring of holes around the fibre core (Figure 10), a core diameter of 3μm, and a length of about 10cm. The chirp parameter is* $\alpha = 0$ *and the initial pump pulse duration is* $\tau_0 = 62 fs$ *for the solid line;* $\alpha = 7 \times 10^{-5} fs^{-2}$ *and* $\tau_0 = 65 fs$ *for the dotted line;* $\alpha = -10^{-4} fs^{-2}$ *and* $\tau_0 = 70 fs$ *for the dashed line.*

depleted under these conditions resulting in a highly efficient generation of a broadband visible light.

Due to dispersion effects, supercontinuum generation in an MS fibre is sensitive to the initial chirp of the laser pulse coupled into the fibre. Changing the initial chirp of the pulse, we were able to noticeably change the efficiency of SC generation process in an MS fibre and to considerably modify the spectral shape of SC emission. The examples of chirp-controlled SC generation in an MS fibre are given in Figure 17.

Dispersion properties of an MS fibre are largely determined by its core–cladding geometry. To demonstrate the influence of the core–cladding fibre geometry on the process of SC generation, we compared the spectra of supercontinua generated in MS fibres with a single core surrounded with a single ring of air holes (Figure 10) and MS fibres with a more complicated core–cladding geometry where the core, in fact, consists of several coupled channels (see the inset in Figure 18). An instructive way of analyzing the dispersion of MS-fibre-guided modes in the latter case is to use the analogy between the properties of modes guided by an array of coupled fibres and electronic wave functions in a polyatomic molecule (Fedotov *et al.* 2002e). Figure 18 shows the spectra of SC emission generated in such photonic-molecule modes of a microstructure fibre. The characteristic size of the central core area of the MS fibre used in these experiments was about 4μm. As can be seen from the comparison of Figures 12 and 15–17, supercontinua with different spectral shapes can be generated by using MS fibres with different core–cladding geometries. Even when pulses with equal energies, durations, and chirps were coupled

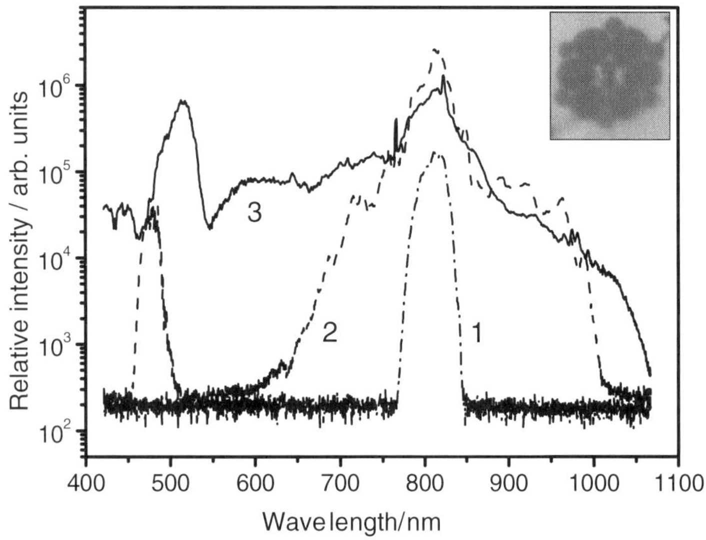

Figure 18. *Spectra of SC emission generated by a 50fs Ti:Sapphire laser pulse with an energy of (1) 0.1nJ, (2) 2nJ and (3) 5nJ in a photonic-molecule mode of a microstructure fibre. The cross-sectional image of the fibre is shown in the inset. The diameter of the central, multiple-core waveguiding structure is 4μm. The fibre length is 1.5m.*

into single-core (Figure 10) and multiple-core (the inset in Figure 18) fibres, substantial differences have been observed in the shapes of SC spectra (cf. Figures 15 and 18), indicating a considerable influence of the fibre design on the process of SC generation. Modification of the microstructure fibre geometry leads to changes in both the effective area and dispersion properties of guided modes. In particular, a decrease in the effective mode area due to the modification of the microstructure geometry may play a very important role, allowing higher power densities of laser radiation to be achieved for the same energies of light pulses coupled into the fibre. However, qualitative changes observed in SC spectra, featuring under certain conditions spectrally isolated emission bands (around 500nm in Figure 18), are most likely related to a modified dispersion of microstructure fibres, leading to an efficient generation of spectral components around 500nm through phase-matched parametric nonlinear-optical processes.

Thus, experimental studies of the mode structure of supercontinuum emission generated by 40fs pulses of 800nm Ti:Sapphire laser radiation in fused silica microstructure fibres have shown that the long-wavelength (720–900nm) and visible (400–600nm) parts of supercontinuum emission can be spatially separated in microstructure-fibre modes. Thus, although multimode phase matching for four-wave mixing processes contributing to supercontinuum generation in a microstructure fibre may make the output supercontinuum beam multimode, single modes of visible and long-wavelength sections of supercontinuum can be extracted with an appropriate spectral filtering. This regime of mode-separated SC generation offers much promise for numerous spectroscopic applications, allowing SC emission to be spectrally sliced in a simple and convenient way, resulting in the formation of radiation with a high spatial quality.

A high spectral and beam-profile quality of mode-separable supercontinua generated by MS fibres and spectrally sliced with a set of optical filters have been demonstrated by spectral measurements and sum-frequency generation experiments, suggesting the possibility of using spectrally sliced SC emission for wave-mixing spectroscopy and time-resolved pump-probe measurements, as well as for seeding optical parametric amplifiers. Our studies have thus demonstrated that MS fibres may serve as sources of broadband emission with a high spectral quality, providing, at the same time, the opportunity of generating radiation with different transverse intensity distributions, allowing various non-collinear geometries of wave mixing to be implemented in a convenient way. These results bring us to a concept of new sources of broadband emission for spectroscopic applications and time-resolved measurements based on supercontinuum generation in MS fibres.

5 Supercontinuum-generating photonic-molecule modes of microstructure fibres

5.1 Design of microstructure fibres: why should we care?

Various designs of microstructure fibres have been proposed since their first demonstrations in early 1970s (Kaiser *et al.* 1973. Some of the remarkable properties of microstructure fibres (such as robust single-mode waveguiding within a broad spectral range (Knight *et al.* 1996) and a high degree of light localization in the fibre core (Fedotov *et al.* 2001a) due to a high index step between the core and the cladding) have been demonstrated to be not very sensitive to the geometry of the cladding and the core of the fibre. In particular, as shown by Monro *et al.* (2000), light remains strongly confined in the fibre core, enhancing nonlinear-optical interactions even in fibres with a random distribution of holes in the cladding. This capability of microstructure fibres to localize light within a very small area of the fibre cross section is due to the high step between the refractive index of the fibre core and the effective refractive index of the cladding—the factor that is also responsible for the radical enhancement of nonlinear-optical processes in tapered fibres (Birks *et al.* 2000, Akimov *et al.* 2002).

In a general situation, however, the design of the fibre is very important for many nonlinear-optical functions of microstructure fibres. In particular, although the effects related to strong light localization show up in microstructure fibres with different geometries, the degree of light localization in the fibre core and, hence, the degree to which nonlinear optical processes are enhanced depend on the air-filling fraction of the cladding, which determines the effective refractive index of the cladding. Microstructure fibres with a photonic-crystal cladding provide new regimes of waveguiding due to a high reflectivity of the cladding around the photonic band gap (Cregan *et al.* 1999) and allow many of the ideas discussed in connection with photonic crystals to be realized. The possibility of designing microstructure fibres with different configurations of the core and the cladding helps to engineer and tailor the dispersion of such fibres (Reeves *et al.* 2002), making these fibres very attractive for nonlinear optics and short-pulse formation and control. A hole-assisted microstructured fibre, comprising a material index profile for waveguiding and air holes for modifying optical properties, has been demonstrated recently (Hasegawa *et al* 2001) to allow an anomalous dispersion larger than that of conventional fibres to be achieved without introducing large optical losses. With a special design of the core and the cladding, a microstructure fibre can be made highly birefringent (Ortigosa-Blanch *et al.* 2000), as demonstrated, for example, for a cobweb photonic-crystal (PC) fibre (Knight *et al.* 2000). This opens the way for polarization control of supercontinuum generation as shown recently by Apolonski *et al.* (2002).

In this section, we will present the results of our supercontinuum-generation experiments performed with microstructure fibres fabricated with the use of the technique that involves stacking capillaries into a preform and then pulling this preform at elevated temperatures (Knight *et al.* 1996, Alfimov *et al.* 2000). Although the fibres were produced with a standard technological procedure, the core–cladding configuration of these fibres supports guided modes whose dispersion properties differ from the dispersion properties of previously reported guided modes in microstructure fibres, thus offering new ways of optimizing nonlinear-optical processes occurring in these modes. Cobweb microstructure fibres employed in our experiments (Figure 1c) have a core with the cross section in the form of a cyclic photonic molecule (Figure 19). Air holes arranged in a two-dimensionally periodic structure in the cladding of this fibre and a larger hole at the centre of the fi-

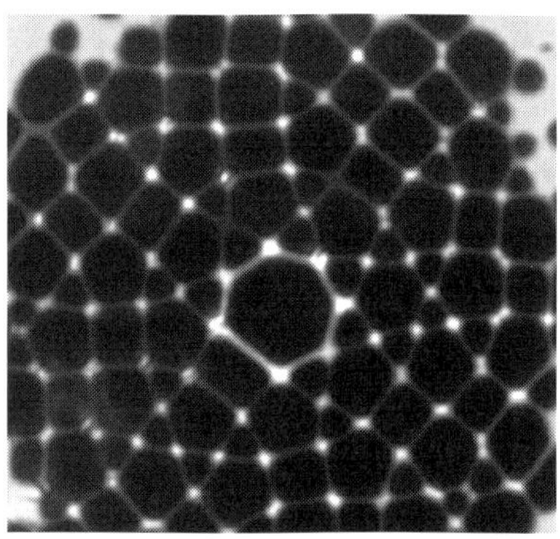

Figure 19. *Cross sectional image of a photonic-molecule microstructure-integrated bundle of fibres with a radius of each fibre in the bundle equal to 2μm and the distance between the neighbouring fibres of 7.4μm.*

bre form a cyclic-molecule-like structure, consisting of an array of small-diameter glass channels linked by narrow bridges, around the central hole (Figure 19). This photonic-molecule (PM) microstructure-integrated bundle of fibres can guide the light through total internal reflection, providing a very high light confinement degree due to the large refractive index step. We will show that fibres of this type offer new solutions to the key problem of nonlinear waveguide optics of very short light pulses, opening new ways to engineer waveguide dispersion. In a broader perspective, our observation of SC generation in microstructure fibres of this type is, to the best of our knowledge, the first nonlinear-optical result obtained with photonic-molecule-like structures. We will demonstrate below that the photonic-molecule analogy provides a very useful insight into the properties of dispersion and waveguide modes in microstructure fibres with a special design of the fibre core, permitting a better understanding of dispersion-tailoring and field-distribution aspects and allowing the potentialities of such fibres in the context of nonlinear-optical applications to be assessed.

5.2 Photonic-molecule modes of microstructure fibres

Supercontinuum generation, as mentioned above, is often a result of the joint action of many nonlinear-optical processes, involving the generation of new spectral components and spectral broadening of these components. The efficiencies of these processes are highly sensitive to the dispersion of the nonlinear medium, which gives rise to the phase mismatch of different spectral components and leads to group-delay and pulse-spreading effects. The possibility of tailoring the dispersion of photonic-crystal fibres by changing their structure is, therefore, another important and very useful property of these fibres, which offers much promise for the enhancement of nonlinear-optical interactions.

Our goal here is to proceed with the discussion of the possibilities offered by considering microstructure fibres with the special configuration of holes shown in Figure 19. This is essentially a hollow-core photonic-crystal fibre design, demonstrated earlier by Cregan *et al.* (1999), but with a very thin wall separating the central hole from the rest of the structure and with a very high refractive index step between this wall (Figure 19) and the adjacent areas, which is achieved by the high air-filling fraction of the microstructure. In contrast to earlier work, we are now interested in the light guided along the narrow ring surrounding the hollow core of this fibre. Thus, instead of air-guided modes, we consider the modes of the cladding of the hollow microstructure fibre. While the waveguiding in the hollow core of a photonic-crystal fibre reported by Cregan *et al.* (1999) is achieved due to the reflective properties of the periodic structure of its cladding within the spectral range around the photonic band gap, it is the total internal reflection that results in the formation of guided modes in the central ring of our microstructure fibre.

A standard approach to the mathematical description of the modes guided by the central ring in the structure shown in Figure 19 would be to treat these modes as cladding modes of a hollow-core fibre. This approach, which works very well in the case of circular fibre claddings (Snyder and Love, 1983), becomes not very practical in our case because of the complicated structure of our waveguiding ring. On the other hand, this waveguiding ring, as is readily seen from Figure 19, can be considered as an array of standard circular fibre cores linked by narrow bridges. As can be realized by looking at Figure 19, light can also be coupled into glass channels that do not belong to the central ring. However, it is also seen from this image that the coupling between the channels forming the central ring of the structure is much stronger than the coupling between any of the central channels and the channels lying beyond the central ring. We were also able to avoid coupling light to any of the channels except for those belonging to the central ring by using objectives with a higher numerical aperture.

The considered configuration of fibre cores linked by bridges is reminiscent of the configuration of atoms linked by chemical bonds in a cyclic polyatomic molecule consisting of identical atoms (a generic diagram of such a molecule is shown in Figure 20), which brings us to an idea of using a photonic-molecule analogy to understand the optical properties of such a waveguiding structure. Below, we will extend this analogy by introducing a polyatomic photonic molecule. This effort will later prove to be quite rewarding by providing us with an illustrative and physically clear model of dispersion properties and mode structure of the considered rather complicated optical fibre.

The main purpose of the analysis performed in this section is to use the photonic-molecule analogy to qualitatively explain the general features of dispersion and mode

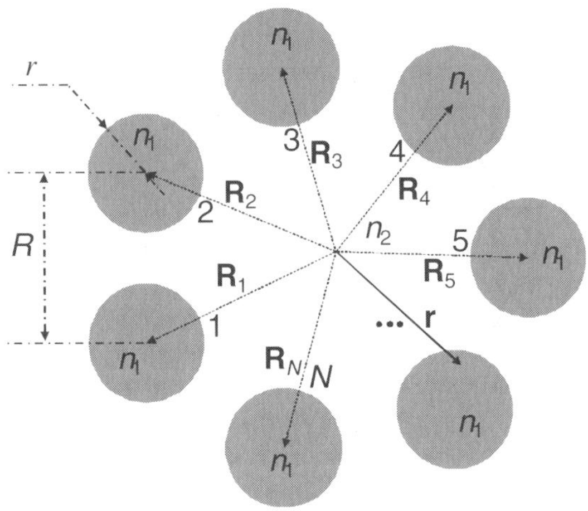

Figure 20. *A photonic-molecule model for an optical fibre where the light is guided along a set of N cores with the refractive index* n_1 *surrounded by a material with the refractive index* n_2 *and linked by narrow bridges with the distance between the neighbouring cores equal to R.*

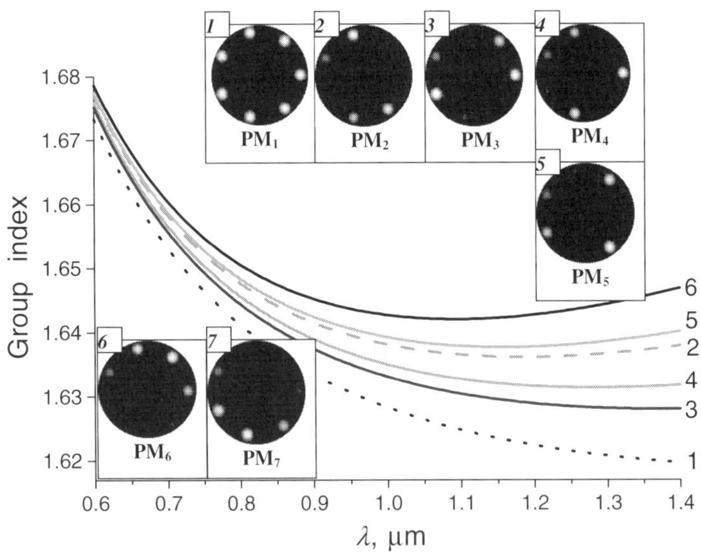

Figure 21. *The group index as a function of radiation wavelength for (1) the material of the fibre; (2) a single isolated fibre from the considered photonic-molecule structure (or an array of identical fibres completely isolated from one another); and (3)* PM_1*, (4)* PM_2 *and* PM_3*, (5)* PM_4 *and* PM_5*, and (6)* PM_6 *and* PM_7 *modes of a seven-core photonic-molecule fibre (*$N = 7$*). The radius of a single fibre in the photonic-molecule fibre structure is 2 μm, the distance between the neighboring fibres in the photonic-molecule structure is 7.4 μm. The insets show light intensity distributions in (1)* PM_1*, (2)* PM_2*, (3)* PM_3*, (4)* PM_4*, (5)* PM_5*, (6)* PM_6*, and (7)* PM_7 *modes of a seven-core photonic-molecule fibre.*

structure of an optical fibre where the light is guided along a set of N channels with the refractive index n_1 (Figure 20) surrounded by a material with the refractive index n_2 and linked by narrow bridges. The fibre channels, forming a photonic molecule in the plane of Figure 20, are made of glass whose wavelength dependence of the group index is given by curve *1* in Figure 21. We start our analysis with a set of N isolated single-mode fibres consisting of a bare core surrounded with air. As long as these fibres remain isolated, the dispersion relation of such an array is degenerate (curve *2* in Figure 21), and this degeneracy is N-fold. As soon as we allow them to be coupled to each other, the degeneracy is partially or completely lifted, similar to polyatomic molecules consisting of identical atoms, giving rise to several dispersion branches (cases 3–6 in Figure 21).

Physical and mathematical aspects of the analogy between a bundle of coupled micro-structure-integrated fibres and a polyatomic molecule should be emphasized. Physically, the action of a refractive-index step on a light field is similar to the influence of a potential jump on an electron wave function in a molecular system. Therefore, the way the distribution of the refractive index in the plane of Figure 20, corresponding to a bundle of coupled fibres, modulates the light field is similar to the way the distribution of potential in a polyatomic molecule modifies the electron wave function. Mathematically, this analogy stems from the similarity of coupled-theory equations for an array of fibres to perturbation-theory equations for the electron wave function in a polyatomic molecule.

In our analysis, we will neglect polarization effects and employ a scalar-wave-equation approximation. Since the strongest mode coupling is achieved for waveguide modes with equal propagation constants in a photonic-molecule fibre with identical cores, we will neglect also the coupling of modes with different mode indices. The modes of the photonic-molecule fibre can be then represented as superpositions of modes of isolated fibres:

$$\Psi(\mathbf{r}) = \sum_n A_n f(\mathbf{r} - \mathbf{R_n}), \tag{11}$$

where $\mathbf{r}$ is the radius vector in the plane of fibre cross section (the plane of Figure 20), $\mathbf{R}_n$ are the coordinates of the nth core centre (Figure 20), A_n and f are the amplitude and the transverse distribution of the field for the nth fibre.

Since only the neighbouring fibres are coupled to each other in our photonic-molecule bundle (Figure 20), the coupled-mode equations for the field amplitudes can be written as

$$\frac{dA_n}{dz} - i\beta A_n - i\alpha(A_{L(n)} + A_{R(n)}) = 0\,. \tag{12}$$

Here, β is the propagation constant for the relevant mode of an isolated fibre and α, the coefficient characterizing mode coupling in n-th and n'-th fibres, is given by

$$\alpha = \frac{\omega^2}{2\beta c^2}\frac{\int \Delta\varepsilon(\mathbf{r}) f(\mathbf{r} - \mathbf{R}_{n'}) f^*(\mathbf{r} - \mathbf{R}_n) d\mathbf{r}}{\int |f(\mathbf{r} - \mathbf{R}_n)|^2 d\mathbf{r}}\,, \tag{13}$$

where ω is the radiation frequency, $\Delta\epsilon(\mathbf{r})$ is the deviation from the unperturbed dielectric constant at a given point with a radius vector $\mathbf{r}$, and

$$L(n) = \begin{cases} n-1 & n > 0 \\ N, & n = 0 \end{cases} \qquad R(n) = \begin{cases} n+1 & n < N \\ 1, & n = N \end{cases}.$$

The propagation constants can now be found from the characteristic equation corresponding to the set of Equations (12). In the general case of arbitrary N, these propagation constants can be calculated with the use of numerical methods. There are several simple analytical solutions, however, that provide a useful physical insight into the dispersion of the photonic-molecule fibre. In particular, a symmetric field distribution,

$$\Psi_1(r) = A \sum_n f(\mathbf{r} - \mathbf{R}_n), \tag{14}$$

where A is a constant, is allowed by Equation (12) for any N. This symmetric mode of our fibre is analogous to a symmetric wave function of a polyatomic molecule or, more generally, a system of identical particles (bosons) in quantum mechanics. Similar to a symmetric wave function of such a quantum system, which remains invariant in time, the symmetric mode of our fibre is characterized by an amplitude A that is independent of propagation coordinate z (as long as guiding losses are negligible). To find the propagation constant for such a symmetric mode, we set $A_n = A$ in Equation (12), which yields

$$\frac{dA}{dz} - i(\beta + 2\alpha)A = 0. \tag{15}$$

This equation defines a guided mode with a propagation constant

$$B_1 = \beta + 2\alpha. \tag{16}$$

Equation (16) shows that mode coupling in the array of fibres results in a renormalization of propagation constants, leading to the increase in the propagation constants in the case of the totally symmetric mode of a photonic-molecule fibre.

An antisymmetric solution

$$\Psi_N(\mathbf{r}) = A \sum_n (-1)^n f(\mathbf{r} - \mathbf{R}_n) \tag{17}$$

is also allowed by Equation (12) for even N. Keeping in mind that $A_{n+1} = A_n$ in the case of this antisymmetric solution, we can rewrite Equation (12) as

$$\frac{dA}{dz} - i(\beta - 2\alpha)A = 0. \tag{18}$$

As can be seen from Equation (18), the propagation constant in the case of the antisymmetric solution is renormalized in accordance with

$$B_N = \beta - 2\alpha. \tag{19}$$

An analogy to this antisymmetric mode, with the amplitudes A_n independent of the propagation coordinate z, can also be found in quantum chemistry and quantum mechanics, where the antisymmetric wave function describing a system of identical quantum particles (fermions) remains invariant in the process of temporal evolution of the system.

By analogy with the classification of modes in a standard fibre (see, e.g., Snyder and Love 1983), we identify the fundamental mode of our photonic-molecule fibre as the mode with the largest propagation constant. The highest value of the propagation constant in the case of our PM fibre is achieved for the symmetric mode. This mode will therefore be referred to as the fundamental mode. We introduce the mode index l to enumerate PM-fibre modes, which will be denoted as PM_l modes, starting with $l = 1$, which corresponds to the fundamental PM mode.

Our model of the fibre structure employed in our experiments (Figure 19) consisted of seven identical glass cores with a radius of $2\mu m$. The refractive index of the cladding was set equal to the refractive index of atmospheric-pressure air, since the air-filling fraction of the structure was very high (see Figure 19). The distance R between the neighbouring cores was estimated as $7.4\mu m$. To estimate the coupling coefficient appearing in Equation (12), we employed the expression for the coupling coefficient, (where λ is the wavelength), from the model of two coupled identical planar waveguides (Yariv and Yeh, 1987). For characteristic geometric sizes of our structure, this model allows the parameter C to be estimated as $0.016\mu m^{-2}$. Only lowest order modes of isolated fibres were included in our calculations. The inclusion of higher order modes will, of course, change dispersion branches of a PM fibre and make the analysis much more complicated. The model that includes only fundamental modes of each elementary fibre, on the other hand, allows the general physical features of dispersion of a PM fibre to be understood without reproducing the fibre dispersion in all the details.

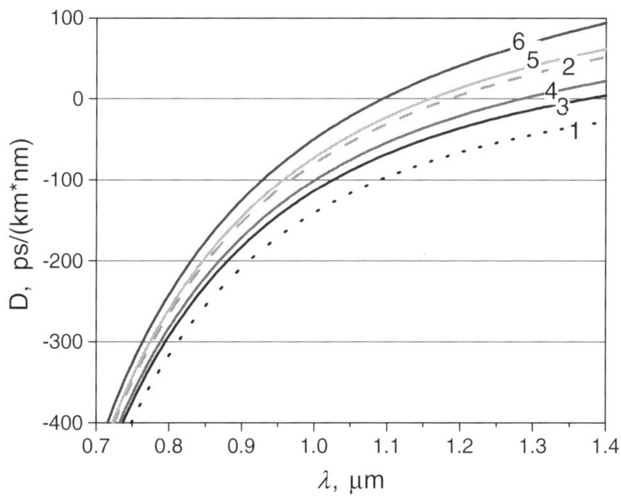

Figure 22. *Group-velocity dispersion of photonic-molecule modes as a function of radiation wavelength for (1) the material of the fibre; (2) a single isolated fibre from the considered photonic-molecule structure (or an array of identical fibres completely isolated from one another); and (3) PM_1, (4) PM_2 and PM_3, (5) PM_4 and PM_5, and (6) PM_6 and PM_7 modes of a seven-core photonic-molecule fibre ($N = 7$). The radius of a single fibre in the photonic-molecule fibre structure is $2\mu m$, the distance between the neighbouring fibres in the photonic-molecule structure is $7.4\mu m$.*

Photonic-molecule modes, as can be seen from Figures 21 and 22, differ from each other by their dispersion properties. This implies that the efficiency of nonlinear-optical processes in these modes should be also different. We will discuss here two factors leading to the enhancement of nonlinear-optical processes in higher order PM_l modes relative to lower order PM modes. These two factors are (i) the larger shift of the zero group-velocity dispersion wavelength, characteristic of higher order PM modes, leading even to the reversal of the sign of the group-velocity dispersion (Figure 22), and (ii) the higher group indices and, consequently, lower group velocities attainable with higher order PM modes (Figure 21). The importance of shifting the zero group-velocity dispersion wavelength and achieving the regime of anomalous dispersion for efficient SC generation has been demonstrated by earlier experiments with both holey (Ranka *et al.* 2000a) and tapered (Birks *et al.* 2000, Akimov *et al.* 2001) fibres. With our fibres, we were able, in fact, to switch between normal and anomalous group-velocity dispersion regimes for laser pulses of 1.2–1.3μm radiation by exciting either the fundamental PM_1 mode or the higher order PM_6/PM_7 modes of the fibre (cf. cases 3 and 6 in Figure 22), leading to quite noticeable changes in the properties of SC generation. It should be noted, however, that although the zero group-velocity dispersion point is a very important parameter for the propagation of ultrashort pulses through a fibre and for SC generation, this point alone cannot determine all the features of SC generation. In particular, knowing only the wavelength of zero group-velocity dispersion, we still cannot assess the group-delay effects (e.g., to find the walk-off length) for distant spectral components arising in the process of SC generation. This shows that the behaviour of the entire dispersion curve is important for understanding and optimizing the generation and propagation of new frequency components within a broad spectral range usually covered by SC emission. Multimode fibres often allow phase matching to be achieved for four-wave mixing processes involving higher order waveguide modes (Naumov *et al.* 2001b). Dispersion of higher order waveguide modes is then also important for SC generation.

To understand, on the other hand, why lower group velocities enhance nonlinear-optical interactions, it is instructive to represent the mean (averaged in time) energy flux w in a guided mode of the electromagnetic field as $P = vwS$, where w is the mean (averaged in time) energy density, v is the group velocity, and S is the effective mode area. For two light beams with equal powers and equal effective mode areas, the ratio of their mean energy densities is then equal to the inverse of the ratio of their group velocities, *i.e.* $w_1/w_2 = v_2/v_1$. This simple consideration explains the idea of using slower guided modes to enhance SC generation in our microstructure fibre.

The lowest values of the group index, as can be seen from the results of numerical simulations presented in Figure 21, are achieved for PM_l fibre modes with smooth profiles. The propagation constant and the group velocity of PM-fibre modes decrease with the growth in the number of extrema in the spatial profile of the PM-fibre mode. Obviously, the number of fibres in a PM fibre bundle define the upper bound for the number of such extrema in the profile of the PM_l fibre mode. The decrease in the group velocity for uneven PM_l fibre modes implies that nonlinear-optical processes should be enhanced for such slow modes. In particular, the efficiency of nonlinear-optical interactions for the PM_6/PM_7 degenerate fibre mode shown in insets *6* and *7* in Figure 21 should be noticeably higher than the efficiency of nonlinear-optical processes for the fundamental, symmetric PM_1 mode, shown in inset *1* in Figure 21. This expectation has been fully justified by the results of our experiments on SC generation in PM fibres, presented in Section 5.4. Generally, the waveguide dispersion, as can be seen from Figure 21, may be changed quite substantially by exciting different PM_l modes of a photonic-molecule fibre. In Section 5.4, we will show that higher order modes of a photonic-molecule fibre can be excited quite easily with OPA radiation, leading to the increase in the efficiency of SC generation. Thus, although the analysis performed in this section is valid, rigorously speaking, only for a bundle of single-mode fibres, it provides very useful physical insights into the possibilities of controlling nonlinear-optical processes in PM fibres through the excitation of different modes of such fibre structures.

It should be noted here that our choice of the basis set of modes for our fibre is, of course, not unique. Other basis sets can be constructed from linear combinations of the PM_l modes specified above. Interestingly and quite importantly, the fundamental PM_1 mode has a full seven-fold rotational symmetry, which coincides with the symmetry of the fibre (in the ideal case of a perfectly symmetric central ring). The other six modes form three pairs of doubly degenerate modes: PM_2 and PM_3, PM_4 and PM_5, PM_6 and PM_7. In each pair, the modes have equal propagation constants and identical dispersion properties (Figures 21, 22). While these three groups of modes can be easily distinguished in experiments, the modes forming a pair are experimentally indistinguishable (PM_2 cannot be distinguished from PM_3; PM_4, from PM_5; and PM_6, from PM_7), as their intensity distributions can be obtained from one another by rotating the intensity pattern around the symmetry axis and their dispersion properties are identical. Moreover, any of these three groups of doubly degenerate modes is sufficient to construct a linear combination of modes that would have the same symmetry as the fibre itself.

5.3 Photonic-molecule modes and white-light emission

The photonic-molecule design of a microstructure fibre proved to be very efficient for supercontinuum generation. We compared the efficiency of SC generation in such fibres with the efficiency of SC generation in standard photonic-crystal fibres (see the insets in Figure 23). The energies of 80–100fs OPA pulses required to achieve spectral superbroadening in PM fibre samples were typically higher (see Figure 23) than the energies needed for white-light generation in PC fibre samples of the same length with the PC-fibre core radius equal to the radii of glass channels forming the photonic-molecule structure in the PM fibre (the insets in Figure 23). This result is quite natural, as the effective mode area in a PM fibre is typically an order of magnitude larger than the effective mode area in a PC fibre with the same pitch of the structure (cf. the insets in Figure 23). At the same time, PM fibres, as will be demonstrated below, offer a simple way of controlling the SC generation process through the excitation of different PM-fibre modes and open new opportunities for evanescent-field nonlinear optics and spectroscopy.

The possibility of controlling the process of SC generation is illustrated by the results of experiments presented in Figure 24. Varying the focusing geometry and shifting the fibre end with respect to the light beam coupled into the fibre, we were able to excite, in fact, all the PM-fibre modes shown in insets 1–7 in Figure 21. We observed efficient SC generation for all these modes excited by 80–100fs pulses of 1.1–1.5μm OPA radiation in 4cm PM-fibre samples (Figure 24). This SC emission was imaged onto an NTE/CCD camera, placed behind a set of neutral density filters. Insets *1* and *2* in Figure 24 show such white-light images of the output end of a 4cm PM fibre. The images appear unfocused in Figure 24, as it is impossible, of course, to achieve optimal imaging conditions for all the colours of SC emitted from the fibre.

In the case of asymmetric modes of the PM fibre (e.g., PM_6 and PM_7 modes; see insets *6* and *7* in Figure 21), SC emission produced much brighter images with the same pulse powers coupled into the fibre than in the case of the fundamental PM_1 mode of the

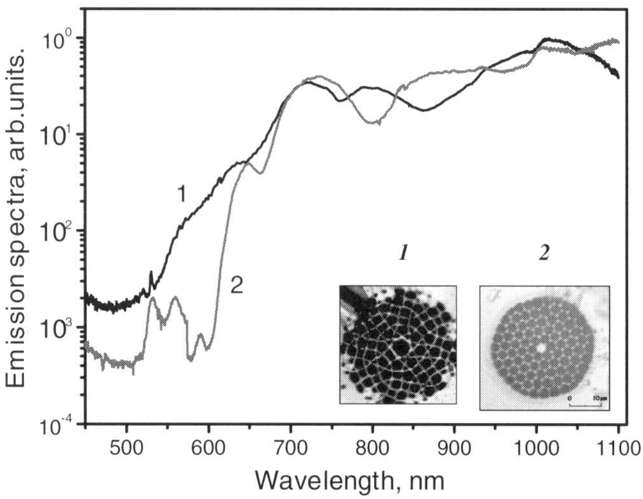

Figure 23. *The spectra of supercontinuum generated in the (1) PM_1 mode of a photonic-molecule fibre and (2) a photonic-crystal fibre. The lengths of photonic-molecule and photonic-crystal fibres were equal to 4cm. OPA radiation pulses with the wavelength of 1.30μm and the energy of (1) 0.2μJ and (2) 5nJ were coupled into the fibres. The initial duration of light pulses is 80fs. The radius of a single fibre in the photonic-molecule fibre structure (inset 1) is 2μm, the distance between the neighbouring fibres in the photonic-molecule structure is 7.4μm. The photonic-crystal fibre (inset 2) has a core radius of 2μm and the air-filling fraction of the cladding equal to 65%.*

same PM fibre (inset *1* in Figure 21). This visual observation was consistent with spectral measurements, which demonstrated that the supercontinuum white light generated in the PM_6 and PM_7 modes had a noticeably broader bandwidth than the supercontinuum generated in the fundamental PM_1 mode (cf. curves *1* and *2* in Figure 24). This observation demonstrates that PM fibres offer the possibility of changing the properties of SC emission by exciting different modes of a PM fibre. The spatial pattern of SC emission at the output of the fibre slightly differed from the distribution of the pump field guided in PM modes. This was due to the fact that different frequency components in the broad spectrum of SC emission were coupled into other PM modes with different coupling efficiencies. Apparently, this effect may have a considerable influence on the process of SC generation in PM modes for longer fibre samples. Since we were able to change the sign of the group-velocity dispersion by exciting different PM modes in our fibres, the initial chirp of laser pulses coupled into the fibre is another important factor influencing SC generation. In both normal and anomalous group-velocity dispersion regimes, it was possible to precompensate the dispersion spreading of the pump pulse in the fibre with an appropriate choice of the initial chirp of the pulse. However, variation of the initial chirp also resulted in changes in the initial pulse duration, which prevented us from performing more systematic studies of chirp-related effects. Therefore, we do not present the SC parameters measured as functions of the initial pulse chirp here.

To demonstrate the reproducibility of the supercontinuum generated in different modes of the photonic-molecule fibre, we first excited the fundamental, PM_1 mode of this fibre

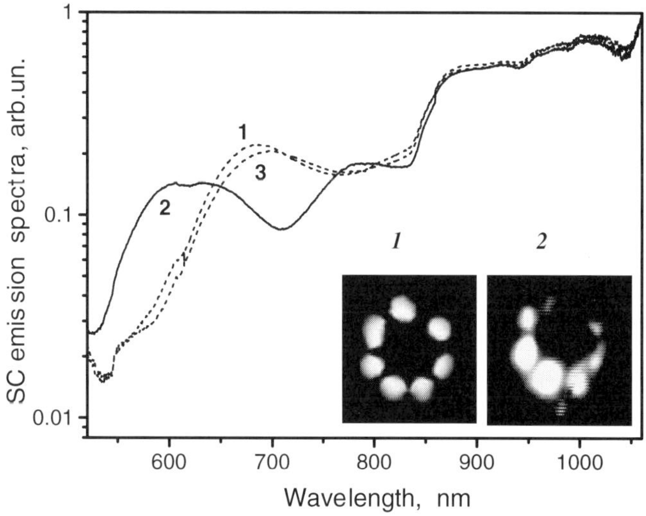

Figure 24. *The spectra of supercontinuum generated in the (1, 3) PM_1 and (2) PM_6/PM_7 modes of a photonic-molecule fibre with a length of 4cm. OPA radiation pulses with the wavelength of 1.30μm and the energy of 0.1μJ are coupled into the fibre. The initial duration of light pulses is 80fs. The radius of a single fibre in the photonic-molecule fibre structure is 2μm, the distance between the neighbouring fibres in the photonic-molecule structure is 7.4μm. Curves 1 and 3 demonstrate that the spectrum of supercontinuum emission recovers after the re-alignment of optical elements coupling OPA radiation into the fibre. The insets show white-light images of photonic-molecule fibres produced with the use of supercontinuum generated in the (1) PM_1 and (2) PM_6/PM_7 modes of these fibres by 80fs 0.1μJ pulses of 1.3μm OPA radiation.*

and measured the spectrum of SC emission at the output of the fibre (curve *1* in Figure 24). Then, the optical system used to couple OPA radiation into the fibre was adjusted in such a way as to excite the PM_6/PM_7 mode of the PM fibre. The excitation of this mode was indicated by the appearance of a white-light emission pattern shown in inset *2* of Figure 24. Spectral measurements were then performed on supercontinuum emitted from the fibre (curve *2* in Figure 24). Then, we re-aligned the beam-coupling optics in such a way as to excite the fundamental, PM_1 mode of the PM fibre again, indicated by the appearance of a white-light emission pattern shown in inset *1* of Figure 24, and measured the spectrum of SC emission again. The result is given by curve *3* in Figure 24. The fact that the spectrum of SC emission recovers after the re-alignment of optical elements coupling OPA radiation into the fibre (curves *1* and *3* in Figure 24 coincide with a high accuracy) demonstrates a rather high reproducibility of SC emission within a single series of experiments.

Thus, although the efficiency of SC generation in the case of photonic-molecule-core microstructure fibres is typically lower than the efficiency of SC generation in single-core photonic-crystal fibres (due to the larger total effective mode area in the case of PM fibres), photonic-molecule fibres offer several important options, which seem to advantageously supplement vast opportunities provided by photonic-crystal fibres. In particular, additional dispersion tunability can be achieved in the case of PM fibres by varying the initial conditions at the input end of the fibre, resulting in the excitation of different PM fibre modes. A promising direction for further studies would be to explore the possibility of coherent control of nonlinear-optical cross-action processes in the fibres of a PM waveguiding structure by using pulses with different temporal and spatial profiles, frequencies, and initial chirps. It would be of interest also to search for new ways to phase-match nonlinear-optical interactions in such fibres, as well as to examine the potential of PM fibres for pulse compression through self- and cross-phase modulation and SC generation. The geometry of PM fibres is also advantageous for a broadband evanescent-field spectroscopy of gases or liquids using SC generation, as well as for the laser guiding of atoms.

Thus, analysis of dispersion properties and the mode structure of a cyclic photonic molecule shows that the created photonic-molecule fibres is not only an interesting object for studying fundamental aspects of photon-localization modes in micro- and nanostructured matter and developing heuristic analogies with quantum chemistry, but this fibre also offers several new ways of dispersion engineering and switching between the regimes of nonlinear-optical interactions. The dispersion branches, as shown above, are highly sensitive to the coupling constants characterizing the bonds in the photonic molecule. This allows the waveguide dispersion to be controlled by changing parameters of the bridges linking elementary glass channels in our structure. This possibility of dispersion engineering seems to offer much promise for nonlinear-optical applications, allowing phase matching to be improved for various nonlinear-optical processes. This fibre design is also very attractive for multiwave mixing, since pump waves may enter different cores of the central ring-like structure to interact then through the bridges linking these cores. With the holes of our fibre structure filled with nonlinear or laser-active media, the evanescent field of guided modes can be used to generate nonlinear-optical signals and produce the lasing effect. While all these ideas and possibilities still have to be analyzed in greater detail, the results of experiments presented in this paper provide an example of how nonlinear-optical processes can be enhanced and controlled in photonic-molecule microstructure fibres, opening new avenues for applications of fibre-optic components in nonlinear optics and spectroscopy, optical metrology, and biomedicine, as well as offering new opportunities in generating ultrashort light pulses and few-cycle light-field waveforms and in controlling parameters of such pulses and waveforms.

6 Laser pulses in hollow-core photonic-crystal fibres and photonic band-gap guidance

Fibres with a two-dimensionally periodic microstructure (two-dimensional photonic crystal) and a hollow core are one of the most interesting and promising types of microstructure fibres. Such fibres were demonstrated for the first time by Cregan *et al.* (1999). The photonic band gap in the transmission spectrum of a two-dimensional periodic cladding in these fibres provides high reflection coefficients for electromagnetic radiation propagating along the hollow core of the fibre, allowing a specific regime of waveguiding to be implemented. This mechanism of waveguiding is of special interest for telecommunication applications, opening, at the same time, the ways to guide high-power and ultrashort pulses, and allowing the enhancement of nonlinear-optical processes, including high-order harmonic generation, in a gas medium filling the fibre core (Naumov and Zheltikov 2002a). The possibility of using such fibres for laser manipulation of small-size particles was recently demonstrated by Benabid *et al.* (2002).

In this section, we will consider waveguide modes of microstructure fibres with a hollow core and a two-dimensionally periodic cladding. The spectrum of modes guided in the hollow core of these fibres displays isolated maxima, indicating that waveguiding is supported due to the high reflectivity of the fibre cladding within photonic band gaps. The high reflectivity of the cladding within the photonic band gaps may substantially reduce optical losses of guided modes as compared with the modes guided in conventional, solid-cladding hollow fibres, and the group-velocity dispersion of modes guided in the hollow core of such fibres is, of course, much lower than the group-velocity dispersion characteristic of waveguide modes in standard fibres. These properties of hollow-core photonic-crystal fibres make them very attractive for ultrafast photonics applications, as well as for telecommunication applications, high-power laser radiation guiding, laser manipulation and laser guiding of atoms and charged particles, and high-order harmonic generation, and transport of ultrashort laser pulses.

Microstructure fibres are usually fabricated with the use of a preform consisting of a set of identical glass capillaries. Seven capillaries were removed from the central part of the preform for the hollow core of photonic-crystal fibres. The cross-section image of a fibre fabricated by drawing such a preform is presented in Figure 25 (Konorov *et al.* 2002). A typical period of the structure in the cladding of the fibre shown in Figure 25 is about 5μm. The diameter of the hollow core of the fibre is then approximately equal to 13μm. The length of fibre samples employed in our experiments ranged from several centimeters up to 1m.

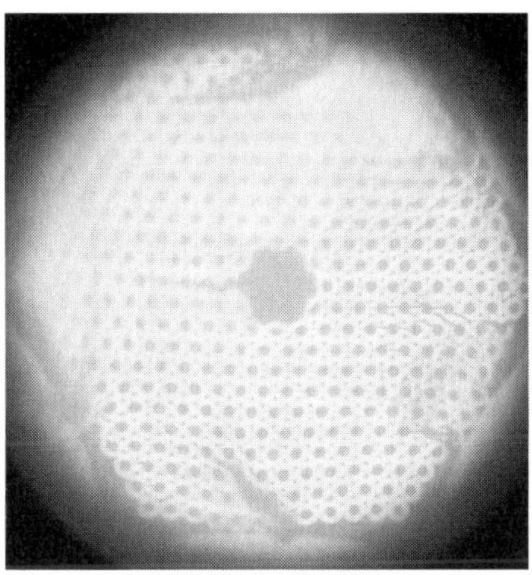

Figure 25. *Cross-sectional image of a microstructure fibre with a two-dimensionally periodic cladding consisting of an array of identical capillaries. This periodic cladding supports guided modes in the hollow core of the fibre due to the high reflectivity of a periodic structure within photonic band gaps. The hollow core of the fibre is formed by removing seven capillaries from the central part of the structure. The period of the structure in the cladding is about 5μm and the core diameter is about 13μm.*

The idea of lowering the magnitude of optical losses in a hollow fibre with a periodically microstructured cladding relative to the magnitude of optical losses in a hollow fibre with a solid cladding is based on the high reflectivity of a periodic structure within photonic band gaps (Yariv and Yeh 1987). In hollow fibres, the refractive index of the core is lower than the refractive index of the cladding. Therefore, the propagation constants of hollow-fibre modes have nonzero imaginary parts, and the propagation of light in such fibres is accompanied by radiation losses. The coefficient of optical losses in hollow fibres scales as λ^2/a^3 (Marcatili and Schmeltzer 1964), where λ is the radiation wavelength and a is the inner radius of the fibre. Such a behaviour of the magnitude of optical losses prevents one from using hollow fibres with very small inner diameters in nonlinear-optical experiments. Our estimates show that the magnitude of radiation losses for the fundamental mode of a hollow fibre with a fused silica cladding and an inner radius of 6.5μm may reach 20cm^{-1} for 0.8μm radiation, which, of course, imposes serious limitations on applications of such fibres. Radiation losses can be radically reduced in the case of hollow fibres with a periodic cladding.

Experimental results confirm the possibility of using hollow photonic-crystal fibres with a core diameter of about 13μm to guide coherent and incoherent radiation. Figure 26 displays the spatial distributions of intensity of incoherent (Figure 26a) and coherent (Figure 26b) radiation obtained by imaging the output end of a hollow photonic-crystal fibre with the above-specified parameters. Optimizing the geometry of coupling of laser radiation into the fibre, we were able to achieve a high degree of light-field confinement in the hollow core of the fibre without losing too much energy through mode excitation in the photonic-crystal cladding (Figure 26a). The spatial distribution of radiation intensity at the output end of the fibre under these conditions corresponded to the fundamental waveguide mode.

(a) (b)

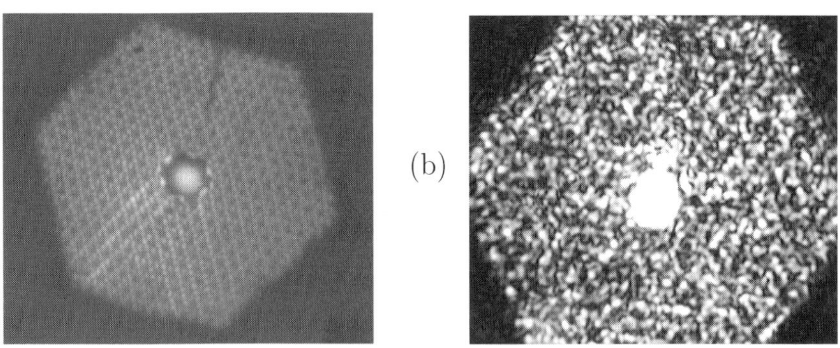

Figure 26. *Radiation intensity distribution in the cross section of a hollow photonic-crystal fibre with a period of the structure in the cladding of about 5μm and the core diameter of approximately 13μm. (a) A waveguide mode is excited in the hollow core with a broad beam of incoherent light. (b) The fundamental waveguide mode of the hollow core is excited with 633nm diode-laser radiation.*

To investigate the spectrum of modes guided in the hollow core of photonic-crystal fibres, we used a diaphragm to separate radiation transmitted through the hollow core from radiation guided by the cladding. The spectra of modes supported by the hollow core of photonic-crystal fibres were measured within the range of wavelengths from 450

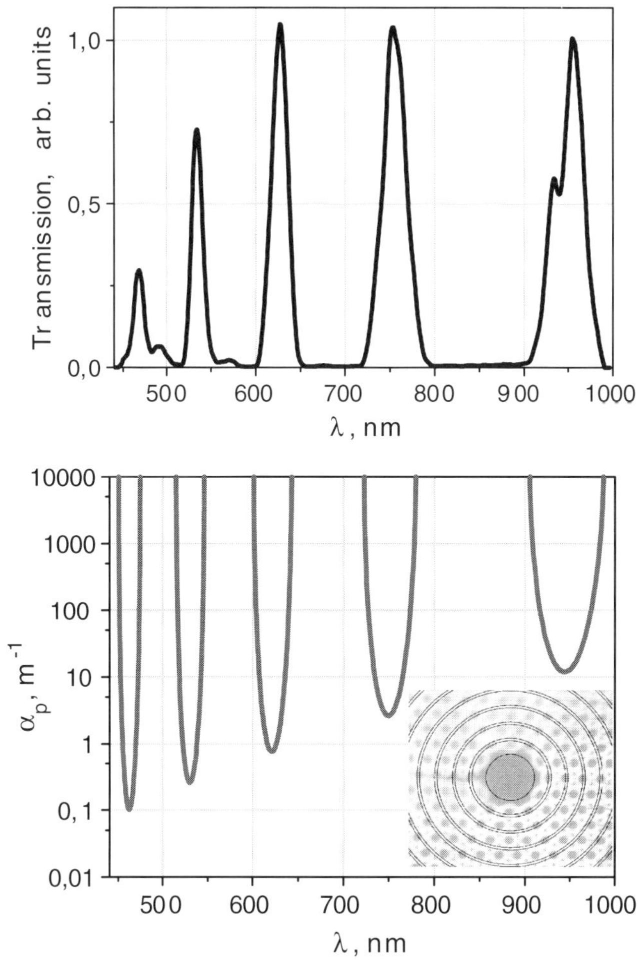

Figure 27. *The spectrum of modes measured for a hollow photonic-crystal fibre with a period of the structure in the cladding of about 5μm and the core diameter of approximately 13μm. (b) The attenuation coefficient of the TE_{01} waveguide mode calculated as a function of the wavelength for a periodic coaxial waveguide (see the inset) with r_0 = 6.5μm, b = 4.3μm, and c = 0.7μm.*

up to 1000nm. These spectra displayed characteristic well-pronounced isolated peaks (Figure 27a). The origin of these peaks is associated with the high reflectivity of a periodically structured fibre cladding within photonic band gaps, which substantially reduces radiation losses in guided modes within narrow spectral ranges. Radiation with wavelengths lying away from photonic band gaps of the cladding leaks from the hollow core. Such leaky radiation modes are characterized by high losses, giving virtually no contribution to the signal at the output of the fibre.

To model the spectrum of guided modes and the spatial distribution of radiation intensity in a hollow photonic-crystal fibre, we employed the model of a periodic coaxial waveguide. A two-dimensional periodic structure of the fibre cladding is replaced within the framework of this model by a system of coaxial glass cylinders (see the inset in Figure 27b) with a thickness $b \approx 4.3$ μm and the inner radius of the ith cylinder equal to $r_i = r_0 + i(b + c)$, where r_0 is the radius of the hollow core (about 6.5μm for our fibre) and c is the thickness of the air gap between the cylinders. The latter parameter was chosen in such a way as to include the air-filling fraction of the fibre cladding (about 14% in our experiments) and was set equal to approximately 0.7 μm for our calculations. In our theoretical analysis, we employed the results of earlier work (Ouyang *et al.* 2001) devoted to the properties of modes in coaxial waveguides.

Figure 27b presents the attenuation coefficient for the TE_{01} mode calculated as a function of the wavelength for a periodic coaxial waveguide with the above-specified parameters. Comparing these results with the experimental data shown in Figure 27a, we find that the model of a periodic coaxial waveguide provides qualitatively adequate predictions for the positions and the widths of spectral bands where the hollow core of a photonic-crystal fibre can guide radiation with virtually no losses. The model of a periodic coaxial waveguide, as can be seen from Figures 28 and 29, also gives a satisfactory qualitative description for radiation intensity distributions in the fundamental (Figure 28) and higher order (Figure 29) waveguide modes of a photonic-crystal fibre. Our theoretical results are consistent on the qualitative level with the predictions of simulations (Broeng *et al.* 2000) performed with the use of a more accurate and more sophisticated model of a hollow-core photonic-crystal fibre.

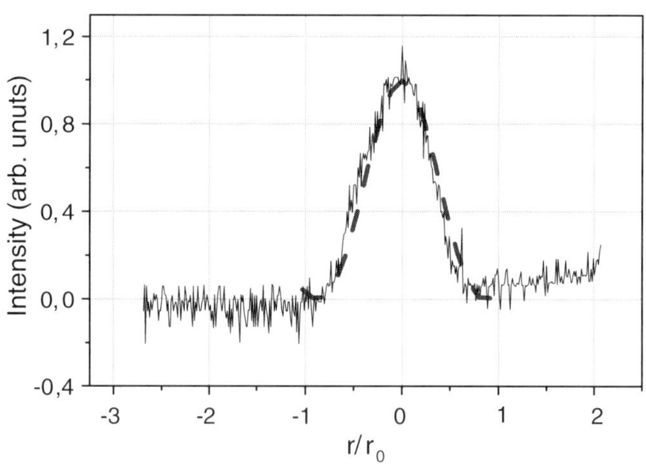

Figure 28. *Transverse intensity distribution of electromagnetic radiation (solid line) measured at the output of a hollow-core photonic-crystal fibre and (dashed line) calculated with the use of the model of a periodic coaxial waveguide.*

The spatial distribution of 633nm diode-laser radiation (this wavelength falls within one of the passbands in Figure 27, corresponding to the guided modes of our fibre) at the output of an 8cm hollow photonic-crystal fibre shown in Figure 29b indicates the existence of multimode regimes of waveguiding around this wavelength. Multimode waveguiding regimes in hollow photonic-crystal fibres can be employed to enhance high-order harmonic generation in nonlinear gases filling the hollow core of photonic-crystal fibres (Naumov and

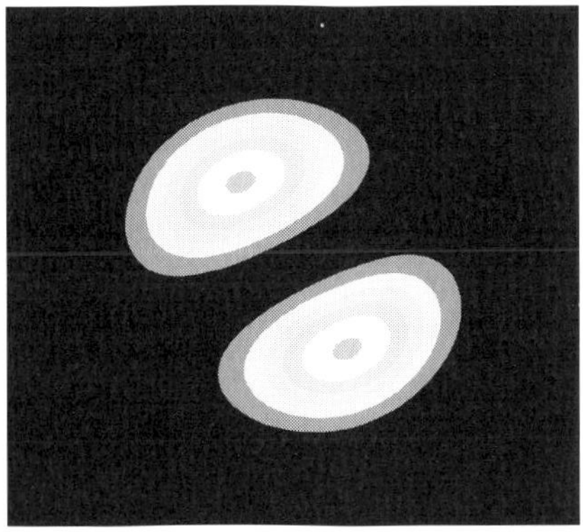

(a)

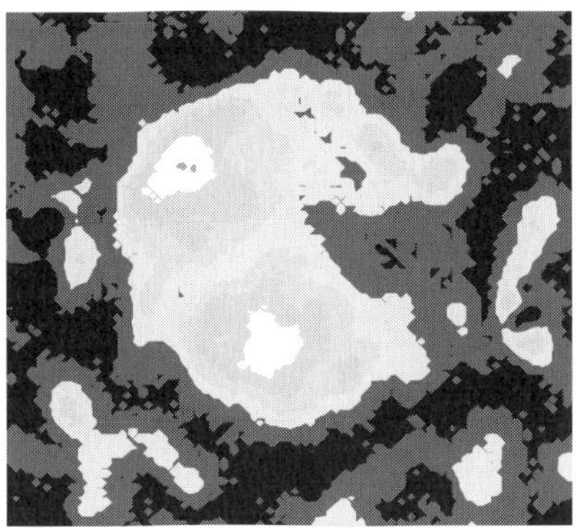

(b)

Figure 29. *Transverse distribution of the electric field squared in a higher order mode of a hollow-core photonic-crystal fibre calculated with the use of the model of a periodic coaxial waveguide. (b) Transverse intensity distributions of electromagnetic radiation measured at the output of a hollow-core photonic-crystal fibre with a higher order waveguide mode of the fibre excited with 633nm radiation of a diode laser.*

Zheltikov 2002a). The waveguide contribution to the mismatch of propagation constants related to guided modes of the pump and harmonic radiation increases with a decrease in the core diameter of a hollow fibre. Our photonic-crystal fibre with a small core diameter is, therefore, characterized by a strong dispersion of guided modes, allowing considerable phase mismatches related to the material dispersion of nonlinear gases to be compensated. This efficient phase-mismatch compensation becomes possible due to the unique properties of hollow photonic-crystal fibres, as the leaky modes guided in hollow fibres with a solid cladding and a diameter of the hollow core of about 13μm would have, as mentioned above, unacceptably high losses.

Since hollow waveguides with a periodic cladding permit radiation losses characteristic of hollow-waveguide modes to be radically reduced, waveguides of this type seem to offer new solutions in guiding high-power laser radiation and enhancing nonlinear-optical processes, including self- and cross-phase modulation, as well as optical harmonic generation and wave mixing. The results of preliminary experiments and theoretical studies (Naumov and Zheltikov 2002b) suggest that polycapillary glass structures and photonic-crystal fibres can guide and focus ultrashort x-ray pulses, including ultrashort field waveforms synthesized from high-order harmonics. Strong waveguide dispersion, inherent in hollow fibres with a small core radius, can be employed to compensate for the initial chirp and, eventually, to compress short x-ray pulses.

The fibres created and studied in this work can be used for the creation of high-sensitivity gas sensors based on linear and nonlinear spectroscopic techniques. The waveguiding regimes attainable with hollow photonic-crystal fibres with a small core diameter allow the amount of gas necessary for spectral analysis to be considerably reduced and permit nonlinear-spectroscopic studies to be performed with low-power laser pulses. In particular, experiments on four-wave mixing in hollow fibres (see, e.g., Naumov *et al.* 2001b) are usually carried out with capillaries with a typical inner diameter on the order of 100 μm (the use of capillaries with smaller inner diameters leads to the fast growth of optical losses). The fibres considered in this paper would allow comparable levels of nonlinear signal to be achieved with 60 times less powerful laser pulses. The spectra of modes guided by these fibres seem optimal for laser frequency conversion through stimulated Raman scattering.

7 Concluding remarks

The format of this chapter allows me to give just a brief introduction into the ultrafast optics of photonic crystals, providing only a general impression of how photonic crystals can help to control the parameters of short light pulses and light beams. However, the examples considered above demonstrate very clearly that photonic crystals can be considered as promising elements for modern optical technologies, permitting considerable conceptual progress in designing compact frequency converters and pulse compressors, new types of waveguides, and elements for local photoexcitation and high-resolution microscopy. Remarkable properties of field localization, confinement, and enhancement characteristic of photonic crystals can be employed to increase the efficiency of nonlinear-optical interactions of short light pulses and to guide laser beams. By properly engineering the photonic-crystal lattice, one can often construct a potential required for the solution of many problems of classical and quantum electrodynamics, including the laser guiding of atoms and the entanglement of quantum states. These and many other useful applications of photonic crystals call for the development of new technologies for the fabrication of three-dimensionally periodic structures with a photonic band gap in the visible range and elaboration of new efficient methods for the analysis of temporal and spatial properties of light fields in such structures.

We have shown that the whole catalog of nonlinear-optical processes involving ultrashort pulses can be radically enhanced with the use of microstructure and photonic-crystal fibres. Fibres of this type may provide a high degree of light confinement in the fibre core, simultaneously offering broad opportunities for engineering the dispersion of guided modes. Due to their remarkable properties, microstructure fibres make nonlinear optics accessible even to unamplified femtosecond pulses. Enhancement of the broad class of nonlinear-optical phenomena, accompanying the propagation of femtosecond pulses in microstructure fibres, makes it possible to generate a supercontinuum emission starting with nano- and even subnanojoule energies of laser radiation. The range of applications of supercontinuum generation in microstructure fibres is now rapidly expanding, leading to revolutionary changes in optical metrology, opening new horizons in optical coherence tomography, and suggesting new solutions for the creation of compact and practical sources of broadband emission based on supercontinuum generation in microstructure fibres.

Acknowledgments

The material of this chapter is based on the results of our research group at M.V. Lomonosov Moscow State University in collaboration with several research groups from different countries. I am particularly grateful to A.B. Fedotov, D.A. Sidorov-Biryukov, S.O. Konorov, A.N. Naumov, D.A. Akimov, O.A. Kolevatova, and A.A. Podshivalov, of M.V. Lomonosov Moscow State University; M.V. Alfimov and A.A. Ivanov, of the Center of Photochemistry, Russian Academy of Sciences; V.I. Beloglazov, N.B. Skibina, A.V. Shcherbakov, and L.A. Mel'nikov, of Technology and Equipment for Glass Structures Enterprise, Saratov; Yu.N. Kondrat'ev, V.S. Shevandin, and K.V. Dukel'skii, of S.I. Vavilov State Optical Institute, St. Petersburg; D. von der Linde, Ping Zhou, and A.P. Tarasevitch, of the University of Essen; R.B. Miles, of Princeton University; J.W. Haus, of Dayton University; T.A. Birks, W.J. Wadsworth, and P.St.J. Russell, of the University of Bath; D. Chorvat Jr., I. Bugar, and D. Chorvat, of the International Laser Center in Bratislava; and the late Valery B. Smirnov, of St. Petersburg State University. I am most happy to have this precious opportunity to collaborate with these brilliant scientists and excellent research groups.

The research described in this chapter was supported in part by the President of Russian Federation Grant no. 00-15-99304, the Russian Foundation for Basic Research project no. 00-02-17567, the Volkswagen Foundation (project I/76 869), CRDF Award no. RP2-2266, the European Research Office of the US Army under (Contract no. N62558-02-M-6023), and the "Fundamental Metrology" Federal Science and Technology program of the Russian Federation.

References

Agrawal G P, 1989, *Nonlinear Fibre Optics* (Academic Press, Boston).

Akhmanov S A and Nikitin S Yu, 1997, *Physical Optics* (Clarendon, Oxford).

Akimov D A, Fedotov A B, Podshivalov A A, Zheltikov A M, Ivanov A A, Alfimov M V, Bagayev S N, Pivtsov V S, Birks T A, Wadsworth W J, and Russell P St J, 2001, *JETP Letters* **74** 460.

Aközbek N, Scalora M, Bowden C M, and Chin S L, 2001, *Optics Communications* **191** 353.

Alfano R, ed., 1989, *The Supercontinuum Laser Source* (Springer-Verlag, New York).

Alfimov M V, Zheltikov A M, Ivanov A A, Beloglazov V I, Kirillov B A, Magnitskii S A, Tarasishin A V, Fedotov A B, Mel'nikov L A, and Skibina N B, 2000, *JETP Letters* **71** 489.

Apolonski A, Povazay B, Drexler W, Wadsworth W J, Knight J C, and Russell P St J, 2002, *Journal of Optical Society of America B* **19** 2165.

Armstrong J A, Bloembergen N, Ducuing J, and Pershan P S, 1962, *Physical Review* **127** 1918.
Benabid F, Russell P St J, and Knight J C, 2002, *Conference on Lasers and Electro-Optics (CLEO'2002)*, Long Beach, May 19–24, 2002, CMJ5.
Bennett P J, Monro T M, and Richardson D J, 1999, *Optics Letters* **24** 1203.
Birks T A, Knight J C, and Russell P St J, 1997, *Optics Letters* **22** 961.
Birks T A, Wadsworth W J, and Russell P St J, 2000, *Optics Letters* **25** 1415.
Bloembergen N and Sievers A J, 1970, *Applied Physics Letters* **17** 483.
Brabec T and Krausz F, 2000, *Review Modern Physics* **72** 545.
Broeng J, Barkou S E, Søndergaard T, and Bjarklev A, 2000, *Optics Letters* **25** 96.
Coen St, Chau A H L, Leonhardt R, Harvey J D, Knight J C, Wadsworth W J, and Russell P St J, 2001, *Optics Letters* **26** 1356.
Cregan R F, Mangan B J, Knight J C, Birks T A, Russell P St J, Roberts P J, and Allan D C, 1999, *Science* **285** 1537.
Diddams S A, Jones D J, Jun Ye, Cundiff S T, Hall J L, Ranka J K, Windeler R S, Holzwarth R, Udem T, and Hänsch T W, 2000, *Physics Review Letters* **84** 5102.
Dumeige Y, Vidakovic P, Sauvage S, Sagnes I, Levenson J A, Sibilia C, Centini M, D'Aguanno G, and Scalora M, 2001, *Applied Physics Letters* **78** 3021.
Fedotov A B, Koroteev N I, Loy M M T, Xiao X, and Zheltikov A M, 1997, *Optics Communications* **133** 587.
Fedotov A B, Zheltikov A M, Tarasevitch A P, and von der Linde D, 2001a, *Applied Physics B* **73** 181.
Fedotov A B, Giammanco F, Naumov A N, Marsili P, Ruffini A, Sidorov-Biryukov D A, Zheltikov A M, 2001b, *Applied Physics B* **72** 575.
Fedotov A B, Ping Zhou, Kondrat'ev Yu N, Bagayev S N, Shevandin V S, Dukel'skii K V, Smirnov V B, Tarasevitch A P, von der Linde D, and Zheltikov A M, 2002a, *Journal of Experimental and Theoretical Physics*, in press.
Fedotov A B, Ping Zhou, Tarasevitch A P, Dukel'skii K V, Kondrat'ev Yu N, Shevandin V S, Smirnov V B, von der Linde D, and Zheltikov A M, 2002b, *Journal of Raman Spectroscopy* **33** no. 11/12.
Fedotov A B, Naumov A N, Haus J W, Miles R B, Sidorov-Biryukov D A, Chigarev N V, and Zheltikov A M, 2002c, *Journal of Optical Society of America B* **19** 1162.
Fedotov A B, Naumov A N, Zheltikov A M, Bugar I, Chorvat D Jr, Chorvat D, Tarasevitch A P, and von der Linde D, 2002d, *Journal of Optical Society of America B* **19**, 2156.
Fedotov A B, Naumov A N, Bugar I, Chorvat D Jr, Sidorov-Biryukov D A, Chorvat D, and Zheltikov A M, 2002e, *IEEE Journal of Selected Topics in Quantum Electronics* 8 665.
Fejer M M, Magel G A, Jundt D H, and Byer R L, 1992, *IEEE Journal of Quantum Electronics* **28** 2631.
Gaeta A L, 2002, *Optics Letters* **27** 924.
Golovan L A, Zheltikov A M, Kashkarov P K, Koroteev N I, Lisachenko M G, Naumov A N, Sidorov-Biryukov D A, Timoshenko V Yu, and Fedotov A B, 1999, *JETP Letters* **69**, 300.
Harris S E and Sokolov A V, 1998, *Physical Review Letters* **81** 2894.
Hartl I, Li X D, Chudoba C, Rhanta R K, Ko T H, Fujimoto J G, Ranka J K, and Windeler R S, 2001, *Optics Letters* **26** 608.
Hasegawa T, Sasaoka E, Onishi M, Nishimura M, Tsuji Y, and Koshiba M, 2001, *Optics Express* **9** 681.
Hentschel M, Kienberger R, Spielmann Ch, Reider G A, Milosevic N, Brabec T, Corkum P, Heinzmann U, Drescher M, and Krausz F, 2001, *Nature* **414** 511.
Herrmann J, Griebner U, Zhavoronkov N, Husakou A, Nickel D, Knight J C, Wadsworth W J, Russell P St J, and Korn G, 2002, *Physical Review Letters* **88** 173901.

Holzwarth R, Udem T, Hänsch T W, Knight J C, Wadsworth W J, and Russell P St J, 2000, *Physics Review Letters* **85** 2264.
Joannopoulos J, Meade R, and Winn J, 1995, *Photonic Crystals* (Princeton University Press, Princeton).
Kaiser P, Marcatili E A J, and Miller S E, 1973, *Bell System Technology Journal* **52** 265.
Knight J C, Birks T A, Russell P St J, and Atkin D M, 1996, *Optics Letters* **21** 1547.
Knight J C, Broeng J, Birks T A, and Russell P St J, 1998, *Science* **282** 1476.
Knight J C, Arriaga J, Birks T A, Ortigosa-Blanch A, Wadsworth W J, and Russell P St J, 2000, *IEEE Photonic Technology Letters* **12** 807.
Knight J C and Russell P St J, 2002, private communication.
Konorov S O, Fedotov A B, Kolevatova O A, Beloglazov V I, Skibina N B, Shcherbakov A V, and Zheltikov A M, 2002, *JETP Letters* **76** 341.
Le Blanc S P and Sauerbrey R, 1996, *Journal of Optical Society of America B* **13** 72.
Marcatili E A J and Schmeltzer R A, 1964, *Bell System Technology Journal* **43** 1783.
Monro T M, Bennett P J, Broderick N G R, and Richardson D J, 2000, *Optics Letters* **25** 206.
Naumov A N, Zheltikov A M, Fedotov A B, Sidorov-Biryukov D A, Tarasevitch A P, Ping Zhou and von der Linde D, 2001a, *Journal of the Optical Society of America*, **B 18** 811 .
Naumov A N, Giammanco F, Sidorov-Biryukov D A, Fedotov A B, Marsili P, Ruffini A, and Zheltikov A M, 2001b, *JETP Letters* **73** 263.
Naumov A N and Zheltikov A M, 2002a, *Quantum Electronics*, **32** 129.
Naumov A N and Zheltikov A M, 2002b, *Laser Physics* **12**, no. 11.
Naumov A N, Fedotov A B, Zheltikov A M, Yakovlev V V, Mel'nikov L A, Beloglazov V I, Skibina N B, and Shcherbakov A V, 2002, *Journal of Optical Society of America B* **19**, 2183.
Nazarkin A, Korn G, Wittmann M, and Elsaesser T, 1999, *Physical Review Lett*ers **83** 2560.
Ortigosa-Blanch A, Knight J C, Wadsworth W J, Arriaga J, Mangan B J, Birks T A, and Russell P St J, 2000, *Optics Letters* **25** 1325.
Ouyang G, Yong Xu, and Yariv A, 2001, *Optics Express* **9** 733.
Paul P M, Toma E S, Breger P, Mullot G, Augé F, Balcou Ph, Muller H G, and Agostini P, 2001, *Science* **292** 1689. Petit S, Talebpour A, Proulx A, and Chin S L, 2000, *Laser Physics* **10** 93.
Pezzetta D, Sibilia C, Bertolotti M, Haus J W, Scalora M, Bloemer M J, and Bowden C M, 2001, *Journal of Optical Society of America B* **18** 1326.
Ranka J K, Windeler R S, Stentz A J, 2000, *Optics Letters* **25** 25.
Ranka J K, Windeler R S, and Stentz A J, 2000, *Optics Letters* **25** 796.
Reeves W H, Knight J C, Russell P St J, and Roberts P J, 2002, *Optics Express* **10** 609.
Scalora M, Bloemer M J, Manka A S, Dowling J P, Bowden C M, Viswanathan R, and Haus J W, 1997, *Physical Review A* **56** 3166.
Snyder A W and Love J D, 1983, *Optical Waveguide Theory* (Chapman and Hall, New York).
Tempea G and Brabec T, 1998, *Optics Letters* **23** 762.
Wadsworth W J, Ortigosa-Blanch A, Knight J C, Birks T A, Mann T P M, and Russell P St J, 2002, *Journal of Optical Society of America B* **19** 2148.
van der Ziel J P, Ilegems M, Foy P W, and Mikulyak R M, 1976, *Applied Physics Letters* **29** 775.
Yablonovitch E, 1993, *Journal of Optical Society of America B* **10** 283.
Yariv A and Yeh P, 1977, *Journal of Optical Society of America B* **67** 438.
Yariv A and Yeh P, 1987, *Optical Waves in Crystals* (Wiley, New York).
Zheltikov A M, 2000, *Physics Uspekhi* **170** 1203.
Zheltikov A M, Tarasishin A V, and Magnitskii S A, 2000, *Journal of Experimental and Theoretical Physics* **91** 298.
Zheltikov A M, 2002, *Physics Uspekhi* **172** 743.

Ultrafast spectroscopy studies of semiconductors

Jaroslav Kovac

International Laser Center and Slovak Technical University, Slovak Republic

1 Introduction

A detailed knowledge of carrier scattering mechanisms and relaxation processes is an essential basis for the development of ultrafast semiconductor opto-electronic devices. Recent years have witnessed a great deal of interest in artificially structured low-dimensional semiconductor hetero-structures. Both fundamental interactions and possible device applications have been addressed on ultra-short time-scales. Quantum confined systems may take the form of quantum wells (QW) (Pelouch *et al.* 1992, Sun *et al.* 2000), quantum wires (QWR)(Sung *et al.* 1996, Lienau *et al.* 1999) and more recently quantum dots (QD) (Watabe *et al.* 1996, Sosnovski *et al.* 1998). Ultrafast relaxation processes and quantum capture in the low-dimensional structures are important, not only for the basic physics involved, but for applications in devices such as lasers and photodetectors. Time resolved optical experiments are a major source of experimental data on the dynamics of free carriers in semiconductors. Optical spectroscopy using ultrashort pulses provides direct insights into these processes occurring on time scales between about 10^{-14}s and 10^{-10}s (Diels 1996). Many physically interesting relaxation processes take place in a few picoseconds or less. Most experimental studies have concentrated on interband excitations at photon energies close to, or higher than, the fundamental bandgap of the semiconductor. The dynamical properties of excitons and/or electron-hole pairs in the band-to-band continuum have been investigated and an excellent review is available (Shah 1996). However, a variety of elementary excitations occur in the photon energy range below the fundamental bandgap. Intraband (free carrier), inter-valence band, and inter-subband transitions are determined by a single type of carrier, that is, electrons or holes, and can serve as a probe of the dynamics of single component plasma. Details of this topic are reviewed by Elsaesser and Woerner (1999).

This chapter reviews progress in this field. After a brief introduction into generation and recombination processes in semiconductors, ultrafast spectroscopy techniques are described. The second part of the chapter is devoted to experimental results including pump-and-probe and ultrafast time resolved luminescence studies in various semiconductors and low-dimensional structures.

2 Semiconductors: basic concepts

2.1 Energy bands in semiconductors.

The basic description of the electronic properties of a semiconductor is given by its band structure, that is, the variation of electron energy E with the wave-vector $\mathbf{k}$. The electron energy $E(\mathbf{k})$ can be restricted to the first Brillouin zone (Ebeling 1993). Figure 1(a) shows the energy band behaviour for GaAs, including the features relevant to optoelectronic properties (Kalt 1995). The profiles of various bands are shown, starting from the centre of the first Brillouin zone Γ up to the intersection with the zonal boundary points L and X. GaAs has a direct band gap located at the centre of the Brillouin zone. The energy band gap, E_g, is temperature dependent and at room temperature, $E_g = 1.42$eV. The Γ valley is spherically symmetric with an approximately parabolic energy dispersion in relation to wave vector $\mathbf{k}$. The electron band structures near the X and L indirect minima are highly anisotropic. The minimum of the conduction band (CB) and maxima of the valence bands (VB) lie at $\mathbf{k} = 0$. The valence states arising from the $\mathrm{p}_{3/2}$ atomic levels give rise to heavy hole (hh) and light hole (lh) bands. The $\mathrm{p}_{1/2}$ states give the split-off valence band which lies at a lower energy determined by spin-orbit coupling.

The energy band gaps of semiconductors range from very small values in semiconductors such as InSb to several eV for semiconductors like ZnS and GaN. Energy band gaps of some binary III–V semiconductors are shown in Figure 1(b) as a function of lattice constant. All semiconductors can be classified either as direct gap with the valence band maximum and conduction band minimum at the same $\mathbf{k}$ vector in the Brillouin zone, or indirect gap in which these extremes occur at different points in the Brillouin zone. The optical properties of direct band gap semiconductors differ considerably from those having indirect band gaps. The continuous and dotted lines in Figure 1(b) indicate direct and indirect band gaps, respectively, for ternary compounds created by alloying two binary semiconductors.

In undoped semiconductors the valence bands are completely filled with electrons at absolute zero, whereas the conduction bands are completely empty. To investigate various optical properties such as rate of absorption or emission and how electrons and holes distribute themselves within the various bands, we need to know the density of states in terms of energy by using the energy-momentum relation in parabolic bands. The effect of energy and momentum conservation can be conveniently illustrated by measuring optical absorption as a function of photon energy (Hunsperger 1995). The strongest absorption in semiconductors occurs when an electron in a given energy level absorbs a photon and makes a transition to a higher energy state. Optical transitions are subject to certain selection rules where the energy and the momentum ($\mathbf{p} = \hbar\mathbf{k}$) of the electron and photon must be conserved. If only photons and electrons are involved $E_i + E_{\mathrm{ph}} = E_f$ and

(a)

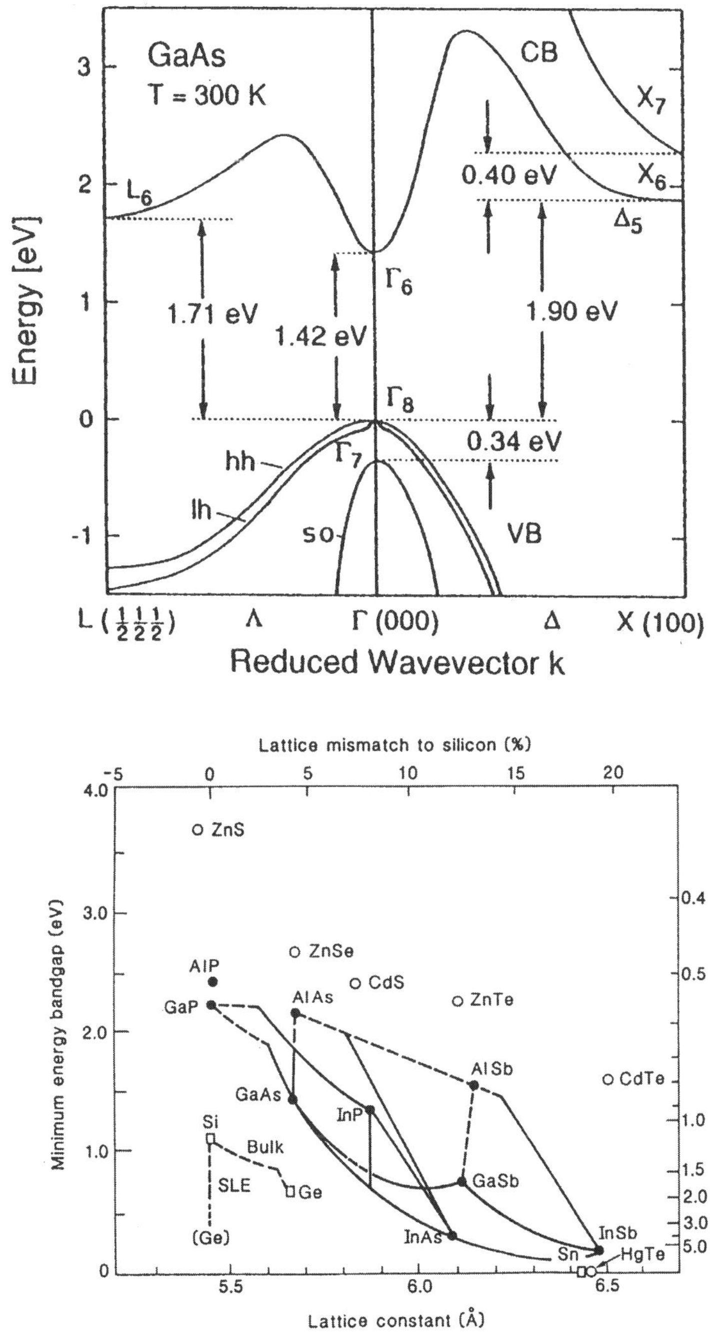

(b)

Figure 1. *(a) Plot of the band structure of GaAs, and (b) energy band gaps of some binary III-V semiconductors as a function of lattice constant.*

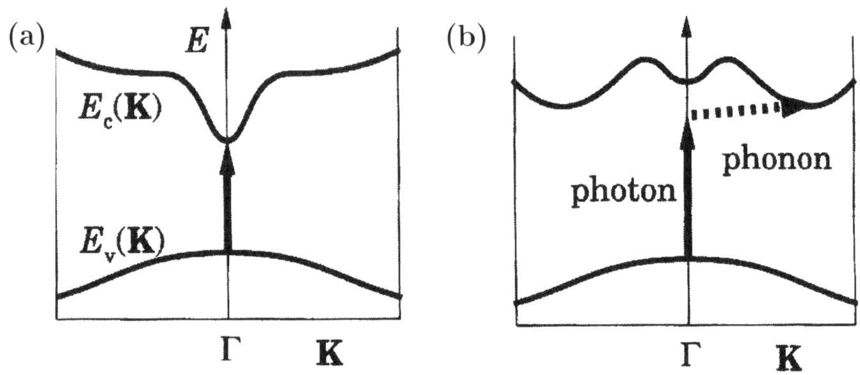

Figure 2. *(a) Direct transition in direct gap (b) direct transition in indirect gap and (c) indirect transition in indirect gap semiconductor (Hunsperger 1995).*

$\mathbf{k}_i + \mathbf{k}_{\mathrm{ph}} = \mathbf{k_f}$ (where the subscripts i and f refer to initial and final states and E_{ph} is the photon energy). The momentum of the photon $\mathbf{p}_{\mathrm{ph}}$ is very small so that $\mathbf{k}_i = \mathbf{k}_f$ is a good approximation. The transition appears on the $E(\mathbf{k})$ diagram as a vertical (direct) line in Figure 2 (a). It is possible, however, to have optical transitions in which $\mathbf{k}$ changes if momentum is transferred to, or absorbed from, the lattice vibrations, that is, phonons. The conditions restricting an absorptive transition in which a photon, electron and phonon are involved are given by $E_i + E_{\mathrm{ph}} \pm E_{\mathrm{phonon}} = E_f$ and $\mathbf{k}_i \pm \mathbf{q} = \mathbf{k}_f$ where E_{phonon} and $\mathbf{q}$ are the phonon energy and wavevector. Such a transition is called indirect and appears on the $E(\mathbf{k})$ diagram as shown in Figure 2(b).

Photon generation usually results from electron-hole recombination. The radiative recombination process can be divided into two types, (a) spontaneous emission, in which an electron and hole randomly recombine, and (b) stimulated emission in which an electron and hole are stimulated to recombine by an already existing photon. As in the case of absorption, energy and momentum criteria limit the possible emissive transitions.

Coulomb interactions between electrons and holes have a profound influence on the interband optical properties of semiconductors. A hydrogenic series of exciton bound states occurs just below the fundamental absorption edge. These exciton states may be resolved in absorption or emission particularly at low temperature and can be enhanced at room temperature in quantum confined structures. Coulomb contributions also enhance the optical absorption above the band gap energy in the electron-hole continuum regime. In a bulk semiconductor the exciton eigen-energy, given by $E_n^{3D}(k) = E_g - R_y^*/n_B^2 + \eta^2 K^2/2M$ is characterized by the principal quantum number n. The total exciton wave vector $\mathbf{K} = \mathbf{k}_e + \mathbf{k}_h$ and the dispersion is related to the motion of the centre of mass $M = m_e + m_h$ (Kalt 1996). The excitonic Rydberg energy $R_y^* = e^2 \mu_{\mathrm{ex}} m_0 / 2(4\pi\varepsilon_0)^2 \varepsilon_1 \eta^2$ is calculated in analogy to the hydrogen model, while accounting for the exciton reduced mass μ_{ex} and the modified strength of the Coulomb interaction in the solid by including the real part of the dielectric function ε_1. The excitonic Bohr radius is $a_B^* = 4\pi\varepsilon_0\varepsilon_1\eta^2/m_0\mu_{\mathrm{ex}}e^2$. Many basic properties of excitons are influenced by the nature of the band gap to which they are related. These effects can be nicely illustrated via optical interactions in direct or indirect band gap compound semiconductors (Weisbuch *et al.* 2000, Kalt 1996).

2.2 Quantum confined effects in low-dimensional structures

Since the 1980s much effort has been devoted to the study of two-dimensional (2D) semiconductor quantum wells (QW). Much of the effort has concerned the understanding of the basic quantum mechanical effects in QWs and their potential applications for opto-electronic devices. Epitaxial growth techniques such as molecular beam epitaxy (MBE) (Herman and Sitter 1989) or metal-organic vapour phase epitaxy (MOVPE) (Kovac *et al.* 1999) have been developed to allow controlled growth of semiconductor nanostructures on an atomic scale. If a semiconductor with a smaller band gap (GaAs) is sandwiched between layers of a larger band gap semiconductor (AlGaAs), and if the thickness of the smaller band material is comparable to or smaller than the carrier de Broglie wavelength in the semiconductor, the quantum confined effects in QW become important and lead to profound changes in optoelectronic properties. In these structures the carriers are confined along the growth z direction but are free to move in the x and y directions. The confinement of the carriers leads to a quantization of the eigen-energies into sub-bands in the direction of the restriction while maintaining parabolic dispersion due to the motion in all remaining directions. The energies and dispersion of such sub-bands in 2D systems are illustrated in Figure 3(a) for the case of a GaAs/AlGaAs quantum well. The sub-band energies are shifted with respect to the bulk band gap by the confinement energies $E_{n_z,q}$ (where q =e or h) of the electrons and holes given by $E_{n_z,q} = \eta^2 n_z^2 \pi^2/2m_q L_z^2$, assuming a QW of width L_z and infinite well height where n_z is the quantum number and m_q is the electron or hole effective mass.

Due to the smaller effective mass of the electrons, the quantization in the conduction band is much larger than the quantization in the valence band. The first light hole (lh) level lies typically farther from the valence band edge than the respective heavy hole (hh) level due to the lighter effective mass. If QWs are stacked on top of each other, either a multi-quantum well (MQW) or a superlattice (SL) structure is formed. The difference between these two types is that in the superlattice the wells are so close to each other that the electron wavefunction is not localized within a single QW. Typical absorption spectra are shown in Figure 3(b) for GaAs/AlGaAs MQW structures at low temperature. The absorption follows the step-like density of states in 2D systems with distinct hh and lh exciton resonances at each step.

More recently, great interest has been shown in the lower-dimensional hetero-structures, one-dimensional (1D) quantum wires (QWR) and zero-dimensional (0D) quantum dots (QD). However the lack of technological capability in the fabrication of high-quality 1D and 0D hetero-structures has so far limited the number of fundamental studies and their possible applications. The electronic states in 1D and 0D structures are described in a similar way to 2D. The additional lateral confinement of the wave functions, along the y axis in QWR and along x and y axis in QD, results in a larger confinement energy for the discrete electronic states. For a particle confined in a quantum dot, if L_x and L_y define the size of the semiconductor along x and y, respectively, the discrete set of energy levels can be calculated according to $E_{n_x,n_y,n_z,q} = \eta^2\pi^2/2m_q[(n_x/L_x)^2 + (n_y/L_y)^2 + (n_z/L_z)^2]$, where n_y and n_x are the quantum numbers due to the confinement along y and x, respectively. The above formula comes from the solution of the Schrödinger equation assuming an infinite rectangular confining potential. Integration over the respective momentum directions gives the characteristic $1/\sqrt{E}$ density of states dispersion for QWR's and delta like dispersion for QD's, as shown in Figure 4 (Cingollani 1993).

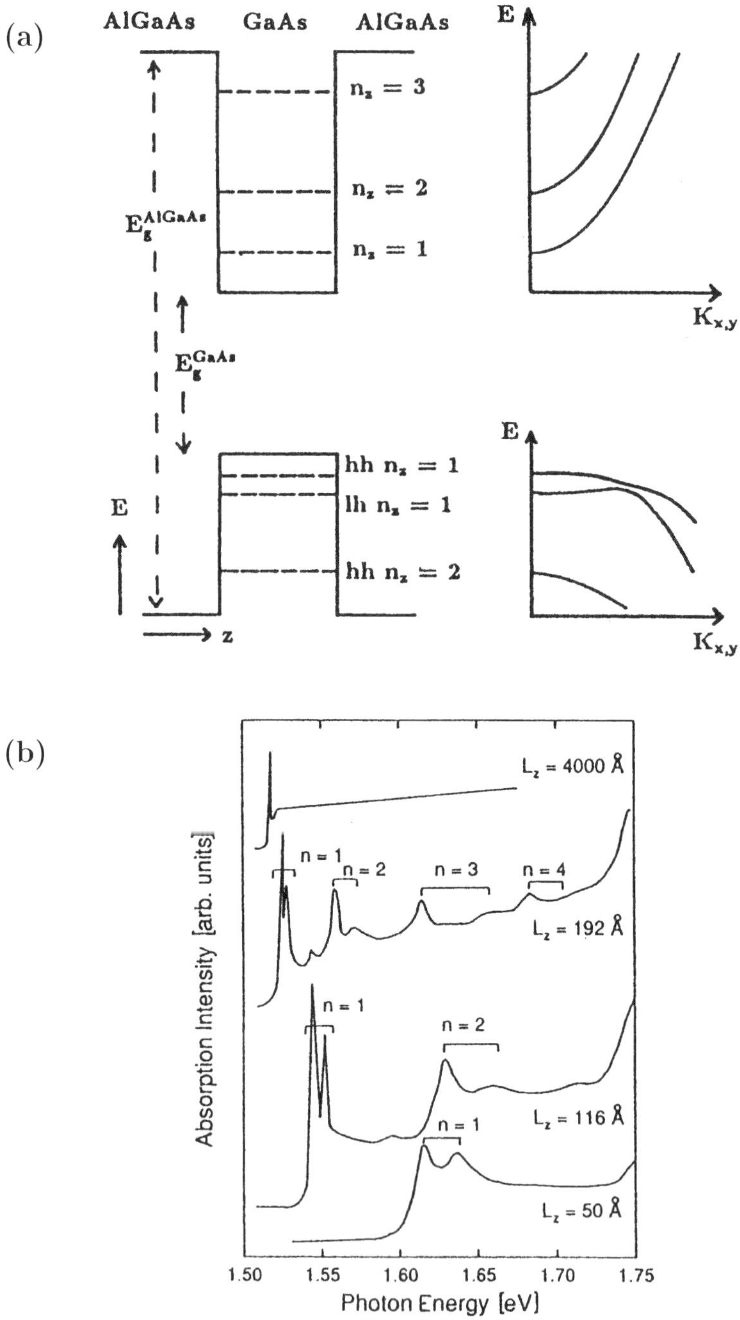

Figure 3. *(a) Confined electron and hole energy levels and subband dispersion in AlGaAs/GaAs QW, and (b) optical absorption in GaAs/$Al_{0.25}Ga_{0.75}Al$ MQW with various well thicknesses (Kalt 1996).*

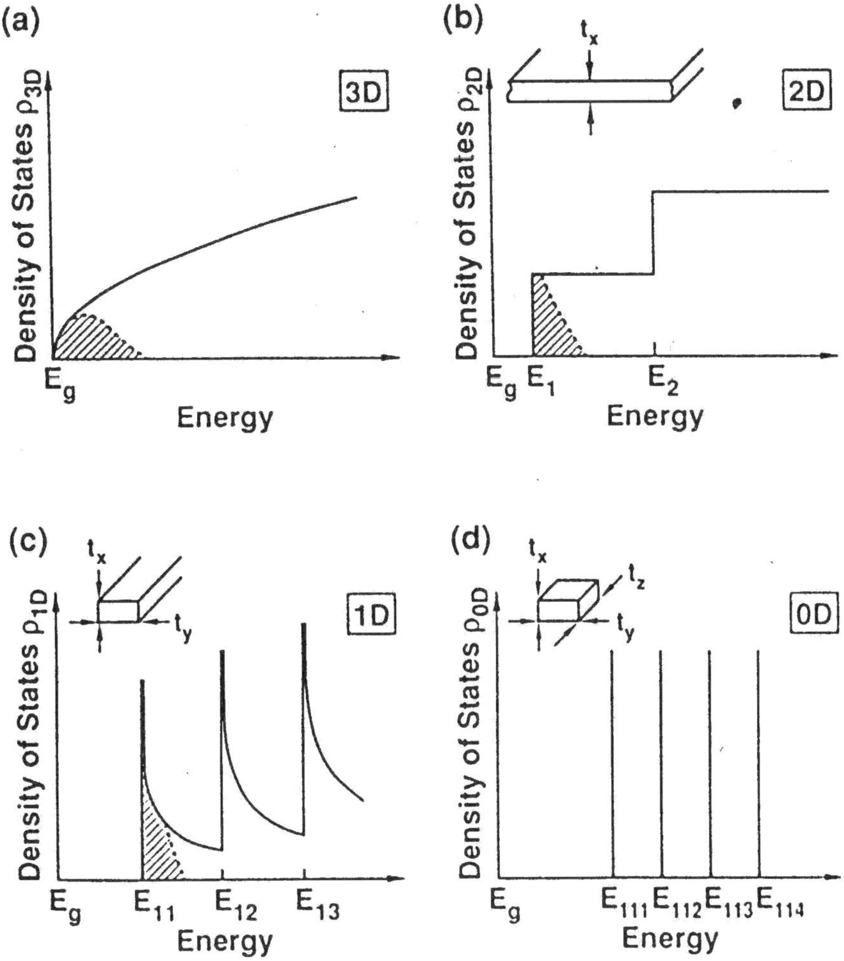

Figure 4. *Shape and density of states in low dimensional semiconductor structures.*

The carrier confinement strongly modifies the properties of excitons in these low-dimensional structures with respect to the bulk. The consequences of exciton confinement in QWs are illustrated by comparison to 3D excitons. The binding energy and radius of the lowest exciton state in bulk (3D) and QW (2D) are $E_{\mathrm{bind}}^{3D} = R_y^*$; $E_{\mathrm{bind}}^{2D} = 4R_y^*$ and $a_1^{3D} = a_B$; $a_1^{2D} = a_B/2$. Thus, the increased binding energy strongly enhances excitonic effects in lowered dimensions (Kalt 1996). The 1D exciton binding energy is expected to be significantly enhanced with respect to the 2D case, but approaches the value of the barrier material in the limit of vanishing wire radius. For the QD case the confinement energies of the carriers are the dominant contribution to the energy shift of optical transitions with respect to bulk material. The properties of excitons in 1D and 0D semiconductor structures are currently less well understood.

2.3 Carrier relaxation and scattering processes

After a semiconductor in equilibrium is excited by an ultrashort pulse, it undergoes several stages of relaxation before returning once again to equilibrium. Simplified diagrams illustrate ultrafast processes in bulk semiconductors in Figure 5(a) and QW in Figure 5(b), following optical excitation above the band gap (Diels 1996). An ultrashort light pulse of frequency ω_l creates electron-hole pairs in states above the band gap. Their mean excess energy is $\Delta E = \hbar(\omega_1 - \omega_{\mathrm{gap}})$, and their initial energy distribution resembles the excitation spectrum. The dynamics of electrons, holes, excitons and phonons are influenced by their interaction with each other, as well as with defects and interfaces of the system. Interaction with phonons plays a major role in exchange of energy and momentum between carriers and the lattice and determines the energy relaxation of photoexcited carriers.

III–V semiconductors such as GaAs are intermediate between the covalent semiconductors like Si and ionic semiconductors like CdS. Polar optical phonons therefore play an important role in carrier-phonon scattering processes. The acoustical phonon lifetimes are relatively long (in the range of 100ps) compared to other relaxation mechanisms (Heyman *et al.* 1996). On the contrary, emission of the longitudinal optical (LO) phonons represents a very fast mechanism (process 2 in Figure 5(b)). The resonant energy for this transition in GaAs is $\omega_{\mathrm{LO}} = 36\mathrm{meV}$. The strength of the electron-optical phonon interaction $\alpha(q)$ is given by Fröhlich's term, $|\alpha(q)|^2 = 2\pi\eta\omega_{\mathrm{LO}}e^2/\Omega\varepsilon q^2$ where $\mathbf{q}$ is the wave vector of the optical phonon and Ω is the volume of a unit cell. The formula implies, that there is a resonance at low q values. That means that the lower the momentum of the phonon, the stronger is the interaction between electrons and phonons and the higher is the rate of LO-phonon transitions (Hvozdara 2000). Carrier-carrier scattering determines the exchange of energy between carriers, and is primarily responsible for the thermalization of the initially non-thermal distribution of carriers (Ulbrich 1973, Leo *et al.* 1988, Suemoto *et al.* 1998, Hamanaka *et al.* 1999). This process is mediated by Coulomb interactions and includes electron-electron, hole-hole and electron-hole scattering. The temperature which can be attributed to the thermalized electronic system can far exceed the lattice temperature. If the energy of the exciton is raised by an amount larger than the binding energy E_b, the bound system decays into a free electron and hole (exciton ionization). Such a process can be induced, for example, by longitudinal LO phonon scattering and typically proceeds on the time scale of about 100fs in bulk materials at room temperature (Diels 1996). In addition to the energy and momentum relaxation processes discussed above, spin relaxation processes can also play an important role in dynamics of photoexcited systems (Shah 1996, Miller this volume). The scattering processes in QW are similar to those discussed above for bulk semiconductors (Figure 5(b)), but the scattering rates are modified by several factors such as different band structure, phonon modes and density of states in these semiconductors. One major difference in carrier-phonon scattering rates results from the fact that only the momentum parallel to the interface planes is conserved.

Four temporally-overlapping relaxation regimes in photo-excited semiconductors above the band gap can be resolved (Shah 1996).

- **Coherent regime** ($\leq$200fs) occurs when an ultrashort laser pulse creates excitation in the semiconductor with a well-defined phase relationship within the excitation and with the electromagnetic field that created the excitation. Any scattering will result in dephasing.

(a)

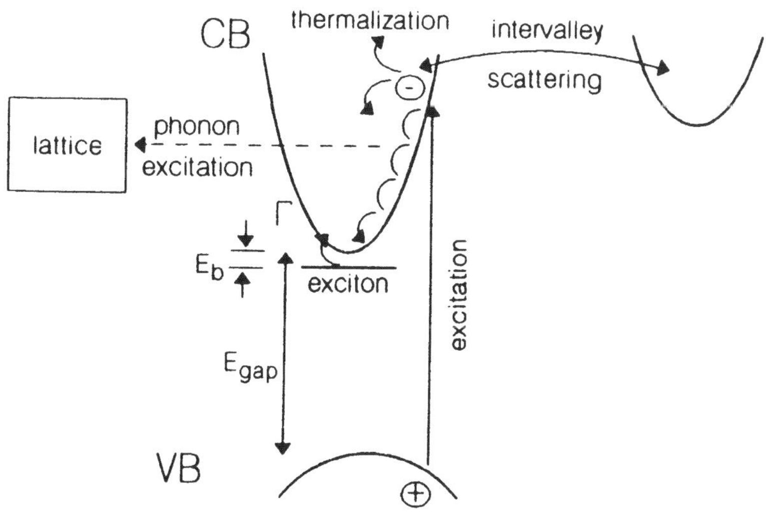

(b)

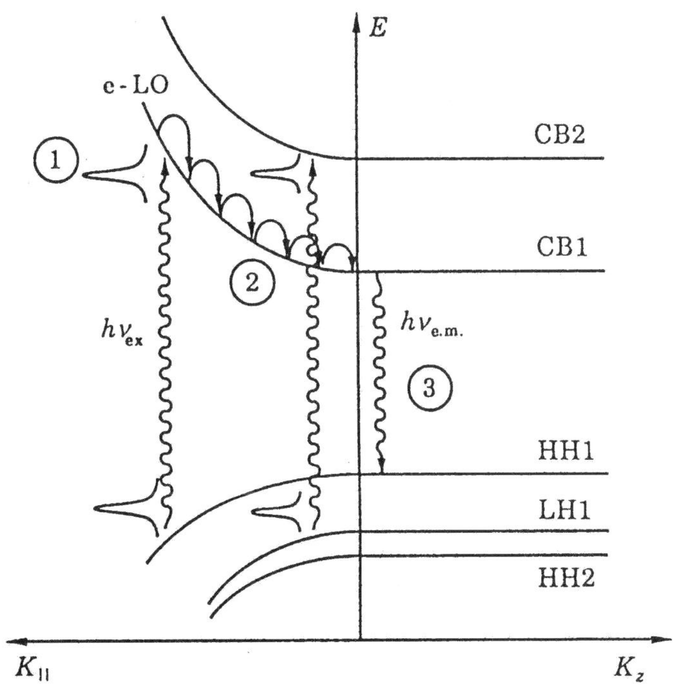

Figure 5. *(a) Ultrafast processes in bulk semiconductors and (b) QW following optical excitation.*

- **Non-thermal regime** (≤2ps) for the case of real excitation, occurs when the distribution of the excitation (free electron-hole pairs or excitons) is very likely to be non-thermal, that is, the distribution function cannot be characterized by a temperature.
- **Hot-carrier regime** (~1–100ps) occurs after carrier-carrier (or exciton-exciton) scattering which is primarily responsible for redistributing the energy within the carrier (exciton) system, leads to the thermalized distribution function of carriers (excitons), that is, a distribution that can be characterized by a temperature. The thermalized electron-hole pairs reach lattice temperature in hundreds of picoseconds through interaction with various phonons in the semiconductor.
- **Isothermal regime** (≥ 100ps) where at the end of the hot-carrier regime, all the carriers, phonons and excitons are in equilibrium with each other, that is, the carrier distribution can be characterised by the same temperature as that of the lattice. These excess electron-hole pairs (or excitons) recombine either radiatively or non-radiatively and return the semiconductor to thermodynamic equilibrium.

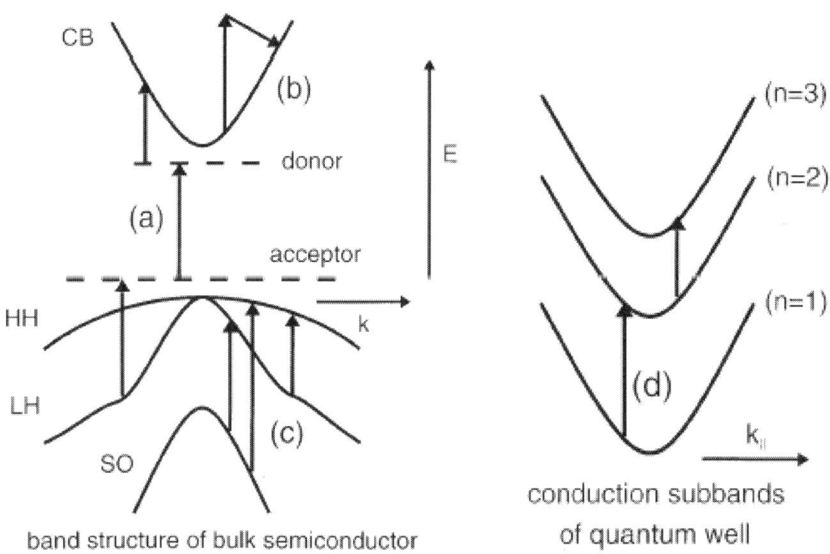

Figure 6. *Optical excitations below the fundamental band gap of doped bulk (left) and QW (right).*

Doped semiconductors can also exhibit absorption at photon energies less than the fundamental absorption edge. Absorbing transitions are depicted in Figure 6 for bulk and low dimensional structures(Elsaesser and Woerner 1999). There are four different types of optical excitations: (a) transitions between electronic states of impurity atoms and vacancies, and from impurity levels into the valence or conduction band continuum; (b) indirect intraband excitations (free carrier absorption); (c) intervalence band transitions of holes; (d) intersubband transitions between valence and conduction subbands in low dimensional structures. Below band gap excitations occur mainly in the infrared spectral range beyond a wavelength of $1\mu m$ (1–$100\mu m$). Intervalence and intersubband excitation create carrier distributions populating different valence bands and subbands, respectively. Population transfer between those bands is part of the relaxation, thermalization and cooling of such nonequilibrium populations towards a quasi-equilibrium carrier distribution and occur on the time scales between about 30fs to several tens of picoseconds.

3 Ultrafast spectroscopy techniques

Progress in ultrafast spectroscopy techniques has been closely related to the development of femtosecond laser sources. Femtosecond pulses are an ideal tool for investigating many physically interesting relaxation processes in semiconductors. The measurement techniques of ultrafast spectroscopy comprise two steps (Hermann *et al.* 1987). The first step consists of exciting the sample to be investigated with an intense ultrashort light pulse, whereby energy is transferred to the sample during the interaction. After the passage of the excitation pulse, a balancing process occurs which can be of a local or non-local nature and as a result the sample strives towards the initial or a changed equilibrium or quasi-equilibrium state. Various kinds of relaxation processes, energy and charge transfer as well as diffusion processes may be involved in these balancing processes. The second step of the ultrashort pulse spectroscopy consists of measuring the temporal evolution of time dependent luminescence intensity, spectral absorptivity or reflectivity. With a suitable choice of wavelength of the exciting radiation as well as detection, these optical methods are very informative, because the various types of physical processes generally produce a time varying absorptivity in certain spectral ranges and are usually accompanied by emission of light with a characteristic spectral distribution.

3.1 Luminescence spectroscopy techniques

Picosecond time resolution luminescence spectroscopy: direct detection

Conventional time-correlated single photon counting methods using a photomultiplier tube are inadequate in luminescence lifetime measurement and rise-time analysis on sub-nanoseconds time scales because temporal resolution is limited to ~1ns and the signal waveform may be distorted by the photoelectron transit time spread. The streakscope used with the streak method has a superior temporal resolution which can be better than 1ps, and furthermore, is free of waveform distortion (Weisbuch *et al.* 2000). Figure 7 shows the operating principle of the streak tube (www.Hamamatsu.com). When luminescence pulses enter the photocathode, they are converted into photoelectrons in proportion to the number of incident photons. Being accelerated by the electrode, these photoelectrons pass through a pair of deflection plates. At the instant they pass through the deflection plates, a high speed sweep voltage is applied to the deflection plates so that the photoelectron's trajectories are swept from top to bottom. The swept photoelectrons are than multiplied in the microchannel plate (MCP) by a factor of 10^4, and reconverted into an optical image by the phosphor screen. The optical image produced on the phosphor screen is called the "streak image". In this way, the time is converted into the spatial (vertical) axis. The time taken for photons to reach the photocathode and the intensity can be determined

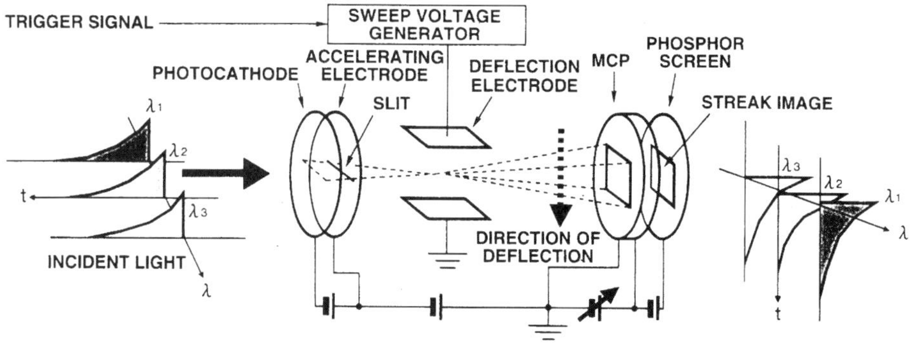

Figure 7. *Operating principle of streak tube.*

by the position and luminance of the streak image. In addition, a streak image can be obtained in which the vertical axis serves as the time axis and the horizontal axis as the wavelength axis if the monochromator is used to focus a spectrum onto the photocathode. The streak images thus obtained are read out by a CCD camera which is optically coupled to the streak tube. Thus a wide variety of luminescence can be measured, ranging from extremely weak to bright intensities.

Femtosecond time resolution: luminescence upconversion spectroscopy

Luminescence upconversion spectroscopy provides a means for measuring a low-level luminescence signal with time resolutions limited only by the laser pulse. This method has been applied to a wide variety of semiconductors. Schematics of the luminescence upconversion setup and sum frequency generation are illustrated in Figure 8 (a,b) (Shah, 1988). The output of an ultrafast laser is divided into two beams and the sample is excited by one of these beams. The luminescence from the sample is collected and focused into a nonlinear crystal. The second beam of ultra-short pulses is suitably delayed and then focused onto the nonlinear crystal to overlap the luminescence spot. The angle between the two beams and the nonlinear crystal are adjusted for phase-matched sum-frequency generation $\omega_s = \omega_L + \omega_G$ (where ω_L and ω_G are the frequency of the luminescence and the gate pulses). The sum-frequency signal is generated only during the gating laser pulse width as illustrated in Figure 8(b). The intensity I_S of the sum frequency is proportional to the product $I_L I_G$ of the intensities of the luminescence and of the gate pulse. Time resolved luminescence spectra can be obtained by scanning the relative delay between the two laser beams. Since the time resolution is obtained at the nonlinear crystal, the sum-frequency signal can be dispersed by a spectrometer and detected in a time-integrated manner by photon counting electronics.

3.2 Pump-probe spectroscopy

The principle of pump-and-probe spectroscopy consists of exciting a sample with a strong pump pulse and subsequently, after a defined adjustable delay time, measuring any

(a)

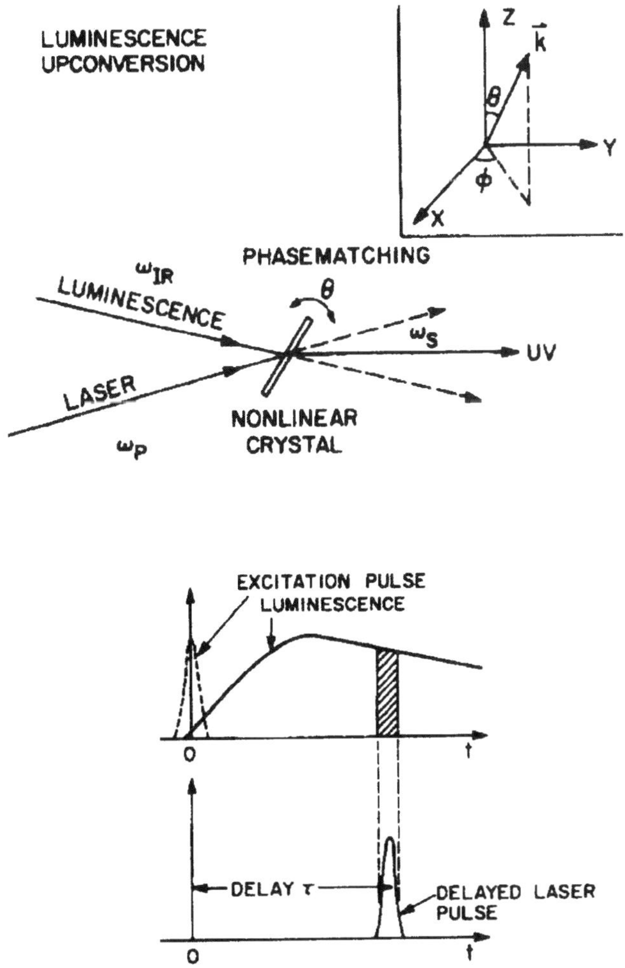

(b)

Figure 8. *(a) Schematics of the luminescence upconversion setup and (b) sum frequency generation during the gating laser pulse (Shah 1988).*

changes of absorption, reflection, amplification or polarization rotation of a probe pulse. The probe pulse can have the same wavelength as the excitation pulse (degenerate pump-probe spectroscopy) or any other wavelength, by which we can probe different transitions of the sample (non-degenerate pump-probe spectroscopy). The probe beam is typically much weaker than the pump, and ideally the spot diameter of the probe at the sample is considerably smaller than that of the pump so that it samples a region of uniform photo-excited density. In pump-probe transmission spectroscopy, one generally measures the change in transmitted probe pulse energy induced by the pump as a function of delay time. The time resolution is determined by the pulse width the delay time between the pump and probe pulses, so that the transmitted probe may be detected by a slow detector, and can be spectrally resolved if desired. The schematic diagram of a pump-and-probe beam spectrometer using single pulse excitation and continuum probe is illustrated in

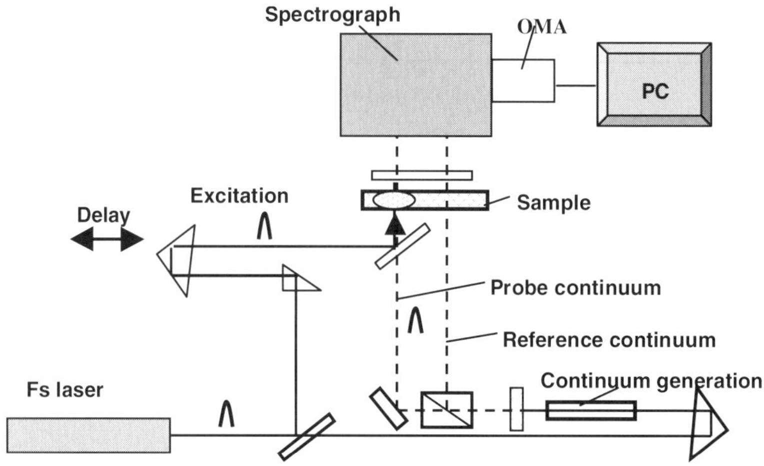

Figure 9. *The schematic diagram of a pump-and-probe pulse spectrometer.*

Figure 9. The femtosecond laser pulse is divided into excitation and probe pulses. The probe laser pulse generates white light continuum and then divides into a probe and reference pulses. The probe pulse passes through the sample in a region that is pumped by the excitation pulse, whereas the reference pulse passes through an unaffected part. After the sample, the excitation light is absorbed by a filter, while the probe continuum and reference continuum are dispersed spectrally in a spectrograph and recorded with temporal integration in different traces by means of an optical multichannel analyzer (OMA). The data are typically presented in the form of normalized Differential Transmission (DT) plots given by $\Delta T/T_0=(T-T_0)/T_0$; the change in transmission $\Delta T=(T-T_0)$ induced by the pump pulse is divided by the transmission of the probe in the absence of the pump, T_0.

4 Ultrafast spectroscopy of bulk semiconductors

When a direct band gap semiconductor such as GaAs is excited by above-band-gap optical radiation, electrons are excited out of the valence band into conduction band states. As a result, filling of the states in the conduction band can change the absorption spectrum of the medium. At wavelengths near the excitation, a local decrease, or spectral hole appears. Disappearance of the hole corresponds to the time it takes electrons to scatter out of their initial states by carrier-carrier or carrier–phonon scattering. The rates and types of initial scattering that occur depend upon the band structure of the semiconductor, the excess energy of the excited carriers, and the excited carrier density. Figure 10(a,b) shows pump-probe wavelength measurements performed by Wei-Zhu Lin (1988) to investigate the induced time-varying changes in optical absorption that follow excitation in a GaAs thin layer. The samples were excited up to carrier density $n \sim 10^{18}\mathrm{cm}^{-3}$, using 35fs pulses tuned to ~1.97eV. The experimental traces display absorption saturation as transient changes in differential transmission $\Delta T/T$ observed at different probe photon energies. The data exhibit a rapid onset of the absorption saturation at all probe energies

(a)

(b)

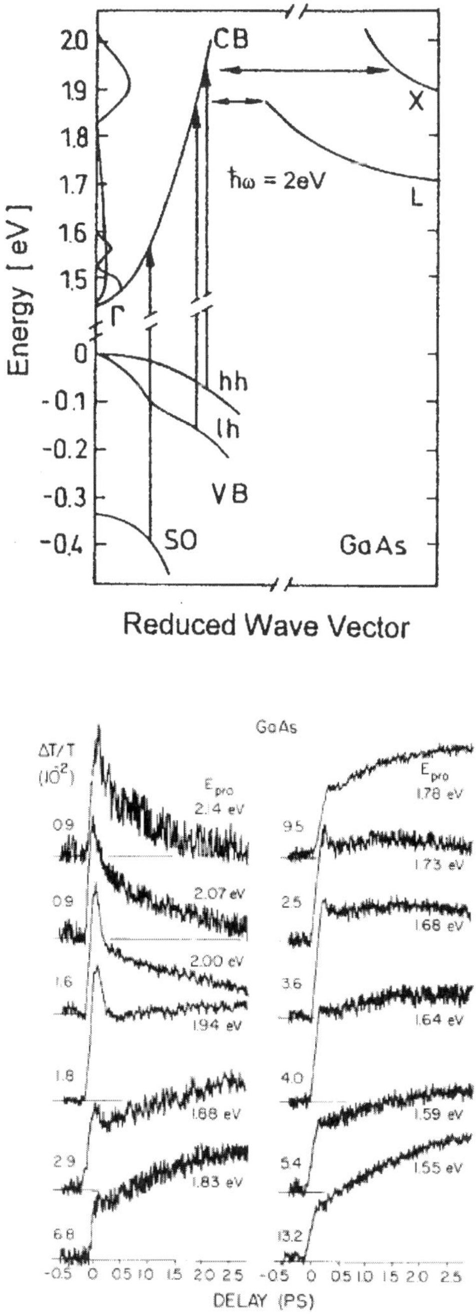

Figure 10. *(a) Schematic of band structure for GaAs indicating allowed optical transitions for 2eV photons and (b) Transient absorption saturation arbitrarily normalised for different probe energies (Wei-Zhu Lin 1988).*

throughout the band. This indicates that the carriers scatter out of their initial excited states and assume a broad distribution in energy within several tens of femtoseconds. The rapid appearance of carriers throughout the band suggests that, at these carrier densities, carrier-carrier scattering plays an important role in the initial scattering process. The picosecond process of the absorption saturation observed at different probe energies is consistent with a cooling of the carrier distribution via carrier-phonon interactions on a 1–2ps time scale. Measurements performed between 1.73 and 1.68eV exhibit a rapid absorption saturation transient which results from carriers generated by pump transitions from the split-off band. The amplitude of the saturation increases as probe energies corresponding to split-off transitions approach the bottom of the band. For probe energies less than 1.76eV, the split-off transition is no longer allowed, and the excited electron distribution is probed only by transitions from the heavy and light hole bands. The rapid initial transient observed for probe photon energies ranging from 2.14 to 1.88eV indicates the presence of an absorption saturation hole for photon energies near the incident pump energy. The decaying absorption saturation for photon energies of 2eV and greater result from a decreased occupancy of the energy states higher in the band. In addition, intervalley scattering effects may contribute to the rising absorption saturation signal observed at photon energies corresponding to transitions which probe near the bottom of the X and L satellite valleys, see Figure 1(a) (Wei-Zhu Lin 1988).

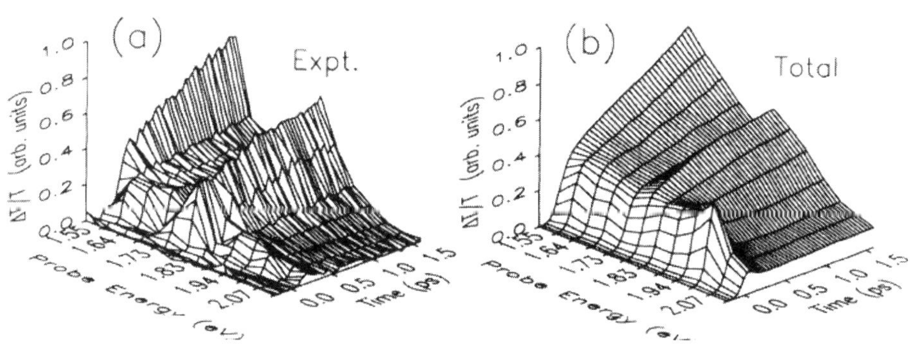

Figure 11. *(a) Experimental DT spectra (Ref in Bailey,1990) and (b) calculated DT including both electrons and holes (Bailey 1990).*

Excellent agreement between pump-probe experiments and theory was reported by Bailey (1990), where the time-dependent distribution function from Monte Carlo simulation were used to evaluate the differential transmission spectroscopy (DTS). Results for the calculation of differential transmission as a function of probe delay and energy are shown in Figure 11 along with the experimental results. Calculations included the contribution of electrons and holes in the heavy, light, and split-off valence bands. The individual contributions to the differential transmission at 1.55eV are shown in Figure 12. In the figure, c and v indicate whether the contribution is for conduction or valence band and hh or lh denotes the optical transitions from heavy or light hole bands. The initial rise in the differential transmission at low energy is also due to the rapid redistribution of holes. The simulations match pump-probe data over a large range of energies, and thus establishes the reliability of the model (Bailey 1990).

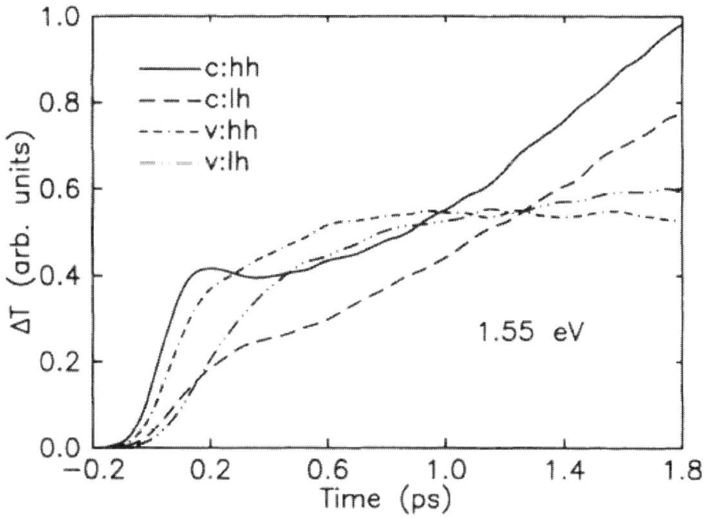

Figure 12. *Relative contributions to the DTS in electron and hole bands (pump 2eV, probe 1.55eV, (Bailey 1990).*

Ultrafast relaxation dynamics of light and heavy holes in GaAs following femtosecond valence-to-conduction band excitation, have been reported in the literature (Ganikhanov *et al.* 1998), using two-colour femtosecond spectroscopy. Synchronized femtosecond pulses in the near-IR (pump 815nm) and mid-IR (probe at 3.28 and 3.87μm) for probing the light- and heavy-to-split-off hole transitions were employed, as shown in Figure 13(a).

Experiments were performed on a 400μm thick bulk pure GaAs sample ($n<10^{14}$cm^{-3}) that was anti-reflection coated in the mid-IR (3–4μm) range. The experiments show that the hole relaxation is generally very fast, as expected. The initial relaxation times are less than 75fs, and a spectral hole burning effect persisting up to 300fs (both hh and lh) as seen in Figure 13(b). The results suggest that carrier-carrier and optical phonon scattering, in particular polar optical-phonon scattering, are the primary processes in the initial redistribution of heavy and light holes (Ganikhanov *et al.* 1998).

The non-thermal photo-excited electrons and holes interact among themselves and with other carriers present in the semiconductor. They achieve a distribution function characterized by a temperature, in times of the order of 100fs for typical densities present in most femtosecond experiments. Initially, the temperatures of the electrons and the holes may be different, but a common temperature T_c for the carrier system is typically achieved in times of picoseconds (Shah 1996). Carrier–phonon interaction is one of the most fundamental scattering process in semiconductors and plays a role in many different aspects of semiconductor physics.

Hot-carrier relaxation has been investigated by Pelouch et al. (1992) for bulk GaAs in the high carrier density limit. Using a time-resolved luminescence up-conversion technique with less than 80fs temporal resolution, carrier temperatures were measured in the 100fs to 1ns range. The sample consisted of a 400nm thick, nominally undoped, GaAs layer

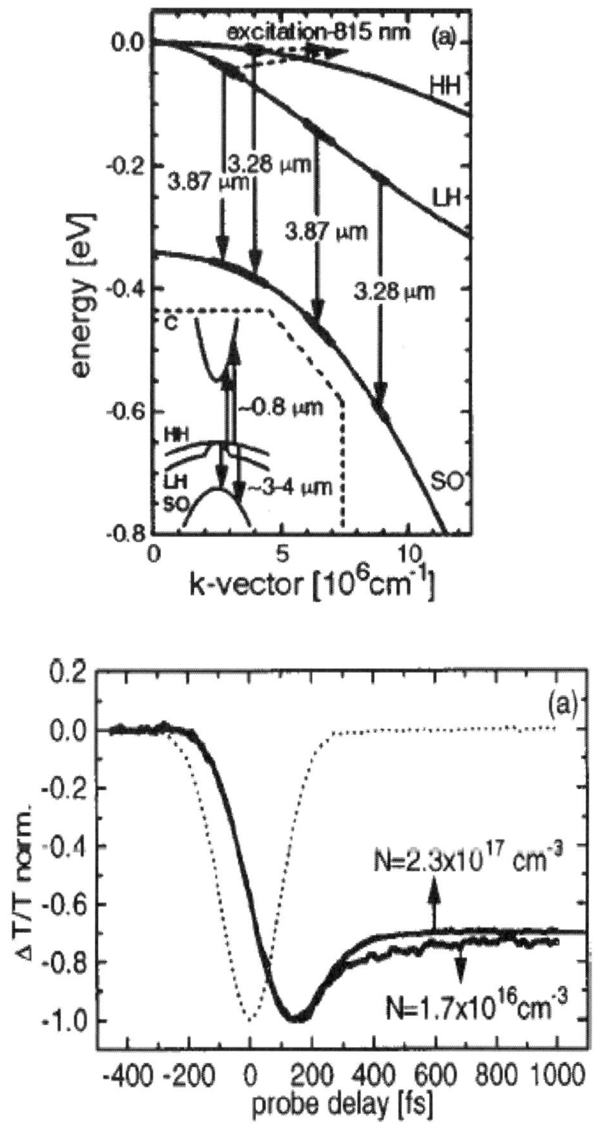

Figure 13. *(a) Valence band energies in GaAs (inset: pump and probe wavelengths) and (b) normalized DT signals at probe wavelength 3.28μm for heavy to split-off hole transition (dotted line cross correlation trace (Ganikhanov et al. 1998).*

surrounded by 200nm AlGaAs layers transparent to the 1.97eV excitation. Time resolved luminescence spectra were recorded at room temperature at incident pump powers of 25, 12.5, and 5mW as shown in Figure 14. The corresponding carrier density was calculated from the values of the quasi-Fermi levels as 1×10^{19} (squares), 5×10^{18} (circles) and 2×10^{18} cm^{-3} (triangles) in Figure 14. Electron temperatures were determined by fitting the high-energy tails of the spectra. Figure 14 displays a plot of the PL spectrum at 100fs for the

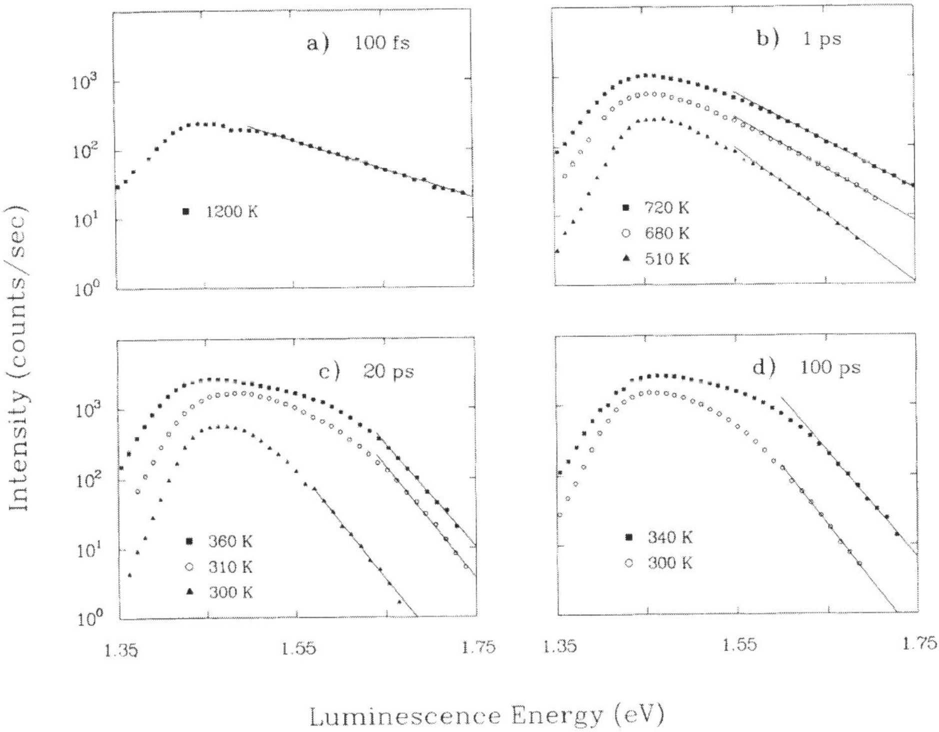

Figure 14. *Time-resolved luminescence spectra for GaAs at room temperature and each excitation power. Inset: carrier temperature fits for each carrier density at delay times (a)100fs, (b) 5ps, (c) 20ps and (d) 100ps (Pelouch et al. 1992).*

highest carrier density that shows that the initial temperature is 1200K. The evolution of the carrier distribution at time delays of 1, 20 and 100ps for each excitation power show the hot-carrier cooling in the 100ps range near lattice temperature. The results were compared with hot-carrier relaxation in GaAs/AlGaAs QWs where a slower cooling rate was found for QWs than for bulk GaAs at carrier densities greater than $2 \times 10^{18} \mathrm{cm}^{-3}$ (Pelouch *et* al. 1992).

5 Studies of low-dimensional structures

5.1 Carrier relaxation in quantum wells (QW)

In reduced dimensionality systems, the effect of quantum confinement leads to formation of several sub-bands in the conduction and valence bands, and also leads to the confined phonon modes for the acoustical and optical phonons. The initial electron and hole distribution functions created by an ultrashort pulse can be rather different in QW and bulk semiconductors. The experimental situation for non-thermal carrier distributions in AlGaAs/GaAs quantum wells has been demonstrated by the experiments of Knox *et al.* (1986). Samples of undoped QW's with 9.6nm GaAs wells and 9.8nm $Al_{0.3}Ga_{0.7}As$ barriers were excited above the lowest interband transition with 100fs laser pulses. A broadband 50fs probe pulse was used to measure the transmission changes induced by this pump pulse. An excitation density of $2 \times 10^{10}cm^{-2}$ was estimated from the transmission change. Figure 15(a) shows the differential transmittance spectra of the sample at room temperature and nominally zero delay for a pump centred at (a) 1.537eV, (b) 1.509eV, and (c) 1.496eV. An increase in the transmittance is seen in the energy range near the pump energy, which is indicated by an arrow for each case. The transmission change on the pump energy is clear evidence of a non-thermal carrier distribution at nominally zero delay. The temporal evolution of a non-thermal distribution for a pump centred at 1.509eV is presented in Figure 15(b). As time evolves the region of higher transmission shifts to lower energy and broadens due to hole-burning. The transformation from a peaked non-thermal distribution function into a Maxwell-Boltzmann-like distribution function occurs in about 200fs. This time is a measure of the carrier-carrier scattering rate responsible for the thermalization of the carrier distribution function (Shah 1996).

The effects of electron-phonon scattering processes on the energy relaxation of photo-excited hot electrons in AlGaAs/GaAs QW, have been investigated by Leo *et al.* (1988). The dependence of energy loss rate on well width and doping was investigated. The MQW structures consisted of 20 undoped GaAs (9nm) layers, clad by $Al_{0.34}Ga_{0.66}As$ barrier layers grown by MBE. The samples were excited by 4ps pulses from a mode locked dye laser and the time resolved luminescence spectra measured using a streak camera. The luminescence intensity varied with photon energy $\hbar\omega$ as $I \sim \exp\{-\hbar\omega/kT_c\}$. The carrier temperature determined the high energy slope and the carrier density the width of the spectra. The temperature T_c of the hot electron-hole plasma was determined from the slope of the luminescence intensity as shown in Figure 16(a) for different delay times and an excitation density of $9 \times 10^{12}cm^{-2}$.

Figure 16(b) depicts cooling curves for different excitation densities. The cooling data were analyzed by comparison with curves calculated for energy loss rates by different scattering mechanisms as a function of the carrier temperature at low carrier densities as shown in Figure 17(a). The solid line is the sum of four interactions; for temperatures about 45K, polar optical scattering with LO phonons is dominant, while at low carrier temperatures, the acoustic deformation potential scattering is most important. The polar-optical energy loss rate reduction parameter α, was determined as shown in Figure 17(b). For higher sheet excitation densities in MQW this factor describes the reduction of the experimental energy loss compared to theory. At low intensities, the energy loss rate is close to that expected for a simple model but there is a sharp reduction in the energy-loss rate at high excitation density. Comparison of hot-carrier relaxation in QW and bulk GaAs has been investigated by time resolved luminescence (Pelouch *et al.* 1992). The results show that hot-carrier cooling is different in QW and bulk GaAs at high carrier densities. The cooling rates in QW structures are similar for low carrier concentrations but significantly slower than that of bulk greater than $2\times10^{18}cm^{-3}$ —in agreement with the results of Leo *et al.* (1988).

Heating and cooling of a pure electron plasma in modulation doped MQW after resonant intersubband excitation was investigated by Elsaesser and Woerner (1999). The experiments used n-type modulation doped $Ga_{0.47}In_{0.53}As$ (8nm)/$Al_{0.48}In_{0.52}As$ (14nm)

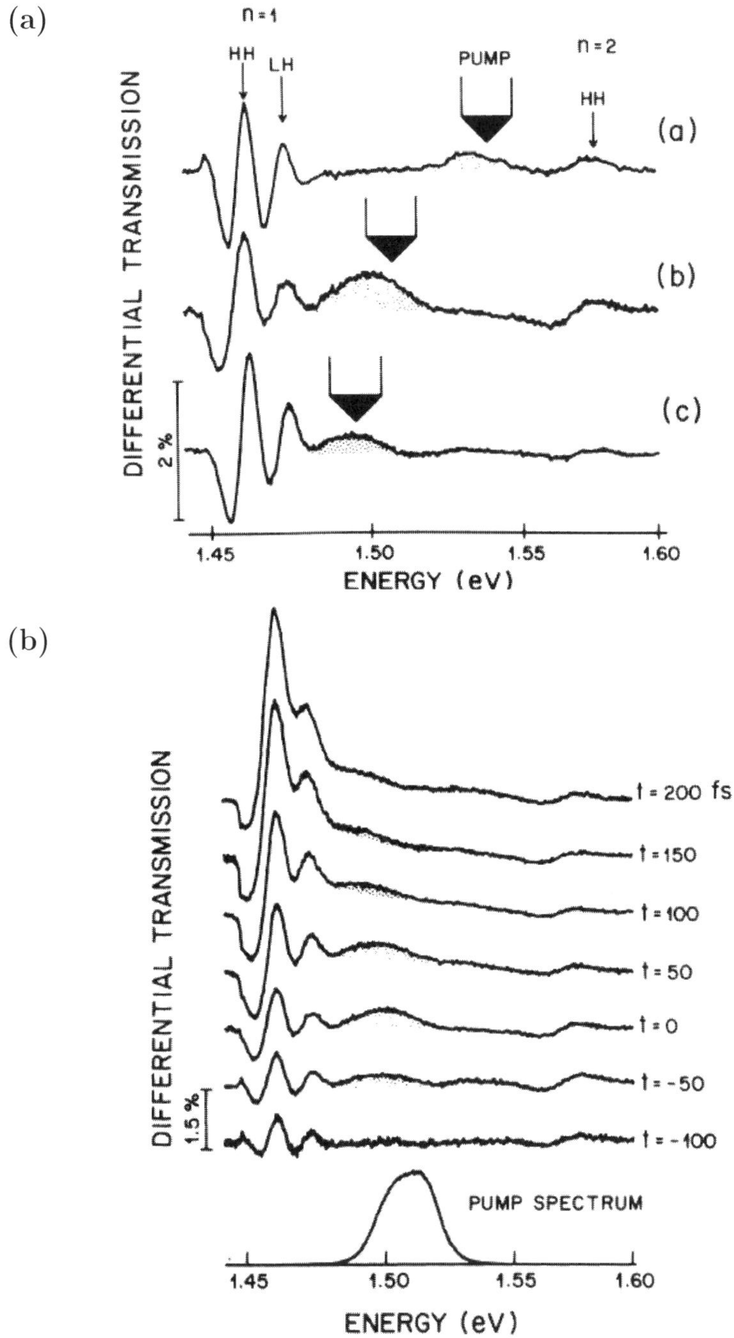

Figure 15. *(a) Differential transmission spectra at nominally zero delay and (b) time dependent DT spectra for a pump centred at 1.509eV (Knox et al. 1986).*

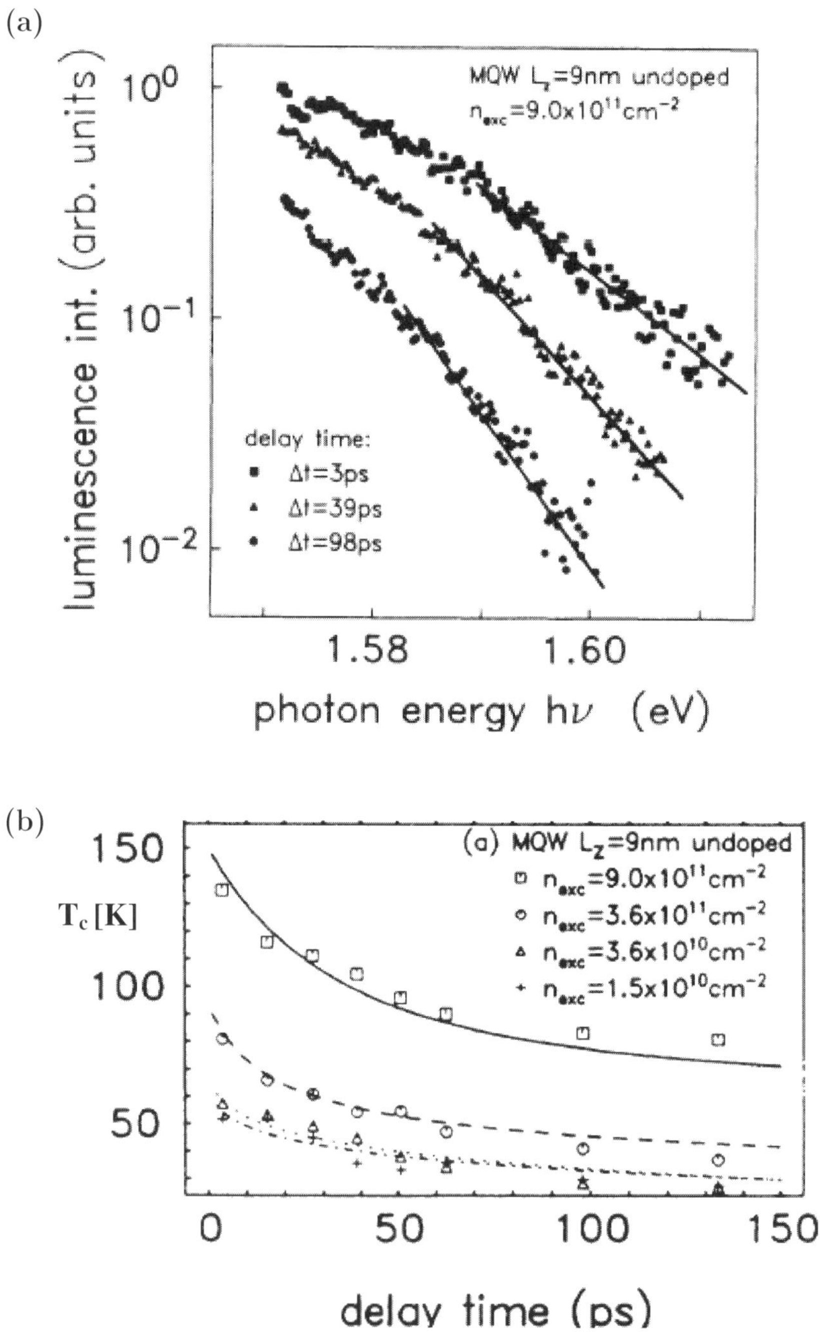

Figure 16. *(a) Transient high-energy luminescence tail of MQW at various delay times and (b) cooling curves for different excitation densities (Leo et al. 1988).*

(a)

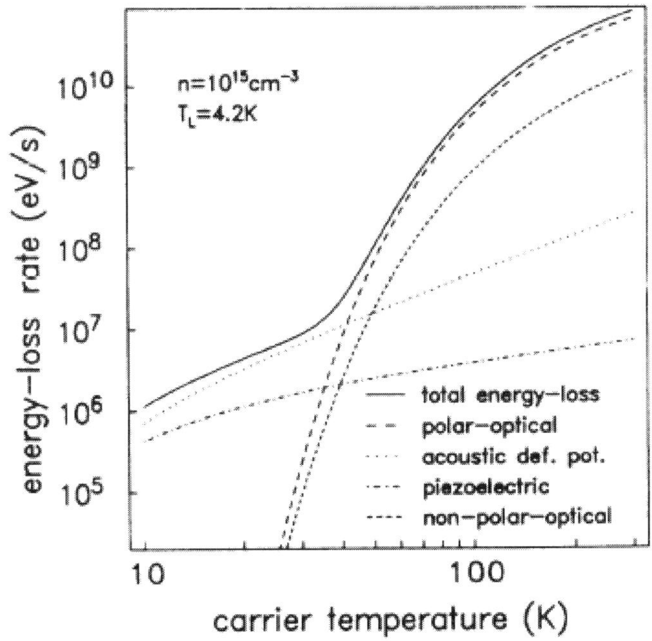

(b)

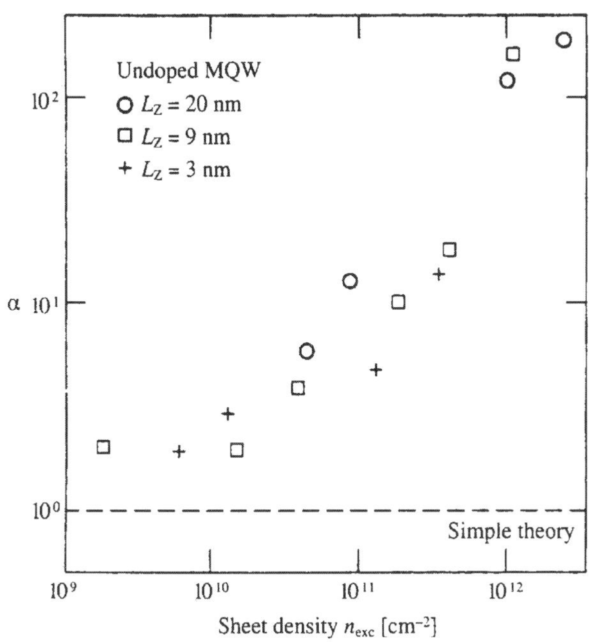

Figure 17. *(a) Energy loss rate per carrier due to different scattering mechanisms as a function of carrier temperature and (b) polar-optical energy loss rate reduction factor as a function of sheet excitation density (Leo et al. 1988).*

QW samples grown by MBE on InP substrates. For such a quantum well width, the inter-subband transition between the n_1 and n_2 conduction subbands is located at mid-infrared photon energies and much higher than the energy of LO phonons. Low temperature (T_L=10K) differential transmission spectra of the interband transition (probe pulse E_{pr} =1.141eV) after excitation of n_1 electrons to the n_2 subband (E_{ex} = 0.2eV) is shown in Figure 18(a). Transient bleaching, which rises within the time resolution of experiment and decays within 1ps is clearly resolved. The bleaching is caused by the transient electron population of the n_2 conduction subband, leading to a blocking of the interband transitions. Transient absorption spectra recorded in the range of the stationary n_1 absorption edge around 0.913eV give insight into the dynamics of n_1 electrons as shown in figure 18(b). The transmission changes are plotted as a function of photon energy of the probe pulses for delay times in the range of 0.2–2ps.

After intersubband excitation (0.2ps) a transmission decrease from the onset of n_1 interband absorption up to the Fermi edge of the initial distribution is observable. The inelastic Coulomb scattering between the cold and back-scattered carriers during the redistribution, leads to a transfer of cold electrons to states above the initial Fermi level corresponding to the formation of a high energy tail. After a time delay of about 2ps, this distinctly non-thermal distribution has evolved into a hot Fermi distribution with similar amplitudes for enhanced absorption and bleaching. The thermalization process creates a hot plasma of n_1 electrons characterized by a carrier temperature. Carrier cooling involves optical and acoustic phonon scattering and represents the final relaxation stage by which the initial electron distribution in the n_1 subband is restored. For the investigated samples cooling of the hot n_1 electron plasma down to the lattice temperature occurs on the longer time scale up to 25ps. This experiment demonstrates a concept of all-optical control of ultrafast processes (Elsaesser and Woerner 1999). Thus femtosecond infrared spectroscopy of below-band-gap excitations provides highly detailed information on carrier dynamics in bulk and nanostructured materials to provide an understanding of fundamental physical interactions underlying ultrafast processes.

5.2 Carrier relaxation in quantum wires (QWR)

One-dimensional electron systems (1D) are predicted to exhibit unique electronic and optical properties (Liennau *et al.* 1999; Maciel *et al.* 1999). Quasi-two-colour femtosecond pump-probe spectroscopy and near -field scanning optical microscopy were combined to study carrier dynamics in QWR by Lineau *et al.* (1999). QWR's with a lateral width of 50nm and thickness of up to 13nm embedded in a 6nm GaAs QW were grown on patterned GaAs (311)A substrates (Figure 19(a)-inset). Sub-40fs pulses were used in two different pump and probe configurations, near-field pump/near-field probe and far-field pump/near-field probe,with a spatial resolution of 200nm. In both cases, the probe light reflected from the sample was detected in the far-field. The reflectivity was measured at room temperature as a function of spatial probe position and energy. The reflectivity change peaked around 1.47ev for a probe delay of 10ps, Figure 19. Probing at energy 1.47eV (QWR) a rise in probe reflectivity was observed within the time resolution of 200fs followed by constant signal amplitude at longer times, Figure 19(b), for far-field excitation. This relaxation involves a complex scattering process between high-lying QWR states and the states at the bottom of the QWR. From the efficiency of these relaxations Lineau *et*

(a)

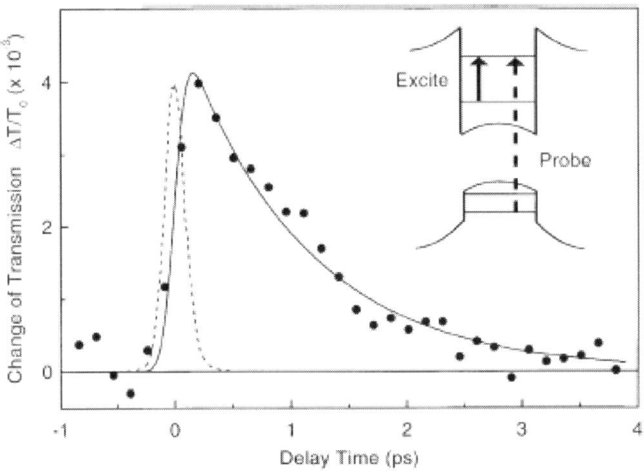

(b)

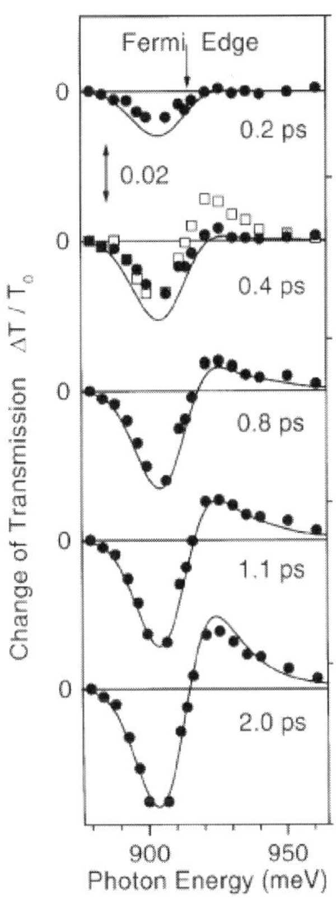

Figure 18. *(a) Time-resolved transmission change of the n_2 interband transition after excitation to the n_1 subband and (b) transmission changes as a function of energy for different delay times in GaAs MQW at low temperatures (Elsaesser and Woerner 1999).*

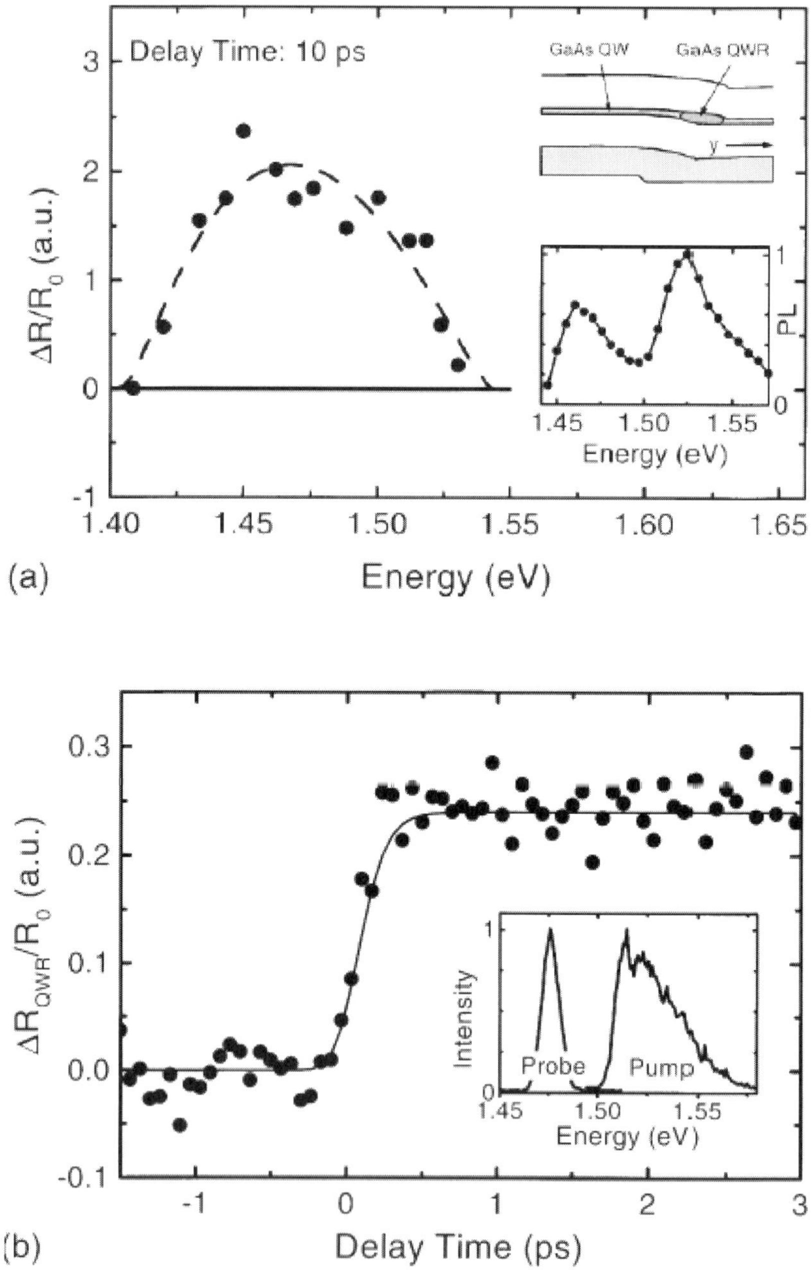

Figure 19. *(a) Pump induced QWR reflectivity change at delay time 10ps (insets: sample structure and room temperature PL spectrum). (b) Temporal variation of the pump induced QWR reflectivity (inset: spectra of pump and probe laser) (Lineau et al. 1999).*

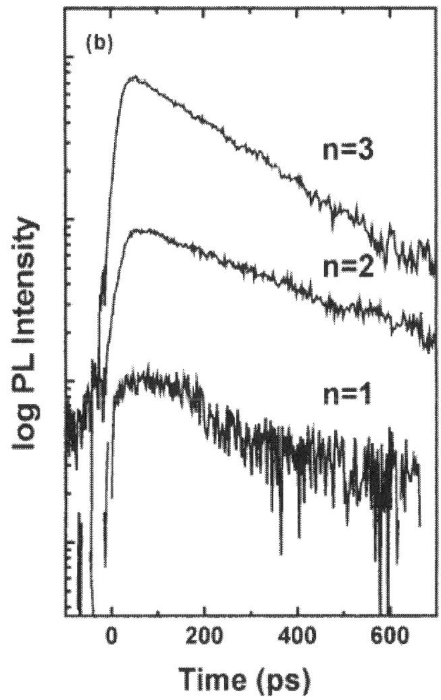

Figure 20. *Time-resolved PL decay times measured at the QWR subband peaks (Maciel et al. 1999).*

al. (1999) concluded that, in addition to carrier-LO phonon scattering, carrier-carrier scattering plays a dominant role.

For far-field excitation with a 30μm spot size a homogeneous carrier distribution is generated along the wire and has no effect on the reflection signal. This was different for near field excitation through the fibre probe. The pump and probe pulses were spatially overlapping and localized into a small spot (~300nm). Thus migration of carriers out of the excitation volume leads to decrease in the local carrier concentration and carrier migration contributes to decay of the pump induced reflectivity on a 10ps time scale. By combining spatial and temporal resolution, the ultrafast relaxation of carriers and carrier migration have been resolved.

The studies of QWR structures show slower relaxation times than for QW materials. In the work of Maciel *et al.* (1999) modulation-doped v-groove QWR's grown by MOVPE on pre-patterned GaAs substrates were investigated. In this case both QWR and SQW were photoexcited. The QW luminescence decayed with lifetimes of several hundred picoseconds, typical of a high quality 2D system, whereas QWR lifetimes were more than an order of magnitude longer. This latter effect may indicate a reduced radiative efficiency for the QWR, which could arise from inhibited hole capture due to the potential barrier separating SQW and QW. Time resolved spectra from QWR's in Figure 20 show the time dependence of the population of carriers in the different sub-bands with the n_3 level decaying most rapidly (~300ps), and the n_1 level being relatively long lived on the nanosecond time scale.

5.3 Carrier relaxation in quantum dots (QD)

Intensive investigations in QD structures have been motivated by promising properties, such as the δ-function-like density of states, and by the perspective of realizing QD devices such as low-threshold lasers. The physics of carrier dynamics is of particular importance for devices because of a 'phonon bottleneck' effect (Bockleman and Bastard 1990, Benisty 1991, Sosnowski *et al.* 1998, Pérone *et al.* 2000). A drastic slow-down of the relaxation process due to the discrete nature of the energy levels was predicted. Recently Sosnowski *et al.* (1998) studied the carrier relaxation under low-carrier-density conditions in self-organized $In_{0.4}Ga_{0.6}As$ QD's by means of time-resolved differential transmission (DT) measurements. The sample investigated was grown by MBE and consisted of four layers of $In_{0.4}Ga_{0.6}As$ QD's in the centre of a 0.2μm-thick GaAs layer with a 1μm $Al_{0.3}Ga_{0.7}As$ carrier confinement layer above and below the GaAs. Atomic-force microscopy revealed a dot density of 5×10^{10}cm^{-2} with a base width of 14nm and height of 7nm. In the PL spectra of this sample the transitions at 1.248eV (e_1h_1) and 1.33 (e_2h_2) were resolved. Differential transmission measurements at 10K used 85fs pump pulses from a Ti:sapphire regenerative amplifier, while probe pulses were obtained by filtering the continuum beam. Figure 21(a) shows DT signals measured using pump pulse centred at 920nm (into the n_2 states) and probe wavelengths of 980 and 923nm, corresponding to the e_1h_1 and e_2h_2 transitions in QD's. The pump pulses produced an estimated carrier density of 10^{10}cm^{-2} in the n_2 state, corresponding to one electron-hole pair for every 20 QD's.

These measurements show that the injected carriers leave the n_2 states (decay at 923nm) with the same time constants as they arrive at the n_1 states (risetime at 980nm). The n_2 to n_1 relaxation time was fitted with two exponential components 0.6 and 5.2ps. The fast component was attributed to phonon mediated hole relaxation and the slower to electron relaxation, which was much more rapid than expected. Electron-hole scattering involving discrete dot states might be ruled out as a possible relaxation mechanism, because the separation of the electronic levels (58meV) is much larger than the LO phonon energy of 30meV. Figure 21(b) shows the normalized DT signal measured with 800nm pump pulses (injection into the GaAs barrier) at probe wavelengths of 980 and 925nm. When carriers are photo-injected into the GaAs barrier region, relaxation into the n_2 and n_1 states of the dot occurs with time constants 2.8 and 8ps, respectively. These measurements indicate that, even at low carrier densities relaxation is more rapid than expected for electrons and indicates the phonon bottleneck is not effective for this system.

Time-resolved luminescence measurements of relaxation dynamics in self-ordered nanometer scale InAs/GAAs QD, using high-excitations were investigated by Bogani *et al.* (1996). The dots were grown by MBE and consist of three monolayers of InAs. From AFM pictures the formation of cone-like structures with base diameters on the order ~25nm were observed. A picosecond Nd:YAG excitation laser (532nm) with a time-correlated single photon counting system were used for time-resolved measurements at 10K. Typical decay times of the order of ~700ps were found at the fundamental transition, decreasing progressively to about 100ps at the highest energy peak as shown in Figure 22. Correspondingly, the risetimes (in brackets) decreased from about 300ps to ~30ps. From the energy difference between adjacent peaks in the spectra, an average spacing of the order of 40–50meV is deduced for the excited state levels in these structures. The observed results indicate a cascade mechanism towards the lower energy states, on shorter than expected time scales of at most a few hundred picoseconds.

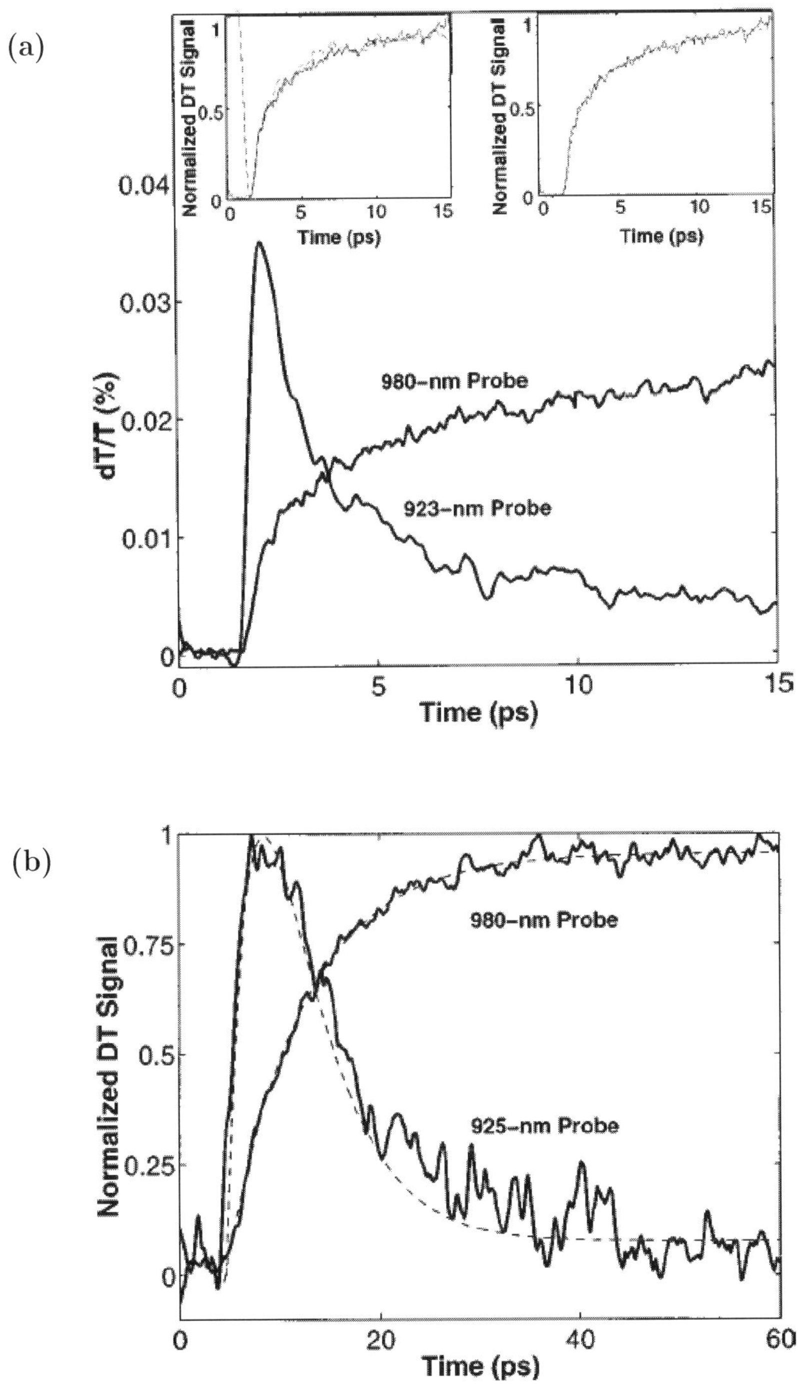

Figure 21. *(a) DT signal at 920 nm pump and (b) DT signals at 800nm pump wavelength as a function of time (Sosnowski et al. 1998).*

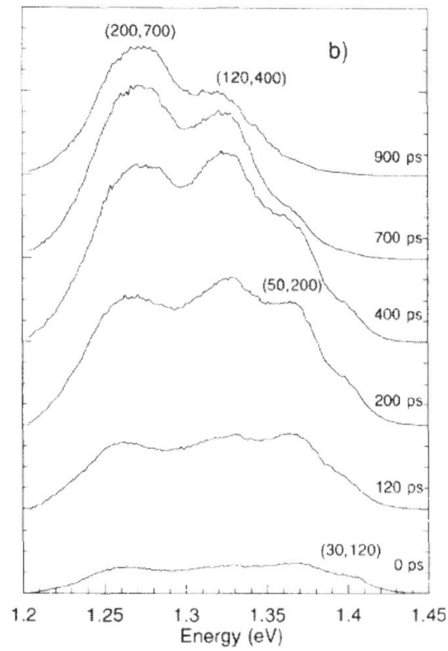

Figure 22. *Time-resolved spectra of QD sample for increasing delays to excitation (Bogani et al. 1996).*

6 Conclusions

In this chapter I have shown that ultrafast spectroscopy experiments are a major source for obtaining experimental data on the dynamics of free carriers in semiconductors and low-dimensional structures. Ultrafast spectroscopy provides direct insights into relaxation processes for electron-hole plasmas in semiconductors occurring on time scales in the range 10^{-14} and 10^{-10} s. Femtosecond pulses are an ideal tool to investigate many physically interesting relaxation processes which strongly influence both optical properties and carrier transport in semiconductors. The investigation of hot-carrier relaxation in the high-carrier-density limit is important in order to understand the physics governing 3D and 2D hot carrier relaxation and may have important consequences in the development of semiconductor devices. Most studies of ultrafast relaxation dynamics of free carriers in semiconductors have been focused on the dynamics of electrons. Precise measurement of hole dynamics are generally very difficult because of the high phonon scattering rates which require time resolutions on the order of 20fs. A variety of elementary excitations occur in the energy range below the fundamental bandgap and can serve as a probe of the dynamics of single component plasmas in semiconductors and low dimensional structures. The relaxation dynamics in 0D (QD) qualitatively differ from higher dimensional systems. The discrete density of states leads to severe energy conservation problems for scattering of 0D electrons with optical phonons (phonon bottleneck) and also with other 0D electrons which can lead to slow carrier relaxation processes. On the other hand, the number of virtual scattering processes remains large.

Acknowledgment

The author would like to acknowledge support from International Laser Center and Microelectronics Department, Slovak University of Technology in Bratislava.
I would also like to acknowledge many of the ideas discussed in this review that were evolved from several authors and laboratories during the past years.

References

Bailey D W, Stanton C J and Hess K, 1990, *Phys. Rev.* **B 42** 3423.
Benisty H, Sotomayor-Torres C M and Weisbuch C, 1991, *Phys.Rev.* **B 44** 10945.
Bocklemann U and Bastard G, 1990, *Phys. Rev.* **B 42** 8947.
Bogani F, Cabaresi L, Mattolini R, Colocci M, Bosaccchi A and Franchi S, 1996, *Sol. State Electron.* **40** 363.
Cingolani R, 1993, 'Semiconductor Superlattices and Interfaces', *Proceedings of the International School of Physics "Enrico Fermi", Course CXVII,* (North Holland).
Diels J-C and Rudolf W, 1996, *'Ultra short laser pulse phenomena'* (Academic Press, San Diego).
Ebeling K J, 1993, *'Integrated optoelectronics',* (Springer-Verlag, Berlin).
Elsaesser T and Woerner M, 1999, *Physics Reports* **321** 253.
Ganikhanov F, Burr K C and Tang C L, 1998, *Appl. Phys. Lett.* **73** 64.
Hamanaka Y, Nishiwaki D, NoNogaki Y, Fujiwara Y, Takeda Y and Nakamura A, 1999, *Physica* **B 272** 391.
Herman M A and Sitter H, 1989, *Molecular Beam Epitaxy,* (Springer Series in Material Science 7, Springer-Verlag, Berlin.)
Herrmann J and Wilhelmi B, 1987, *'Lasers for Ultrashort Light Pulses',* (Academie-Verlag, Berlin.)
Heyman J N, Unterrrainer K, Graig G, Williams J, Sherwin M S, Campman K, Hopkins P F, Gossard A C, Murdin B N and Langerak C, 1996, *Appl. Phys. Lett.* **68** 3019.
Hunsperger R G, 1995, *'Integrated Optics: Theory and technology',* (Springer-Verlag, Berlin.)
Hvozdara L, 2000, dissertation; *'Unipolar GaAs/AlGaAs mid-infrared lasers for spectroscopic applications',* p 21,(Technischen Universitat Wien, Institut fur Festkorperelectronik, Wien)
Kalt H, 1996, *'Optical Properties of III–V Semiconductors',* (Springer-Verlag, Berlin.)
Knox W H, Hirlimann C, Miller D A B, Shah J, Chemla D S and Shank C V, 1986, *Phys. Rev. Lett* **56** 1191.
Kovac J, Kuna L, Gottschalch V, Benndorf G, Gerhardt M, Pudis D, Jakabovic J and Skriniarova J, 1999, *Proc. EW MOVPE VIII* p 409 (Prague).
Leo K, Ruhle W W and Ploog K, 1988, *Phys.Rev.* **38** 1947.
Lienau Ch., Emiliani V, Gunther T, Intonti F, and Elsaesser T, 1999, *Physica* **B 272** 96.
Lin W-Z, Schonlein R W, Fujimoto J G and Ippen E P, 1988, *IEEE J. of Quant. Electron.* **24** 267.
Maciel A C, Kim J, O'Sullivan E D, Ryan J F, Schwarz A, Kaluza A, Hardtdegen H, Schäppers Th, Meertens D, Dieker Ch and Lüth H, 1999, *Physica* **B 272** 101.
Pelouch W S, Ellingson R J, Powers P E, Tang C L, Szmyd D M and Nozik A J, 1992, *Phys. Rev.* **B 45** 1450.
Péronne E, Lampin J F, Alexandrou A, Gauthier-Lafaye O, Julien F H, Brault J and Gendry M, 2000, *Physica* **E 7 2000** 151.
Shah J, 1988, *IEEE J. Quant. Electron.* **24** 276.

Shah J, 1996, *'Ultrafast Spectroscopy of Semiconductors and Semiconductor Nanostructures'* (Springer series in Solid-State Sciences, 115, Springer-Verlag, Berlin.)

Sosnowski T S, Norris T B, Jiang H, Singh J, Kamath K and Bhatttacharya P, 1998, *Phys. Rev.* **B 57** R9423.

Suemoto T, Nansei H, Tomimoto S and Saito S, 1998, *J. Luminesence* **76** and **77** 141.

Sun K W, Sun C K, Wang J C, Wang S Y, and Lee C P, 2000, *Sol. State Comm.* **115** 329.

Sung C Y, Norris T B, Zhang XK, Lam Y L, Vurgaftman I, Singh J and Bhattacharya P K, 1996, *Solid State Electronics* **40** 751.

Ulbrich R, 1973, *Phys.Rev.* **B 8 5719.**

Watabe H, Nagamune Y, Sogawa F and Arakawa Y, 1996, *Solid State Electronics* **40** 537.

Weisbuch C, Benisty H and Houndre R, 2000, *Journal of Luminescence* **85** 271.

Ultrafast semiconductor all-optical processing devices for telecommunications applications

Jean-Louis Oudar

CNRS, Marcoussis, France

1 Introduction

Ultrahigh-bandwidth optical telecommunication systems are the subject of active research as a result of the rapid growth of the internet. Further investigations for reaching even greater transmission capacity are needed in the field of optical transport, so as to propose new services to the users. To attain high transmission and processing capacity the speed of optical demultiplexers, repeaters, routers and other optical control devices needs to be increased. A key device in all-optical systems is the ultrafast optical switch. Semiconductors are well-suited materials for realising compact switches based on saturable absorption, and several methods are available to increase the response speed of semiconductor saturable absorbers.

At present the maximum channel bit-rate is currently limited to 10Gbit/s in long-haul transmission systems, as a consequence of various limitations appearing at higher bit-rates. To prepare the next generation of systems possibly operating at 40Gbit/s, optical signal regeneration is carefully examined as a promising solution enabling cost-effective long-haul transmission of 40Gbit/s signals. Optical 3R regeneration refers to three regenerative functions namely Re-amplifying, Re-shaping and Re-synchronising applied to weak, distorted, noisy or jittered data signals. Passing the signal through a saturable absorber device partially realises the second "R". The regenerative role of the perfect

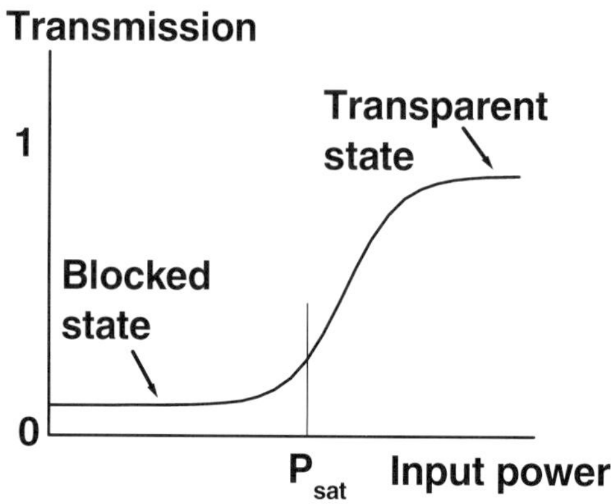

Figure 1. *Transfer function of an idealised saturable absorber.*

saturable absorber device whose transfer function is illustrated in Figure 1 is to be transparent for high incident power levels but opaque for low levels of the optical signal, hence contributing to cancel noise accumulation in "zero" bits.

In the following, we review recent advances in the realisation of semiconductor saturable absorber devices. In Section 2, we focus on the materials aspects, that have led to the controlled fabrication of fast quantum well saturable absorbers. A particular design for semiconductor saturable absorber devices, of the vertical access microcavity type, is presented in Section 3. The main advantageous characteristics of such devices are their fully passive and polarisation-insensitive operation. Section 4 is centred on the systems aspects, and includes a description of long distance transmission loop experiments in which a saturable absorber, included in a fully passive optical 2R regenerator, provides enhanced system performances. Such an application assigns a promising outlook to these totally passive and all-optical saturable absorber devices for next generation optical networks.

2 Fast quantum well saturable absorbers

2.1 Excitonic absorption saturation in quantum wells

For all-optical switching, semiconductor materials have raised a lot of interest in the past few years because of their non-linear optical properties associated with absorption saturation. In semiconductors the optical absorption near the band edge decreases significantly with the increase of non-equilibrium carrier or exciton densities. This offers the possibility to realise all-optical absorption modulators and switching devices, where the free carriers or excitons are injected by optical absorption. Furthermore, saturable absorbers based on semiconductor materials are easily integrated and can be used to achieve compact and stable devices.

Basically, self-induced absorption saturation can be understood as a two-step process: (a) the incoming light creates carriers and excitons, and (b) these additional carriers reduce the absorption coefficient near the band edge. As a result, for an incoming light in the band-edge spectral region, the absorbed power saturates at high intensity. In addition, the speed of the effect is limited by the carrier and exciton dynamics at the microscopic level. While the turn-on of the absorption nonlinearity induced by an intense ultrashort light pulse is nearly instantaneous, the turn-off process, i.e. the recovery time after absorption bleaching is essentially governed by the carrier recombination rate.

Ideally, a saturable absorber would have a very low saturation power, and would saturate completely at high intensity, leaving no residual absorption. In semiconductors, excitonic effects have been subject to a number of studies (see for example Rashba and Sturge 1982), because they give rise to sharp, atomic-like resonances, with low saturation intensities. Detailed measurements on GaAs-AlGaAs heterostructures (Gibbs *et al.* 1979) at cryogenic temperatures (10K) have shown that the intrinsic exciton absorption $\alpha_{\rm ex}$ can be modelled by the sum of a small unsaturable background α_B and a dominant term α_X which saturates with intensity I in the same way as an atomic-like two-level system with saturation intensity I_s, that is

$$\alpha_{\rm ex} = \alpha_B + \frac{\alpha_X}{1 + I/I_s}, \tag{1}$$

with $\alpha_X/\alpha_B = 6.65$ and $I_s = 150{\rm W/cm}^2$. These are interesting parameters for potential applications, however in bulk semiconductors such as GaAs the exciton absorption line can only be observed at cryogenic temperatures, which constitutes a major hindrance to their practical use. With quantum well structures, however, the observation of room temperature excitons has attracted a considerable interest, and led to a number of studies of their nonlinear optical properties (Miller *et al.* 1982).

We note that for an atomic-like two-level system, the saturation intensity I_s at resonance for a photon energy $h\nu$ is expressed as

$$I_s = \frac{h\nu}{2\sigma\tau}, \tag{2}$$

where σ is the absorption cross-section and τ is the upper-level population relaxation time. In semiconductors, the quantity 2σ in Equation 2 may be replaced by a phenomenological parameter, the differential absorption $d\alpha/dN$, which is the derivative of the absorption coefficient α (in ${\rm cm}^{-1}$) with respect to the electron-hole pair density N (in ${\rm cm}^{-3}$).

From a theoretical point of view, several effects contribute to the saturation of the exciton line, namely the screening of the electron-hole Coulomb interaction by free carriers, the reduction of the number of potential exciton-forming electrons in the valence and conduction bands (named as phase-space filling), and the exciton line broadening due to the exciton-exciton collisions, so that the physics of the excitonic absorption saturation is intrinsically a many-body problem (for a review, see Chemla and Miller 1985). These effects begin to occur when the spacing between excitons is about an exciton Bohr radius. Therefore a crude estimate of saturation intensity (for excitation at the exciton peak) is obtained by considering the absorbed intensity needed to produce one exciton in each exciton volume for every exciton lifetime. For bulk GaAs at low temperature, this critical exciton density can be estimated to be $10^{17}{\rm cm}^{-3}$ (Gibbs *et al.* 1979), in order-of-magnitude agreement with the corresponding saturation intensity, taking into account an exciton lifetime of 10ns. In quantum wells, one needs to consider the quasi-two-dimensional (2D) nature of excitons, and for quantum wells, the critical density N_s of excitons for the combined effects of phase-space filling and exchange can be expressed in terms of the

2D exciton Bohr radius a_{2D} as $N_s = 0.117/(\pi a_{2D}^2)$ at 300K (Schmitt-Rink *et al.* 1985). Finally it must be noted that the excitonic absorption line in quantum wells is broadened at room temperature by the electron-phonon interaction, so there is some spectral overlap between the exciton line and the edge of the band-to-band absorption continuum. As a result, under resonant pumping at increasing intensities, it is possible to observe the saturation of the band edge absorption once the excitonic peaks have disappeared. The intensity at which the edge saturates is about two orders of magnitude larger than for the exciton, as shown in Figure 2, which demonstrates very clearly the low saturation intensity of room temperature excitons in quantum wells.

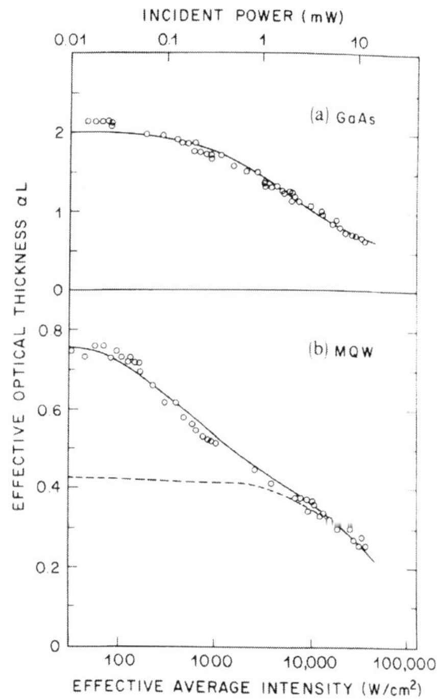

Figure 2. *Comparison of the absorption saturation curves of bulk GaAs and GaAs multiple quantum well structures at room temperature (from Miller et al. 1982).*

Finally, although the most detailed studies of these effects have been performed on GaAs quantum wells, room temperature excitons have also been observed in GaInAs quantum wells grown on InP substrates (Weiner *et al.* 1985), and it was shown that these quantum wells provide high optical nonlinearities at moderate optical powers in the 1.55μm wavelength region useful for telecommunication applications (Weiner *et al.* 1986).

2.2 Speed increase of semiconductor saturable absorbers

As already mentioned, the absorption recovery time after shutting down the saturating optical power is governed by the dynamics of carrier population relaxation. It is essentially determined by the carrier recombination time, which is generally of the order of a few nanoseconds. This constitutes a serious challenge for the fabrication of ultrafast optical devices. Introducing impurities by doping in order to reduce the lifetime tends to smooth out the excitonic absorption peak due to coulomb screening of the electron-hole interaction by the dopant-induced free carriers. To overcome this limitation a number of approaches have been explored.

At the device level, it is possible to increase the recovery speed of saturable absorbers, through the carrier sweep-out induced by strong electric-fields. This approach can be used e.g. in quantum well electroabsorption modulators (Højfeldt *et al.* 2000) where a reverse-bias voltage is used to control the band-edge and excitonic absorption through the quantum-confined Stark effect. However this leads to relatively complex device structures, due to the need for electrodes and external circuitry to operate the device, in spite of the all-optical nature of its saturable absorber functionality.

Defect-induced shortening of the carrier lifetime

At the material level, the speed increase of semiconductor saturable absorbers is classically obtained by increasing the non-radiative recombination rate through the introduction of defects by various means. Doping with compensating impurities, such as iron-doping in InP, introduces deep levels and can reduce the carrier lifetime, to values down to 7ps or less (Marceaux *et al.* 2001). Low temperature (LT) growth has been intensively studied in the GaAs system and has given rather encouraging results (Benjamin *et al.* 1996). LT-GaAs has found applications as a saturable absorber in mode-locked lasers, but other materials with a substantially smaller band gap in the wavelength range around 1.55μm are required for applications in telecommunications. Various studies show that in InP related materials, the growth temperature must be lower than 200°C to significantly reduce the carrier lifetime because of an insufficient excess of deep trap centres. By decreasing the growth temperature, the relaxation time can be reduced but the crystalline integrity is not maintained (Künzel *et al.* 1992). The combination of low-temperature growth and doping can successfully decrease the recovery time and allow one to reach the sub-picosecond region with high optical quality. Be-doping has been reported to reduce the absorption recovery time down to 250fs in LT-InGaAs/InAlAs MQWs (Takahashi *et al.* 1994). Such saturable absorbers have been used as optically controlled gates to regenerate optical data at 10Gbit/s rate (Hirano *et al.* 1998).

Particle bombardment offers an alternative to create trapping/recombination centres and shorten the carrier lifetime. Two different classes of bombardment processes can be distinguished according to the initial kinetic energy of the bombarding particles: impurity-dominated process obtained with particles of low initial energy, and radiation-damaged process obtained with particles of high initial energy. For small initial energy, the particles remain in the material (implantation) whereas for high initial energy they go through the structure (irradiation). As the irradiation induces only intrinsic defects and no impurities, this technique has the great advantage to be efficient for any material, and not sensitive to the chemical nature of the bombarding ion. Proton implantation has been used to reduce the response time of the absorption saturation in the GaAs/AlGaAs down to 150ps without degradation of the excitonic properties (Silberberg *et al.* 1985). Irradiation with high-energy heavy ions has also been recognised as a promising way to control the carrier lifetime in GaAs-based semiconductors (Avrutin and Portnoi 1989).

Recently it has been shown that this approach can be used in the InGaAs/InAlAs system to control the carrier lifetime down to 1.7ps while preserving the excitonic properties (Lugagne Delpon *et al.* 1998).

Pump and probe measurement techniques

The classical way of measuring the dynamics of nonlinear absorption in saturable absorber materials or devices with very high temporal resolution is the pump-and-probe technique. It can be implemented in two basic versions, schematically represented in Figure 3. The simplest version (Figure 3a) is the degenerate configuration, where a single wavelength

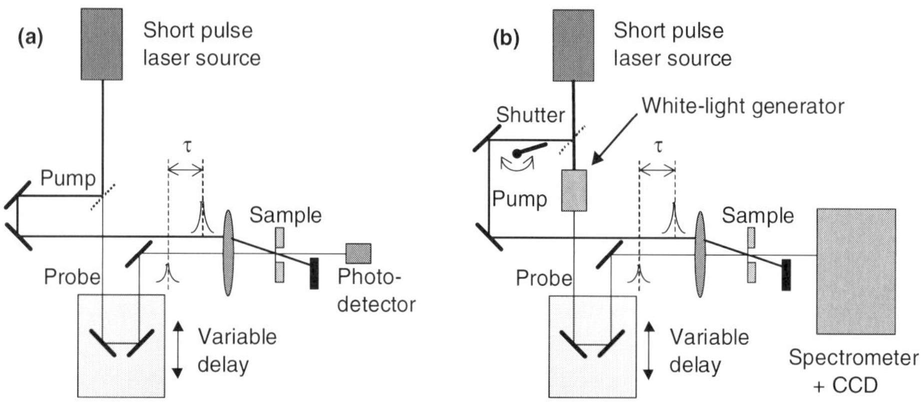

Figure 3. *Schematics of a pump-and-probe experimental set-up in two configurations: (a) degenerate configuration, and (b) non-degenerate configuration.*

is used, for both the pump and probe pulses. Starting with a short pulse laser source (typically a mode-locked laser), the probe pulses are obtained by splitting off a small portion of the laser output power. The remaining, more intense part, is then used as pump pulses. The two beams follow a distinct path, and mirror combinations supported by a mechanical translation stage can provide an adjustable delay line, so as to vary the relative delay between the probe and the pump pulses. After these separate paths, the two pulses are focused at the same point in the nonlinear sample, so that the absorption changes induced by the pump pulse can modify the propagation conditions of the probe pulse. The transmitted (or reflected) probe pulse intensity is then measured as a function of the pump-probe relative delay. The major advantage of this measurement principle is that it provides the dynamics of the sample nonlinear response with a resolution that is limited only by the duration of the short pulse laser source, independently of the temporal response of the photodetector. As a consequence slow photodetectors can be used, as well as lock-in amplifier techniques, to provide accurate measurements of the signal intensities. Most often, the pump and probe beams are perpendicularly polarised, to ease the discrimination between the pump and probe beams, by blocking the residual light scattered from the pump pulse with the help of a crossed polariser in the probe beam detection path.

We note that when the absorption saturation is induced by pulses shorter than the absorption recovery time τ, the degree of saturation depends on the pump fluence (or pump energy density) rather than on its peak power density. As a result, the degree of saturation occurring immediately after the pump pulse scales as the ratio of the pump fluence divided by the saturation fluence E_s of the saturable absorber, this quantity being simply related to the saturation intensity I_s defined in Equation 2, by $E_s = I_s\tau$.

The second basic version is the non-degenerate pump-probe configuration, where the sample nonlinear absorption can be probed at a different wavelength to the pump. While the degenerate configuration is simpler to implement, and directly delivers the relevant information needed to characterise the dynamics of self-induced absorption saturation at a given wavelength, the nondegenerate configuration provides a time-resolved spectrum of the nonlinear absorption process, when the probe pulse wavelength is tuned at a given pump-probe delay. Various experimental set-ups can be used for nondegenerate pump-probe experiments, including the use of two synchronised independently tuneable mode-locked sources. Actually the simplest way to obtain synchronised pulses (with very good timing accuracy) at two different wavelengths is to start from a single powerful short pulse source, and to generate the probe pulses by means of a nonlinear optical process. In the scheme shown on Figure 3b, the probe pulses are produced by a "white light" generator, through self-phase modulation of one part of the laser source. The spectrum of the broadband probe pulse transmitted (or reflected) by the sample is then measured by an optical spectrum analyser for each values of the relative pump-probe delay.

2.3 Carrier trapping and recombination on semiconductor defects

The presence of defects in semiconductors introduces *localised* electronic states, as opposed to the delocalised (Bloch) states of the crystalline material. The energy of some of these localised states may lie inside the material energy bandgap, so they are sometimes called *mid-gap states*. As a consequence the non-equilibrium photogenerated electrons can be trapped by these mid-gap states, with a capture rate that is expected to increase with the defect density. For instance in low-temperature grown GaAs, the additional mid-gap states are created by the excess arsenic incorporated during the material epitaxial growth. These trapping states can have a concentration as high as 10^{20}cm^{-3}, and very fast (subpicosecond) capture times can be obtained (Benjamin *et al.* 1996). With pump-probe dynamical measurements, this capture process gives rise to an initial fast component in the absorption recovery. Very often however this is followed by one or more slower components, the characteristics of which are very sensitive to the precise nature of the defects involved. The physical origin of these slower components is directly linked to the dynamics of the trapped carriers, which may also be influenced by the time-dependence of the material optical excitation itself. When a single isolated ultrashort pump pulse is used, one typically observes an initial fast component (carrier trapping), followed by a slower component, attributed to the population decay of the mid-gap states. This population decay may be described as the recombination of a trapped electron with a hole from the valence band. However when the material is excited by a sequence of highly repetitive strong pulses, one needs to consider more complex phenomena, such as those resulting from the saturation of the trapped carrier population, leading to intensity-dependent recovery times, and also to induced absorption at high pump intensity. A rate equation model (Benjamin *et al.* 1996) has been developed to describe these complex phenomena and obtain quantitative data for the trap emptying times in low-temperature-grown GaAs.

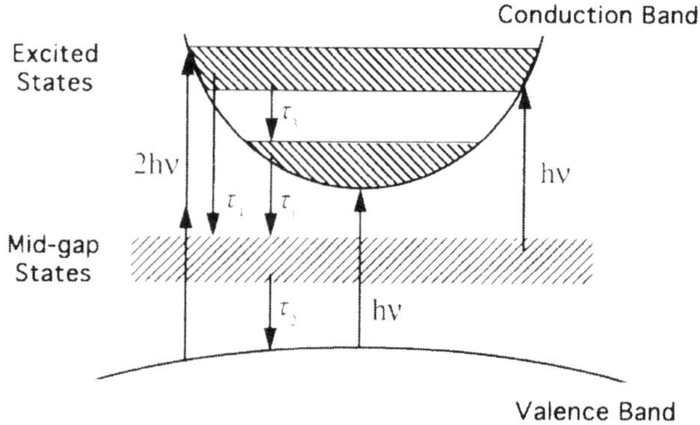

Figure 4. *Band diagram showing the key excitation (absorption of photons with energy hν) and decay processes (with relaxation times τ_1, τ_2, τ_3 and τ_4) that can occur in a direct-gap semiconductor with mid-gap states (reproduced from Smith et al. 1997).*

Figure 4 reproduces the schematic band diagram used by these authors as a basis of this model, which has a sufficient degree of generality to be used in the description of other types of defects and materials.

The effects resulting from the saturation of the trapped carrier population appear when the concentration of photoinduced carriers becomes comparable to the concentration of defects. For example, in low-temperature-grown GaAs with a recovery time of 800fs this value increases up to several picoseconds at high pump intensity (Sosnowski *et al.* 1997). The obvious way to avoid this recombination bottleneck is to use defects that can be liberated very quickly after carrier trapping. Heavy-ion irradiation is very effective to achieve this goal, as this technique produces groups of point defects. Each group of defects can be considered as an individual defect, but with a high number of carrier states, which open up a large number of recombination channels. It has been demonstrated (Stelmakh *et al.* 1998) that absorption saturation in bulk GaAs irradiated by heavy Au+ ions, could keep a subpicosecond recovery time up to excitation densities as high as 1.6mJ/cm^2. Thus the heavy-ion irradiation method seems very attractive for optical processing at high repetition rates, as will be shown in detail below.

2.4 Ultrafast dynamics of heavy-ion and light-ion irradiated InGaAs quantum wells at 1.55μm

The irradiation technique is based on the atom displacements induced on the path of energetic ions through crystalline materials. These atomic displacements create damage responsible for the carrier trapping defects. In semiconductor materials, nuclear collisions between the incident ion and the lattice atoms can displace host atoms, whereas electronic interactions do not create damage but contribute to the ion slowing-down. Both interactions are involved in the mean penetration length that differentiates the implantation from the irradiation.

During a nuclear collision, when the energy transferred by the incident energetic ion to a lattice atom is higher than the minimum energy to produce a displacement, the host atom moves from its original position. However, this displaced atom, to which energy has been transferred by the ion, can in turn displace other atoms when the transferred energy is large enough. The primary displaced atom then produces a cascade of secondary displacements. Altogether, these displacements break the periodicity of the crystalline lattice and create defects in the material. The energy transferred during a nuclear collision increases with the ion mass. The energy density deposited by heavy-ion irradiation on the lattice structure is then relatively high, which favours the formation of extended defects. By contrast, light-ion irradiation deposits a lower energy density and mostly creates isolated displacements along the ion track, leading to the formation of point defects. Such defect morphology distinctions are well illustrated by transmission electron microscopic (TEM) measurements (Mangeney *et al.* 2001). These irradiation-induced defects introduce additional levels in the band gap. Indeed, in highly irradiated samples, subgap absorption can be observed in the linear transmission spectra at room temperature (Mangeney *et al.* 2000b). Measurements have shown that these defects do not disturb the excitonic properties up to irradiation doses as high as 10^{12}cm^{-2} for heavy-ion irradiation in InGaAs/InAlAs MQW structures, as shown in Figure 5.

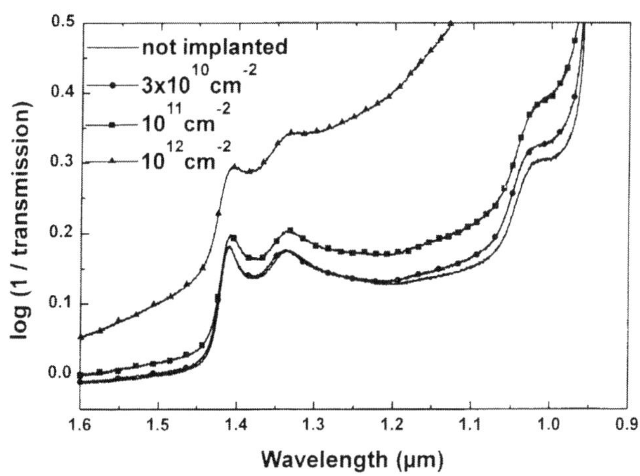

Figure 5. *Evolution of the room temperature transmission spectrum of InGaAs/ InAlAs MQW structures for various Ni+ (100MeV) ion doses (reproduced from Lugagne Delpon et al. 1998).*

At high bombardment energies it is well known that the ions get implanted inside the InP substrate but leave trails of defects along their trajectory in the top layers. The defects may have a high carrier capture rate but have only a small density per unit surface in each quantum well. This can explain the persistence of the exciton line while the lifetime is significantly reduced. For a 50Å well, the exciton diameter is approximately 80Å. The persistence of the exciton line at doses larger than 10^{12}cm^{-2} indicates that one ion creates less than one defect per quantum well.

The carrier lifetime reduction obtained by ion irradiation in semiconductor materials

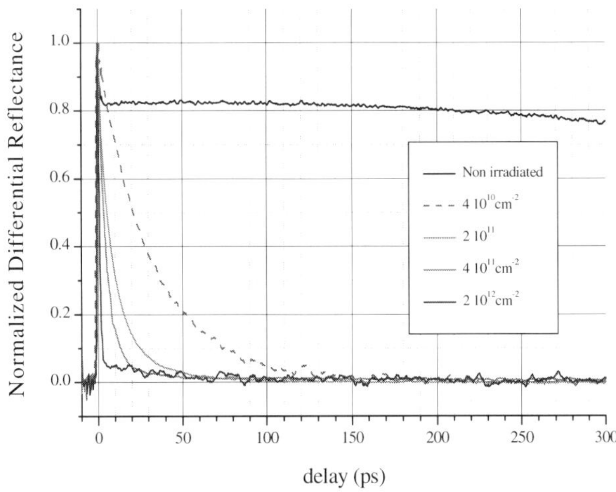

Figure 6. *Differential reflectance vs. probe delay, measured at 23MHz repetition rate on Ni+ (12MeV) irradiated samples with various doses, from 0 to 2×* $10^{12} cm^{-2}$.

can be measured by using a degenerate pump-probe set-up, the principle of which has been described in Section 2.2. Figure 6 shows the standard differential reflectance versus pump-probe delay, measured on InGaAs/InAlAs MQW irradiated by Ni^+ ions. These measurements were performed by using a mode-locked source producing 1ps pulse at 23MHz, at 1.55μm wavelength (Mangeney *et al.* 2000a).

A strong decrease in the absorption saturation relaxation time when the irradiation dose increases is clearly observed. The relaxation follows a mono-exponential decay law with a characteristic time that defines the response time of the saturable absorber device.

More detailed measurements of the 1×10^{11}cm^{-2} irradiated sample, performed with various pump intensities corresponding to photoexcited carrier densities in the region of 1×10^{17}cm^{-3}, showed that the relaxation time was constant ($\sim$12.8ps) within the $\pm$10% experimental accuracy. At the lowest pump energy, the injected carrier density was clearly smaller than the defect density (3.6×10^{18}cm^{-3}) and the dominant process was the capture of free carriers by defect level traps. At the highest pump energy, however, the non-equilibrium carrier density was comparable to the defect density, so the defect level recombination speed was a relevant parameter. A modelling of these measurements by population rate equations involving two consecutive processes, capture and recombination, led to an estimate for the defect level recombination time of less than 2ps. This intensity invariant relaxation time demonstrated the presence of efficient recombination centres in the material, in agreement with what is expected for clusters of point defects (Mangeney *et al.* 2000b).

Response time values extracted from the differential reflectance curves are shown in Figure 7 as a function of irradiation dose for both light and heavy ion irradiation. The response time is inversely proportional to the irradiation dose, reflecting the increasing capture probability with higher defect concentration. Response times as short as 1ps are obtained. Moreover, the response time of saturable absorbers can easily be adjusted from the nanosecond down to sub-picosecond values through the irradiation dose. This

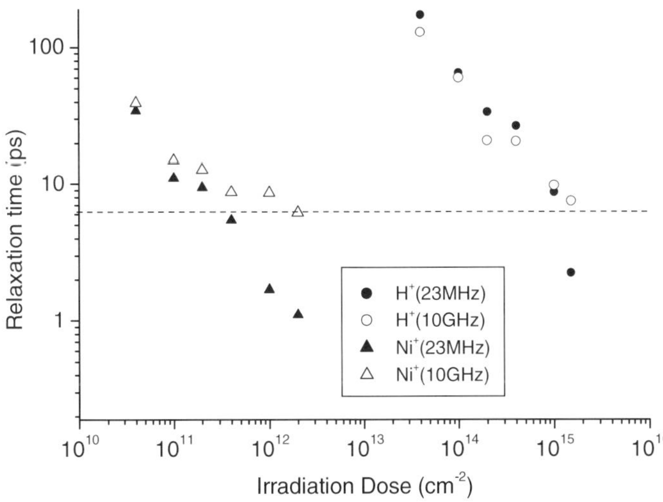

Figure 7. *Variation of the response time on a logarithmic scale with irradiation dose for proton-irradiated and Ni-irradiated InGaAs/InAlAs MQW structures measured at 23MHz and 10GHz repetition rates (Mangeney et al. 2001). The horizontal dashed-line is the resolution limit for the 10GHz data.*

adjustment, obtained with the ion irradiation technique, allows a high flexibility in the use of this material for specific applications.

The capacity of ion-irradiated saturable absorbers to work at high repetition rate can be studied by pump-probe measurements using high repetition rate picosecond optical pulses. This will be described in Section 3.5.

3 Vertical access microcavity saturable absorber devices

In the previous section we discussed the material properties that govern some basic device characteristics, such as the dynamical behaviour, through trapping and recombination, or the wavelength window, through the material band-gap and exciton energy. Other important device parameters, such as the switching power, switching contrast, spectral bandwidth and polarisation properties depend strongly on the device structure itself. Therefore, to some extent, they can be tailored by proper device design, and usually some tradeoffs must be made in the design, depending on the application. In addition, modelling of the device is important, to anticipate the device parameters, and be in a position to make pertinent design choices. In our study, the saturable absorber devices had a microcavity structure, used in the reflection-mode at normal incidence, so that, as opposed to devices with a waveguide structure, they provide a vertical access to the optical signals. The following subsections describe some basic features of microcavities relevant to our devices, a simple approach for the modelling of their optical properties, and some cavity design considerations. Then some details are given on the principle of their fabrication, and on the nonlinear characteristics of fabricated device samples.

3.1 Reflective-type microcavities

The generic structure of a reflection mode vertical microcavity device is schematically represented in Figure 8. It is essentially a compact Fabry-Perot structure used at resonance, where the saturable absorber active layer is sandwiched between two reflecting layer structures, the front and back mirrors, the whole structure being supported by a substrate. The incoming optical signal is sent in vertically at normal incidence. The outgoing optical signal is collected from the same side as the input signal, which clearly separates the optical access (on the upper side) from any other access, thermal, electrical, etc, which can be managed on the lower side.

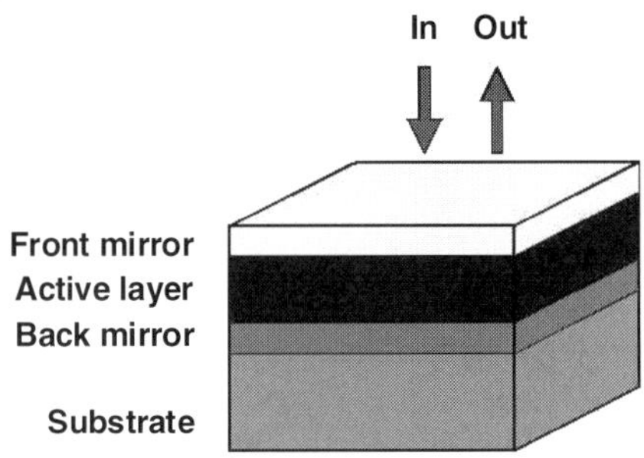

Figure 8. *Structure of a reflection-mode vertical microcavity device.*

As far as device optical parameters are concerned, using such a reflective type microcavity has several advantages. At resonance the intracavity field is enhanced, allowing weaker input signals to saturate the quantum well absorption, and thus leading to a reduction of the switching power. With optical signals at normal incidence on a planar multilayer structure, the device is intrinsically polarisation insensitive, which is an important requirement for processing fibre optical communication signals. Finally, in the reflection mode, a switching device based on a Fabry-Perot structure can have a high on/off contrast ratio, by using interferometric effects to cancel the reflection coefficient at resonance.

Let us consider a classical Fabry-Perot cavity with a front mirror reflectivity R_f, a back mirror reflectivity R_b, a cavity of length L filled with an absorbing material with absorption coefficient α. The reflection coefficient R of this cavity can be written as

$$R = \frac{\left(\sqrt{R_b}e^{-\alpha L} - \sqrt{R_f}\right)^2 + 4\sqrt{R_f R_b}e^{-\alpha L}\sin^2\Phi}{(1-\sqrt{R_f R_b}e^{-\alpha L})^2 + 4\sqrt{R_f R_b}e^{-\alpha L}\sin^2\Phi}, \tag{3}$$

where Φ is the half-roundtrip phase accumulated by the optical beam along its propagation between the mirrors.

At resonance, this phase $\Phi = n\pi$, so that the cavity reflection becomes

$$R_{\text{res}} = \frac{\left(\sqrt{R_b}e^{-\alpha L} - \sqrt{R_f}\right)^2}{\left(1-\sqrt{R_f R_b}e^{-\alpha L}\right)^2}. \tag{4}$$

It will be noticed that this resonant value R_{res} can be exactly cancelled when a specific condition between the cavity parameters is satisfied, namely when

$$R_f = R_b e^{-2\alpha L}. \tag{5}$$

The physical origin of this cancellation property can be understood from the ray diagram displayed in Figure 9 (in the case of oblique incidence for a better clarity). It shows that,

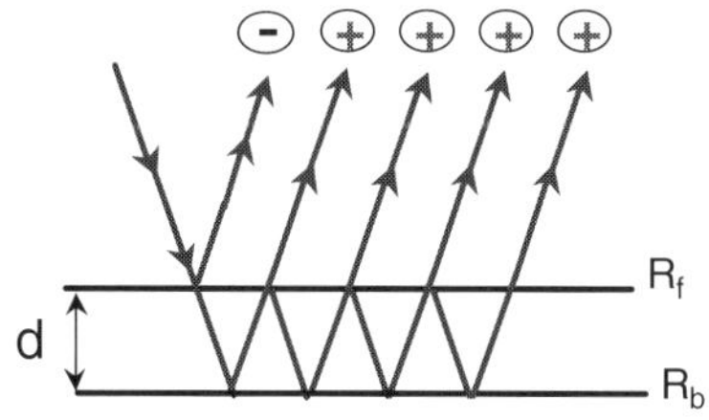

Figure 9. *Ray-diagram showing the sign of the various components, reflected at resonance from a Fabry-Perot cavity.*

at resonance, the ray component directly reflected from the front mirror has an opposite sign to that of all the other ray components that are transmitted by the front mirror, after their propagation back and forth within the cavity. Note that in the general case all these components are complex quantities. However, in the resonant case, the phases of these complex components are either 0 or π, so the use of plus and minus signs in the figure is justified, but this is correct for the resonant case only.

The advantage of this exact cancellation feature has been theoretically discussed in several papers (e.g. Wherett 1984) and has been widely used in the past to obtain a high switching contrast in various reflection-type microcavity devices, such as asymmetric Fabry-Perot electro-absorption modulators (Whitehead and Parry 1989), bistable microcavity devices (Oudar *et al.* 1992), and low-temperature-grown surface-reflection all-optical switches (Takahashi 2001).

For saturable absorber devices, the cavity parameters may be adjusted so that the condition written in Equation 5, sometimes called impedance matching condition (Garmire 1989) is obeyed in the linear regime, when the absorption coefficient is not saturated. In that case, this unsaturated absorption coefficient α_0 should thus be given by

$$\alpha_0 = \frac{1}{2L} \ln \left(\frac{R_b}{R_f} \right) \tag{6}$$

In the nonlinear regime, the absorption coefficient is reduced at increased intensities, and may eventually go to zero in the ideal case of complete saturation. Figure 10 shows schematically the variation of the cavity reflection coefficient, that can be deduced from

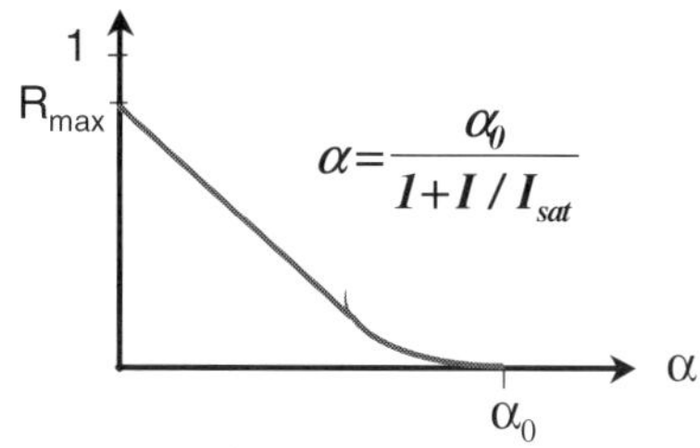

Figure 10. *Schematic variation of the cavity reflection coefficient versus the active material absorption coefficient α.*

Equation 4, when the active material absorption coefficient varies from the unsaturated value α_0 to zero.

When α is zero, we see from Equation 4 that the cavity reflection coefficient reaches a maximum value $R_{\max}$ given by

$$R_{\text{res}} = R_{\max} = \frac{\left(\sqrt{R_b} - \sqrt{R_f}\right)^2}{\left(1 - \sqrt{R_f R_b}\right)^2},$$

We note that in general $R_{\max}$ is smaller than one, but the value of one can be obtained in the limiting case of a back mirror reflectivity of unity. This property is independent of the value of the front mirror reflectivity, provided that the front mirror is lossless, which has been the assumption made in the whole derivation of equations in this section.

3.2 Cavity design considerations

In Section 3.1 we discussed how the device *switching contrast* can be made arbitrarily large, with reflective-type microcavities, through an exact interferometric cancellation of the device throughput in the low intensity regime. The condition for this, expressed in Equation 6, leaves room for the optimisation of other device parameters, such as the *switching power* and the *spectral bandwidth* of the cavity resonance. For a given saturation intensity of the active (quantum well) material, the device switching power may be reduced through the intracavity enhancement factor, which directly depends on the cavity finesse. Continuing with the classical Fabry-Perot model, the cavity finesse f is expressed as

$$f = \frac{\pi\left(\sqrt{R_f R_b}e^{-\alpha L}\right)^{1/2}}{1 - \sqrt{R_f R_b}e^{-\alpha L}} = \pi\frac{\sqrt{R_\alpha}}{1 - R_\alpha}, \tag{7}$$

with

$$R_\alpha = \sqrt{R_f R_b}e^{-\alpha L}.$$

The cavity finesse indicates the effective number of round-trips of light in the cavity, therefore it is a direct measure of the enhancement of the intracavity field at resonance. Of course, as the finesse depends on the absorption coefficient α, for saturable absorber cavity devices, this parameter is intensity-dependent. The consequence of this will be discussed in the next subsection, but at this simple level, it may be noted that an increase in the low-intensity cavity finesse should result in a decrease of the device switching power.

As far as the cavity resonance spectral bandwidth is concerned, the relevant parameter is the cavity Q factor, defined as

$$Q = \frac{\lambda}{\delta\lambda}, \tag{8}$$

where λ is the wavelength at resonance, and $\delta\lambda$ is the resonance full width at half-maximum (FWHM). In order to keep $\delta\lambda$ as large as possible it is important to keep the Q factor as small as possible. As this point it must be noted that the cavity finesse of an empty cavity may also be written in terms of $\delta\lambda$, through the expression

$$f = \frac{\Delta\lambda}{\delta\lambda}, \tag{9}$$

where $\Delta\lambda$ is the free spectral range of the cavity, which directly depends on the cavity optical thickness L_{op}, through the equation

$$\Delta\lambda = \frac{\lambda^2}{2L_{\text{op}}}. \tag{10}$$

As both Q and f are inversely proportional to $\delta\lambda$, they are sometimes considered as not independent from each other. However, if we look at the ratio f/Q, we see that there is some degree of freedom in the cavity design, which is the cavity optical thickness L_{op}. More precisely, we get

$$\frac{f}{Q} = \frac{\Delta\lambda}{\lambda} = \frac{\lambda}{2L_{\text{op}}}. \tag{11}$$

Therefore if we want to design a cavity with a high finesse, to maximise the intracavity field enhancement at resonance, and simultaneously with a small Q factor, to maximise the device spectral bandwidth, we need to work with an f/Q ratio as large as possible, that is, in the *microcavity regime*, where the cavity optical thickness is of the order of the resonance wavelength.

For the ideal case of an empty cavity with two identical mirrors for which the phase of the reflection coefficient is 0 or π (independent of wavelength), the minimum value of L_{op} is $\lambda/2$, so that f/Q is equal to 1. But this is just a specific case, and in the general case one needs to take into account the refractive index dispersion of the intracavity medium, as well as the phase dispersion of the mirror reflection coefficients. The notion of a local free-spectral range (Jewell *et al.* 1988) is then the most useful concept to estimate the finesse of arbitrary microcavities, from the experimental value of $\delta\lambda$. It takes into account the fact that the phase of the reflection coefficient is not constant, and that it may be expanded to first order in wavelength in the vicinity of the cavity resonance. If ϕ is the round-trip phase at a given wavelength λ, the local effective free spectral range in wavelength units is

$$(\Delta\lambda)_{\text{loc}} = \frac{2\pi}{(d\phi/d\lambda)}. \tag{12}$$

In the general case, this phase dispersion may be calculated from the transfer matrix formalism introduced in the next section. In the specific case of a distributed Bragg reflector (DBR) analytical formulas can be derived to compute its value at the DBR Bragg wavelength. For instance, one can define an effective penetration depth for the DBR, defined as the depth (defined from the first DBR interface) where a single mirror interface (with no phase dispersion) should be positioned to give rise to the same phase slope as the DBR itself. Assuming that a fictitious medium of refractive index n is in front of the equivalent mirror, this penetration depth is approximately given by

$$L_{\text{eff}} = \frac{\lambda_B}{4n}\frac{n_L}{n_i}\frac{n_H}{n_H - n_L} \tag{13}$$

for an infinitely thick DBR made of quarter-wave layers of refractive indices n_L and n_H, with light incident from a medium with refractive index n_i (Babic and Corzine 1992). For all practical purposes, one may choose n and n_i as the refractive index n_c of the cavity core. In this case, the cavity optical thickness L_{op} is given by

$$L_{\text{op}} = (2L_c + L_{\text{eff},B} + L_{\text{eff},F})n_c/2, \tag{14}$$

where L_c is the cavity core thickness, and $L_{\text{eff},B}$ and $L_{\text{eff},F}$ are the penetration depths of the back and front mirrors respectively.

From this discussion we see that the best compromise between a large intracavity field enhancement at resonance and a large spectral bandwidth is obtained when the effective cavity optical thickness is kept to a minimum. Monolithic semiconductor microcavities, where the back mirror is grown in the same epitaxial process as the active quantum well structure, although relatively easy to fabricate, do not represent an optimum in this respect, because of a relatively large penetration depth L_{eff}, leading to a small f/Q ratio. Actually, much better performances can be obtained when metallic mirrors are used, because such mirrors have a much smaller phase dispersion than semiconductor distributed Bragg reflectors. As a matter of fact, in the context of the control of spontaneous emission in semiconductor microcavities, it can be shown that a microcavity with two metallic mirrors can even be designed so that the effective L_{op} is smaller than $\lambda/2$ (Abram *et al.* , 1998), which is equivalent to an f/Q ratio larger than 1. However, there is a disadvantage in using a metallic front mirror, because it is absorbing, so that its transmission coefficient is much smaller than $1 - R_f$. This is why a better configuration is that of a metallic back mirror and a dielectric front mirror. With a silver or gold mirror deposited on InP material, a reflection coefficient larger than 93% can be obtained. In addition, a metallic mirror has a better thermal conductivity than a semiconductor Bragg reflector, which helps to reduce the temperature rise in the MQW active layer due to the absorbed optical power. Details of the fabrication procedure will be described in Section 3.4.

In a first generation of such saturable absorber devices, the front mirror was simply the surface of the InGaAs MQW structure, having a front mirror reflectivity of about 30%. With these values of mirror reflectivity, the impedance matching condition (Equation 6) could be achieved with an MQW structure made of 4 groups of 7 InGaAs quantum wells. Each of these groups was located at the antinodes of the intracavity intensity, as schematised on Figure 11. This feature allows to obtain a strong and uniform coupling of the quantum wells with the optical signal, thus keeping to a minimum the effective saturation intensity of the MQW structure.

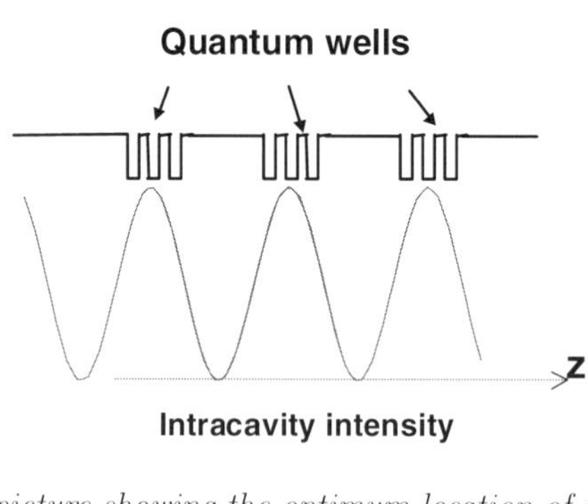

Figure 11. *Schematic picture showing the optimum location of quantum well groups with respect to the intracavity intensity.*

With this cavity structure, one obtains fairly broad resonance lineshapes, as shown in Figure 12. The resonance width is more than 90nm (FWHM). As a result, the devices have a relatively good tolerance to the inevitable fabrication imperfections (Figure 12

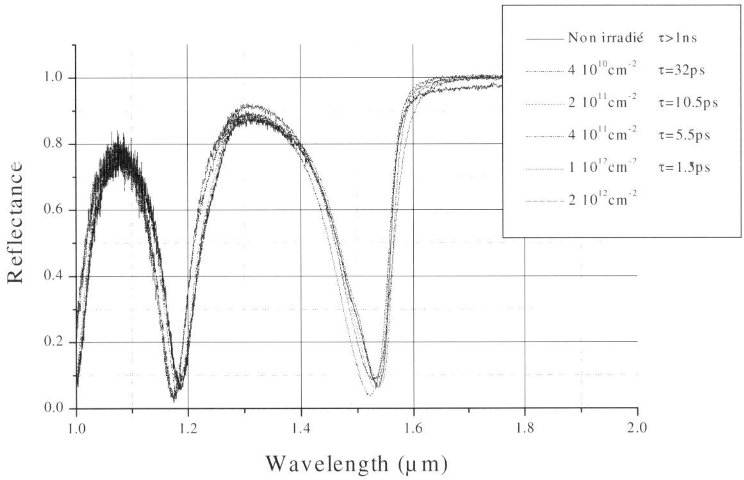

Figure 12. *Reflection spectra of a series of saturable absorber microcavity devices, with different irradiation doses (Ni^+ ions, 12MeV) ranging from 0 to $3\times10^{12}cm^{-2}$.*

displays the superposition of six different microcavity devices with different irradiation doses). Furthermore, the device reflection coefficient in the linear regime has a low value (for high switching contrast) over a large spectral width, compatible with their use in WDM transmission systems.

In this first series of saturable absorber devices the cavity finesse was limited by the low reflection coefficient of the front mirror to a value of 2.2, which did not provide a large intracavity field enhancement at resonance. Therefore, a second generation of such devices was designed, with higher finesse values, ranging from about 8 to 15, consisting of multilayer front mirror structures with reflection coefficients ranging from 70% to 80%, and a single group of absorbing quantum wells. More details on the design parameters and simulation results are given in the next subsection.

3.3 Modelling of linear and nonlinear optical properties

In the two previous subsections, the Fabry-Perot model was used to introduce in a simple way the basic concepts that may be used as guidelines to design improved nonlinear microcavity devices. In spite of its merits, this is an oversimplified model that cannot be used to predict in a detailed way the linear and nonlinear properties of these devices.

To model the linear optical properties of planar multilayer structures, transfer matrix methods provide the basis to build powerful simulation tools of such structures, that may be used for instance to calculate their transmission and reflection spectra, or to compute the intracavity longitudinal variation of the electric field amplitude at a given wavelength. Several versions of them have been developed, that are described in textbooks (Yeh 1988, Born and Wolf 1986, Mac Leod 1986). Let us briefly describe the principle of this method, and give some examples of their use in typical microcavity structures. In the characteristic matrix method each layer k is represented by a matrix M_k written as

$$M_k = \begin{bmatrix} \cos\delta_k & \frac{i}{N_k}\sin\delta_k \\ iN_k\sin\delta_k & \cos\delta_k \end{bmatrix}, \tag{15}$$

where N_k, the effective refractive index for layer k, depends on the incidence angle θ_k of the plane wave in layer k, and on its polarisation state. For s polarisation we have

$$N_k = \sqrt{\frac{\epsilon_0}{\mu_0}} n_k \cos\theta_k, \tag{16}$$

where ϵ_0 and μ_0 are the dielectric and magnetic permittivities of vacuum, and n_k is the refractive index of layer k. For p polarisation it is expressed as

$$N_k = \sqrt{\frac{\epsilon_0}{\mu_0}} \frac{n_k}{\cos\theta_k}. \tag{17}$$

In Equation 15, the parameter δ_k represents the dephasing introduced by layer k. It depends on the thickness d_k of layer k, and on the vacuum wavelength of light λ, through the expression

$$\delta_k = \frac{2\pi}{\lambda} n_k d_k \cos\theta_k. \tag{18}$$

This method allows the definition of an effective index N for a multilayer stack of p layers on a substrate of refractive index N_m, through the equation

$$N = C/B, \tag{19}$$

where B and C are the matrix elements

$$\begin{bmatrix} B \\ C \end{bmatrix} = \prod_{k=1}^{p} \begin{bmatrix} \cos\delta_k & \frac{i}{N_k}\sin\delta_k \\ iN_k\sin\delta_k & \cos\delta_k \end{bmatrix} \begin{bmatrix} 1 \\ N_m \end{bmatrix}. \tag{20}$$

With light incident from a medium of refractive index N_0, the field reflection (r) and transmission (t) coefficients of this stack are then given by

$$r = \frac{N_0 - N}{N_0 + N}, \qquad t = \frac{2N}{N_0 + N}. \tag{21}$$

Using this method, one may also easily compute the electric field amplitude at any point inside the structure. As an example, we consider a model microcavity structure, where the back and front mirrors are Bragg reflector stacks of alternate layers with refractive indices of 2.5 and 3, each of them having a $\lambda/4$ thickness at the nominal wavelength of 1000nm. The back mirror is an 8 period stack with no substrate ($N_m = 1$), and the front mirror is a 2 period stack. The cavity is a $3\lambda/2$ layer of refractive index $n = 3$.

Figure 13 represents for three specific cases the longitudinal variation of the squared electric field amplitude in this structure, over a domain larger than the structure itself. In each case, the superposition of the incident and reflected fields (oriented leftwards and rightwards, respectively) gives rise to standing wave oscillatory patterns, both in the region inside the structure itself (intracavity region), and on the right side of it (incident

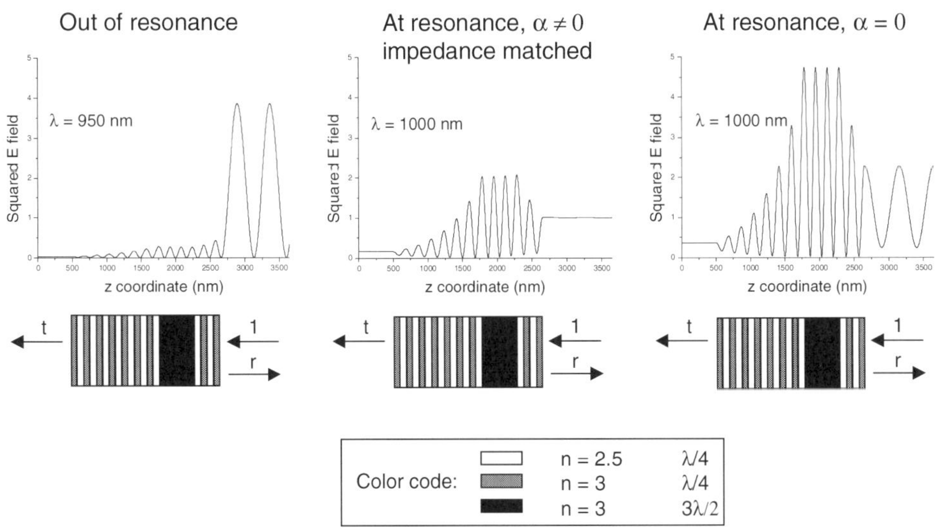

Figure 13. *The upper three curves display the longitudinal variation of the squared electric field in the model structure reproduced below each curve. The incident field comes from the right side of the structure and has an amplitude of one.*

field region). On the left side of the structure, there is only a transmitted (leftward propagating) field, with a constant amplitude.

In the first case (first curve from the left), the wavelength is set out of resonance ($\lambda = 950$nm). The incident field is almost entirely reflected by the structure, so that with an incident field amplitude of one, the squared amplitude of the superimposed incident and reflected field oscillates between 0 and 4. Only a very small amount of the incident field penetrates in the cavity. In the second case the wavelength is exactly at the microcavity resonance ($\lambda = 1000$nm) and the absorption in the cavity layer is adjusted so that Equation 6 is obeyed. In that case the reflection coefficient is cancelled, which is indicated by the absence of oscillations in the incident field region outside the structure. In the third case (last curve from left), the wavelength is still at resonance, but the cavity layer absorption is set to zero. Some light is reflected from the structure. In addition, the intracavity field amplitude becomes larger than the incident field, due to the cavity resonance amplification factor, which is proportional to the cavity finesse. This case is indicative of what occurs at high incident intensity, when the absorption is saturated. We note that, compared to the second case, the cavity resonance amplification factor is larger, because the zero absorption value results in an increased cavity finesse.

In order to model the nonlinear properties of such structures, one has to perform self-consistent numerical calculations, because the variations of the optical field amplitude at a given point inside the structure modify the cavity resonance amplification factor, as already discussed (such a positive feedback is known to give rise to absorptive optical bistability). To this end the following iterative scheme was adopted. For a given incident intensity value, the intracavity field distribution is first computed assuming an unsaturated quantum well (QW) absorption ($\alpha = \alpha_0$). This gives a first determination of the

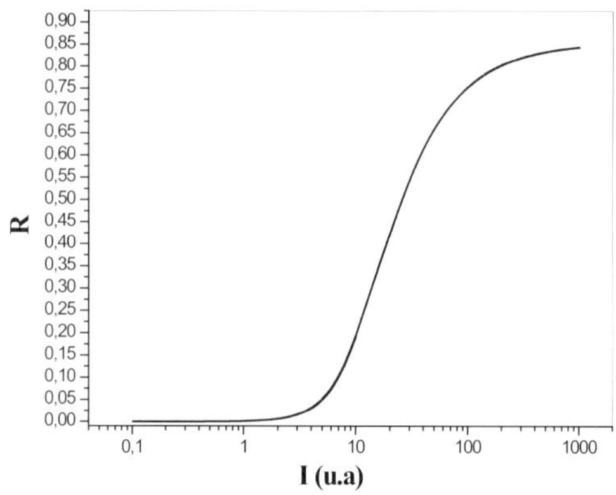

Figure 14. *Variation of the reflection coefficient of structure S28 versus incident intensity in arbitrary units ("u.a." in French).*

local intensity I_{QW} at the QW location. The new value of $\alpha(I_{QW})$ is calculated from the saturation law (Equation 1). From this, a new intracavity field distribution is computed from the matrix formalism, giving a new determination of I_{QW}, and the process is repeated until convergence is attained. Figure 14 gives an example of the reflection coefficient variation with incident intensity that can be obtained with this method.

Such calculations allow us to predict some of the device characteristics, such as the switching power and the switching contrast, for different cavity designs, with different values of the cavity finesse and different numbers of absorbing quantum wells. Table 1 summarises the calculated characteristics of some specific cavity designs studied in our group.

Structure name	Number of QWs	Resonance Width (FWHM in nm)	Finesse	Switching power (in db, relative to structure S28)
S28	28	93	2.2	0
S9	9	35	7.8	−4.4
S7	7	65	8.8	−6.2
S3	3	52	14.6	−8.0

Table 1. *Calculated characteristics of four different cavity structures, with different amounts of quantum wells.*

3.4 Device fabrication

Following the design considerations described in the previous section, we concentrated our efforts on the fabrication of hybrid microcavities that include a metallic (Ag) back mirror, providing a good thermal conductance, associated with a strong optical field confinement.

Figure 15 shows two basic device structures that have been fabricated and studied

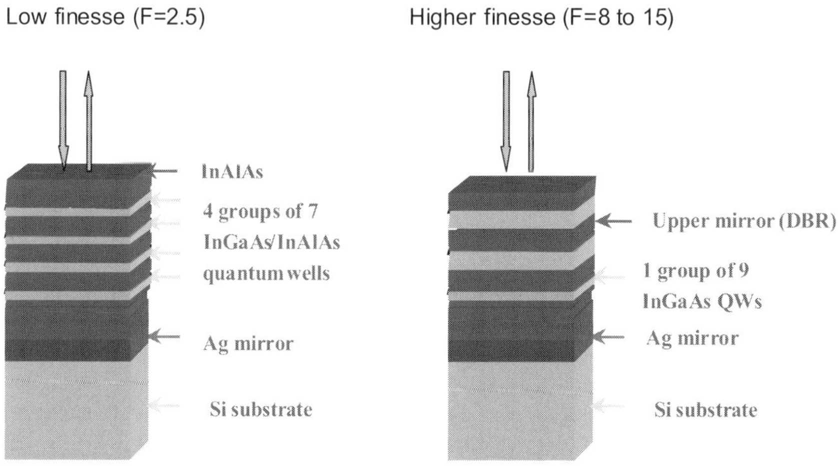

Figure 15. *Schematics of two basic device structures.*

in our group. We started with the low finesse structure depicted on the left part of the figure. In this structure the front mirror is simply the semiconductor surface and the MQW structure consists of 4 groups of 7 QWs, which in this case were 9nm thick InGaAs QWs, separated by 7nm thick InAlAs barriers, each group being located at the antinodes of the intracavity field. This structure is named S28 in Table 1. In a second stage of our studies we designed and fabricated higher finesse structures, as depicted on the right part of Figure 15, where the increased finesse (ranging from approximately 8 to 15) was obtained by the introduction of a front mirror (with reflection coefficient ranging from 0.65 to 0.80) and the use of only one group of absorbing QWs, their number ranging from 3 to 9, (structures S3 to S9 in Table 1). In this case the InGaAs QWs are separated by 7nm thick InP barriers.

In order to obtain this type of device structure with a buried metallic mirror, it is necessary to transfer the metal-coated MQW structure onto a new substrate, and then to remove the initial InP substrate on which the MQW structure was epitaxially grown. Such substrate removal can be performed by mechanical thinning followed by selective etching, using sacrificial etch-stop layers. To make our devices, the following fabrication steps were performed. After the growth of the epitaxial structure, the wafer was cleaved into smaller pieces (typically $7{\times}7mm^2$) for further processing. These samples were ion-irradiated, using the Aramis implanter, at the University of Orsay. Then a silver film was deposited on top of the irradiated structure, to form the metallic mirror. The transfer on a new (Si) substrate was performed through a metallic bonding process, based on Au-In solid-liquid inter-diffusion (SLID). The advantage of this bonding process is that it is performed at relatively low temperature (200°C), and that, once the bond is formed, it can sustain much higher temperatures, up to 450°C (Bernstein 1966, Lee *et al.* 1993). In addition we found that it provides a good device uniformity, both in terms of the resonant wavelength and of the device insertion loss. After bonding, the initial substrate was removed, and the dielectric front mirror was deposited, when needed.

3.5 Nonlinear characteristics and high bit rate operation

In order to evaluate the ability of our heavy-ion saturable absorber devices to operate at high repetition rates, a series of pump-probe measurements was performed, with an actively mode-locked source delivering 5ps pulses at 10GHz repetition rate, with a pump energy of 0.1pJ, at 1.55μm wavelength (Mangeney *et al.* 2000a). The measurements are shown in Figure 16. The relaxation time measured for the 10^{11}cm^{-2} irradiated samples

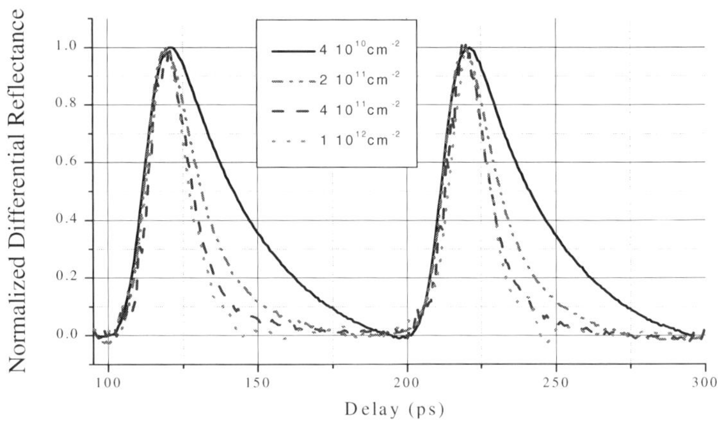

Figure 16. *Normalised differential reflectance of several heavy-ion-irradiated device samples, measured in a pump-probe experiment at 10 GHz (Manganey et al. 2000). There is no degradation of the switching speed occurred at the high repetition rate.*

was 12.8ps, that is, essentially the same as the value measured at 23MHz. For the two faster samples the differential reflectance signal essentially followed the 5ps pump pulse, indicating that the relaxation time was shorter than the pulse duration. It then clearly appears that no degradation of the switching speed occurred at this high repetition rate.

Figure 17 shows the reflectance of the saturable absorber device measured as a function of the incident pulse energy at low repetition rate (23MHz). The on/off contrast ratio between the reflectance at high (above 100pJ) and low (below 0.1pJ) pulse energies is found to be as high as 16:1. Indeed, the reflectance is varied from ~0.03 to 0.45. The switching energy, defined as the energy required for a 10dB switching contrast is ~8pJ. This energy is rather independent of the irradiation dose for values up to 10^{12}cm^{-2}.

For an amplitude modulated signal incident on the device, the contrast enhancement factor can be predicted from the characteristic presented in Figure 17 (Mangeney *et al.* 2000c). If I and $I(H)$ are respectively associated with the low and high intensity levels, this factor is equal to $[I_a(H)/I_a(L)][I_b(L)/I_b(H)]$, where the subscripts a and b correspond to the reflected and incident signals, respectively. Figure 18 shows the computed curves of the contrast enhancement factor versus the incident pulse energy for various contrast ratios of the incident signal. As can be seen in this figure, there is an energy domain over which the contrast enhancement factor remains close to the optimum. This domain weakly depends on the contrast ratio of the incident signal. All-optical discrimination experiments, performed on modulated signals have confirmed these extinction ratio enhancement values, as also shown in Figure 18 (experimental points).

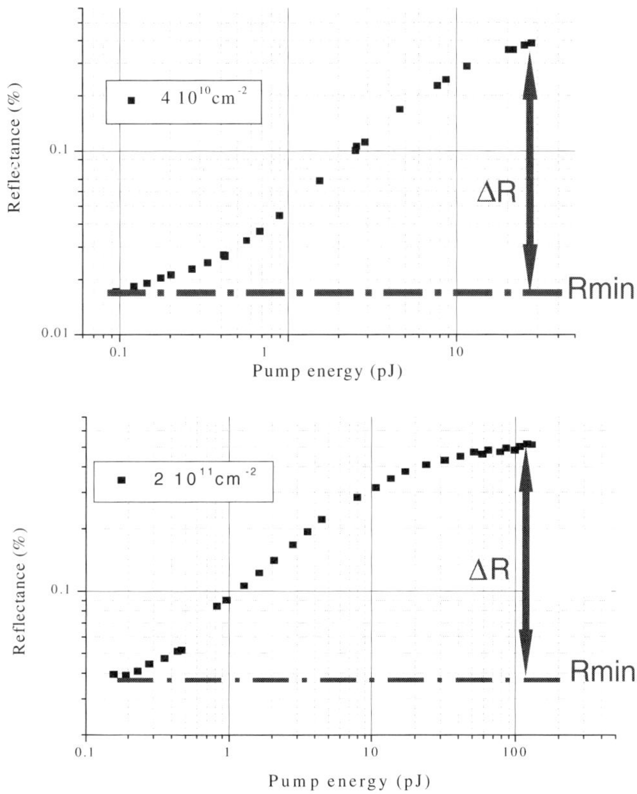

Figure 17. *Reflectance of a saturable absorber device versus incident pulse energy (experimental values: $R_{\min}$ <0.04; $\Delta R > 0.40$; E_{sat} ~5pJ).*

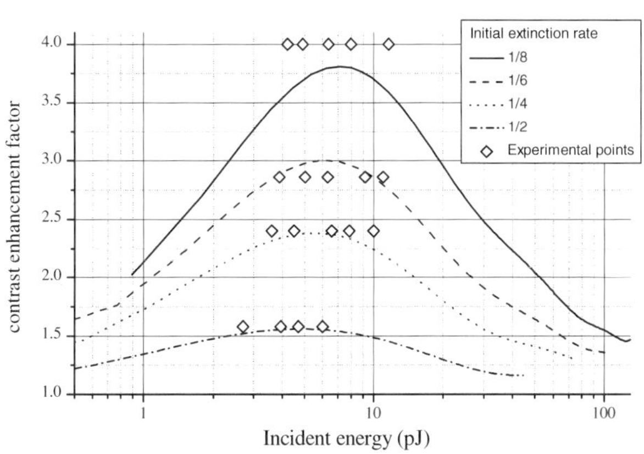

Figure 18. *Extinction ratio enhancement factor versus incident pulse energy for various contrast values of the incident signal. Curves are calculations derived from the results of Figure 17. Dots are experimental results.*

The discrimination properties of this device have been analysed at high frequency using return-to-zero (RZ) signals (see page 255 for details of RZ format) with a regular (0101 $\cdots$) bit pattern at 40Gbit/s, with a purposely degraded extinction ratio (Leclerc *et al.* 2000). The incident optical signal, of 8dBm average power, was focused with a spot size of 4μm ($1/e^2$ diameter). The reflected signal, collected by a circulator, was observed using a streak camera of 2ps temporal resolution. Figure 19 shows the pulse train before and after reflection on the saturable absorber device, showing a clear increase in the extinction ratio, with an average contrast improvement of 2.5dB.

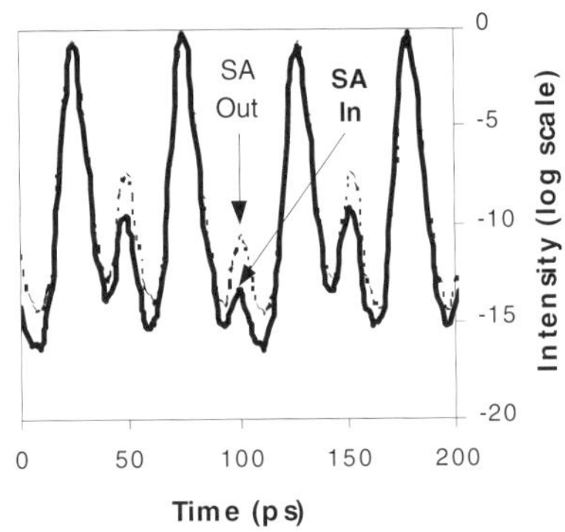

Figure 19. *Example of extinction ratio enhancement measured on modulated signals at 40Gbit/s.*

Finally, Figure 20 compares the transfer curves of two devices, with different finesse values. The $f = 2$ curve was obtained with a device structure of the type named S28 in Table 1, while the $f = 9$ curve was obtained with a device of the S0 type. An improvement of the switching energy of 3dB is deduced from this measurement, to be compared with the expected improvement of 4.4dB, as shown in Table 1.

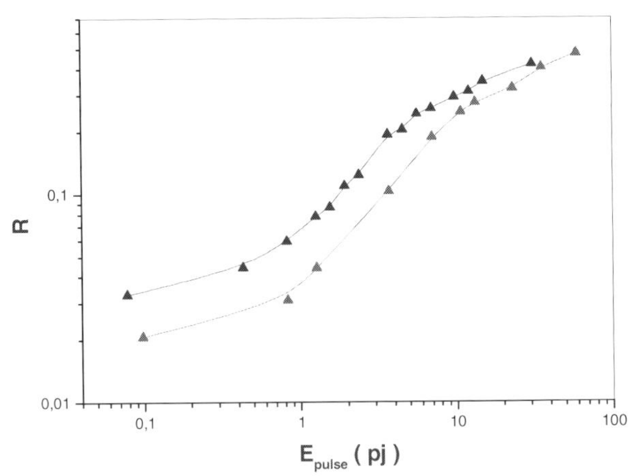

Figure 20. *Comparison of the transfer curves for two devices with different cavity finesse values. Upper is F=9, lower is F=2.*

3.6 Fibre coupling optimisation

Up to this point we have only considered the device behaviour under illumination by a plane wave at normal incidence. In a real situation the device is usually illuminated by a focused beam, with typically a gaussian intensity profile. As the absorption saturation process depends on the light intensity (i.e. power density per unit area), the saturation power should in principle be reduced under tight focusing conditions. Microlensed fibres may be used to couple the incident and reflected signals onto the microcavity device, as depicted on Figure 21.

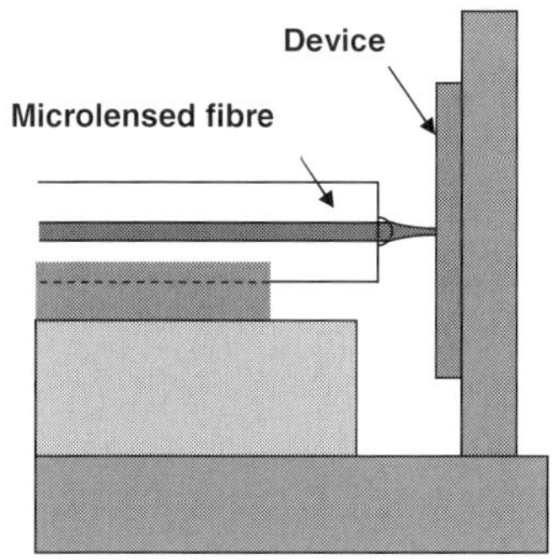

Figure 21. *Schematic view of the device-to-fibre coupling geometry.*

The microlensed fibre can focus the incident signals to small spot sizes, of diameter ranging from 2 to 5μm. However the smaller spot size is not necessarily the better, for the following reason. Under tight focusing, diffraction takes place, and tends to spread out the beam during its propagation back and forth between the cavity mirrors. This well-known phenomenon may limit the effective number of round-trips that contribute to build-up the cavity resonance. One way to assess quantitatively the importance of this effect is to consider the gaussian incident beam as a superposition of plane waves of various incidence angles. Expressing the cavity response to this plane wave superposition allows to calculate the modified resonance lineshape in the diffraction regime.

To this end, we describe here a simple model, which allows one to obtain a closed-form solution to describe the behaviour of a high Q factor resonator illuminated by a gaussian beam, based on the following approach. The spectral response of a Fabry-Perot is given by the well-known Airy function, assuming that the resonator is illuminated by a plane wave. The gaussian beam is described as a superposition of plane waves with different transverse wave vectors. We assume that the input light beam propagates in the z-direction perpendicular to the cavity mirrors, and that the minimum beam diameter $2w_0$ (beam waist) is located on the plane of the top mirror surface. The laser beam intensity I_0 on the beam axis is related to the optical power P by $P = \pi w_0^2 I_0/2$. The intensity versus the beam radius $\rho = \sqrt{x^2 + y^2}$ is given by

$$I(\rho) = I_0 \exp\left[-\frac{2\rho^2}{w_0^2}\right] = \frac{2P}{\pi w_0^2} \exp\left[-\frac{2\rho^2}{w_0^2}\right]. \tag{22}$$

In far field conditions the beam divergence is $\delta = \lambda/\pi w_0$ and the angular intensity distribution $I(\theta)$ is given by

$$I(\theta) = \frac{2P}{\pi\delta^2} \exp\left[-\frac{2\theta^2}{\delta^2}\right]. \tag{23}$$

Due to the Fabry-Perot angular dispersion, the resonance wavelength λ_r depends on the angle θ_i inside the active medium and is expressed as

$$\lambda_r = \lambda_r(0)\cos\theta_i = \lambda_0 \cos\theta_i. \tag{24}$$

For small external angles θ the expression becomes

$$\lambda_r(\theta) = \lambda_0\left[1 - \frac{\theta^2}{2n^2}\right], \tag{25}$$

where n is the average refractive index for the microcavity medium.

For the plane wave case the Fabry-Perot resonance is given by the Airy function. In the high finesse condition the resonance can be approximated by a Lorentzian lineshape of width $\delta\lambda$ (FWHM), given by

$$L_\theta(\lambda) = \frac{1}{1 + \left[\frac{2}{\delta\lambda}(\lambda_r(\theta) - \lambda)\right]^2}. \tag{26}$$

We define a resonance lineshape $L_p(\lambda)$ for the intracavity power, which is proportional to the integral over θ (in the two-dimensional transverse plane) of the resonantly enhanced angular intensity distribution $I(\theta)$

$$L_p(\lambda) = \frac{1}{P}\int_0^\infty I(\theta)\, L_\theta(\lambda)\, 2\pi\theta\, d\theta. \tag{27}$$

This leads to

$$L_p(\lambda) = \frac{2}{\delta^2}\int_0^\infty \frac{2\theta\exp(-2\theta^2/\delta^2)\, d\theta}{1 + \left[\frac{2(\lambda_0 - \lambda)}{\delta\lambda} - \frac{Q\theta^2}{n^2}\right]^2}, \tag{28}$$

where $Q = \lambda_0/\delta\lambda$ is the cavity Q-factor.

We can express $L_p(\lambda)$ in terms of two dimensionless parameters: $x = 2\theta^2/\delta^2$ and $y = 2(\lambda_0 - \lambda)/\delta\lambda$. The power resonance lineshape is now written in a compact form (Oudar *et al.* 1994) as

$$L_p(y) = \int_0^\infty \frac{e^{-x} dx}{1 + (y - px)^2}, \tag{29}$$

with

$$p = \frac{Q}{2}\left(\frac{\lambda}{\pi n w_0}\right)^2. \tag{30}$$

We note that when w_0 becomes infinite, p vanishes, and we get the Fabry-Perot plane wave response at normal incidence.

To calculate the expected lineshape for the reflection spectra, we simply need a minor adjustment. We note that the plane wave intensity reflection coefficient R of a Fabry-Perot cavity can be written as $R = R_{\max}[1 - CL(y)]$, where $L(y)$ is the resonance lineshape, normalised such that $L(0) = 1$, and C is the contrast ratio. At resonance R goes through a minimum value $R_{\min} = R_{\max}(1 - C)$. For the present case of a gaussian illumination beam, we now express the power reflection coefficient spectrum as

$$R_p(y) = R_{\max}\left[1 - CL_p(y)\right]. \tag{31}$$

Figure 22 displays the evolution of the resonance lineshape in dimensionless units, numerically computed from Equation 31, (with $R_{\max} = C = 1$) for several values of the parameter p, which scales as $Q\lambda^2/w_0^2$. At increased values of p we observe three different effects: an asymmetric lineshape, a blue shift of the resonance minimum, and a reduction of the resonance amplitude. This reduction can be interpreted in terms of a diffraction loss. For oblique plane wave components light tends to escape from the beam axis during the multiple reflections within the cavity. This limits the effective number of round-trips involved in the resonance and reduces its amplitude.

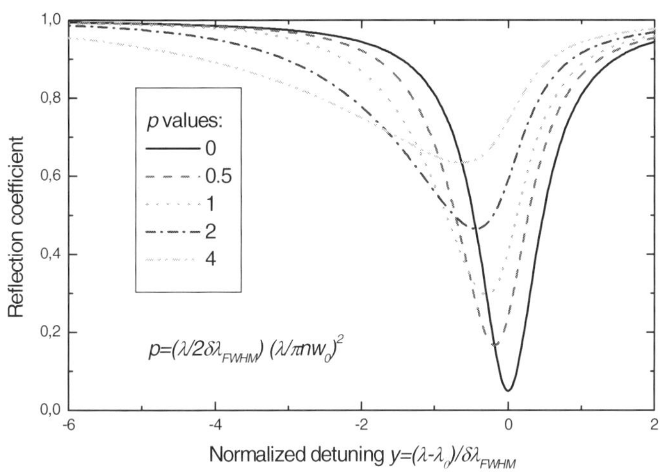

Figure 22. *Evolution of the resonance lineshape for various values of the reduced parameter p defined by Equation 30.*

These effects have important consequences on the behaviour of nonlinear microcavity devices. The reduction of the resonance amplitude at small spot sizes means that higher incident intensities are needed, so that there is no advantage, in terms of beam power, in decreasing the spot size below a critical value. In addition, and even more importantly, the minimum reflected power at resonance increases for small spot sizes, so that the switching contrast is degraded.

The above calculation clearly shows the importance of the reduced parameter p to evaluate the effect of tight focusing on the resonance lineshape. However this analytical result was obtained in the high finesse approximation, while the devices considered here are typically low finesse cavities. Essentially, the high finesse approximation was made by introducing a simplified Lorentzian lineshape for the cavity resonance (Equation 26). Departure from this simplified lineshape comes not only from the usual approximation $\sin^2\Phi \approx \Phi^2$ in the Fabry-Perot equation (see Equation 3 in Section 3.1), but also from the variation of the MQW absorption with wavelength, which can be quite significant over the broad cavity resonance. As a result it can be useful to check the predictions of the analytical model by numerical calculations of specific cases of interest. This can be performed through a generalisation of the above described method. It consists in computing the power reflection coefficient of the device, subject to a gaussian illumination beam of half-angle divergence δ through the expression

$$R(\lambda) = \frac{\displaystyle\int_0^{\pi/2} \theta\, R_\lambda(\theta)\, \exp\left[-2\theta^2/\delta^2\right]\, d\theta}{\displaystyle\int_0^{\pi/2} \theta\, \exp\left[-2\theta^2/\delta^2\right]\, d\theta} \tag{32}$$

where $R_\lambda(\theta)$ represents the variation with incident angle θ of the intensity reflection coefficient R (in the plane-wave regime) at a given wavelength λ. The quantity $R_\lambda(\theta)$ can be computed exactly for a given microcavity structure with the matrix formalism introduced in Section 3.3, and for each wavelength λ the precise absorption value $\alpha(\lambda)$ can be used. An example of the spectral response computed from this equation is displayed in Figure 23, for the microcavity structure S9, for various values of the beam radius w_0.

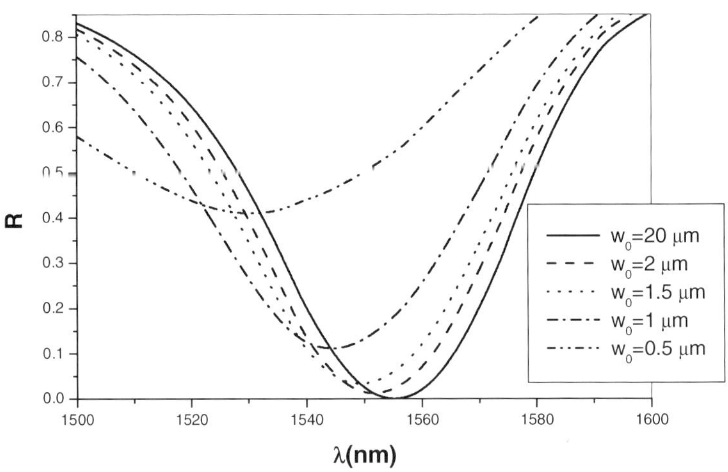

Figure 23. *Spectral response of microcavity S9, for various beam radius w_0 values.*

It can be seen from Figure 23 that the reflection coefficient at resonance remains smaller than 1% for $w_0 = 2\mu$m, while it reaches ~3% for $w_0 = 1.5\mu$m, and 11% for $w_0 = 1\mu$m. From this it can be concluded that a beam radius $w_0 = 2\mu$m represents a good compromise, allowing one to obtain a high intracavity intensity through focusing, and to keep the minimum reflection coefficient at resonance below 1%. We note that, for this precise cavity structure ($Q = 52$, $n = 3.3$), the reduced parameter for $w_0 = 2\mu$m is $p = 0.15$, consistent with the results displayed on Figure 22.

4 System application of ultrafast microcavity saturable absorber devices

4.1 Basic features of signal propagation in long-haul optical transmission systems

Optical communication systems, which have been made technically and economically viable with the development of low loss optical fibres and of diode lasers at appropriate wavelengths have an enormous information capacity due to the high frequency of the carrier wave used to carry the information. At 1.5μm wavelength, the carrier frequency is 200THz, so a 1Tbit/s communication channel can in principle occupy a bandwidth that is of the order of only 1% of the carrier frequency. It is this enormous potential bandwidth that is the driving force behind the world-wide development of lightwave systems. An introduction to the basic concepts used in optical communication systems and an in-depth account of the physical basis needed to understand this complex technology can be found in the textbook of Agrawal (1992).

Historically, the development of optical transmission systems in the telecommunication area has been driven by the need to transmit high bandwidth signals over long distances, either for the terrestrial "backbone "network, or for long-haul point-to-point submarine links. In this section we shall briefly introduce some basic features of signal propagation in long-haul optical transmission systems, and introduce some current technical terms used to describe their characteristics.

For long distance links the communication systems consist of several sections connected to each other by *optoelectronic repeaters*, where the optical signal is first converted into an electrical signal, then regenerated in electronic circuits, and converted back to an optical signal for propagation in the next fibre section. The maximum propagation distance between repeaters is limited by two factors: the fibre loss, which is of the order of 0.3dB/km at 1.3μm, and 0.15dB/km at 1.55μm for state-of-the-art present-day optical fibres, and by the phenomena of chromatic dispersion, which results in a different propagation delay for the various spectral components of the high bandwidth optical signals.

Chromatic dispersion

The concept of chromatic dispersion can be introduced by considering the expansion of the optical guided-wave propagation constant β as a power series of the electric field frequency ω around a particular reference frequency ω_0.

$$\beta(\omega) = \frac{\omega}{c} n(\omega) = \beta_0 + \beta_1 (\omega - \omega_0) + \frac{1}{2}\beta_2 (\omega - \omega_0)^2 + \frac{1}{6}\beta_3 (\omega - \omega_0)^3 + \dots \qquad (33)$$

where $n(\omega)$ is an effective refractive index for the propagating mode. The various coefficients β_i in this expansion have the following physical interpretation. The guided-wave phase velocity is simply given by $v_\phi = \omega_0/\beta_0$, while the group velocity of a wave packet propagating in the fibre is given by $v_g = 1/\beta_1$. As a result of the higher order terms, the different spectral components of a propagating short optical pulse travel at slightly different group velocities and become desynchronised at the fibre output, giving rise to a pulse broadening. The coefficient $\beta_2 = d^2\beta/d\omega^2$ is known as the group velocity dispersion parameter. In optical communication systems, it is customary to express the spectral width of optical signals in wavelength units (using the relation $\omega = 2\pi c/\lambda$), so that it has been found practical to describe the chromatic dispersion of optical fibres by the dispersion parameter D, defined by

$$D = \frac{d}{d\lambda}\left(\frac{1}{v_g}\right) = -\frac{2\pi c}{\lambda^2}\beta_2 \,. \tag{34}$$

We note that $1/v_g$ may be described as the group delay per unit fibre length, so that when expressing $1/v_g$ in ps/km and λ in nm, the dispersion parameter D is expressed in ps/(nm km).

The first and second generation commercial systems were designed to operate at 0.85μm and 1.3μm wavelength, respectively. The advantage of the 1.3μm wavelength is that it combines a low propagation loss, with a cancellation of the dispersion parameter, as shown in Figure 24 for the standard monomode fibres (SMF). However the chromatic dispersion of fibres can be adjusted to some extent, by a careful design of the guiding fibre core index profile. For instance it has been possible to shift the zero dispersion wavelength to the value of 1.55μm, in the so-called dispersion-shifted fibres (DSF). This has allowed a systematic trend towards using the 1.55μm range for long distance transmission systems, giving rise to the third generation optical communication systems, which became commercially available in 1990.

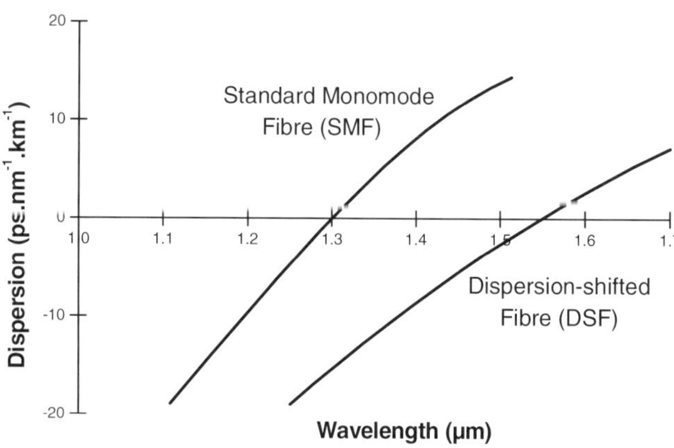

Figure 24. *Variation of the dispersion parameter as a function of wavelength for standard monomode fibres and dispersion-shifted fibres.*

Multichannel transmission

In order to make use of the intrinsically large bandwidth of optical fibres and overcome the speed limitations of electronics, the most recent fibre-optic communication systems transmit multiple channels over the same fibre. There are two basic ways to implement such a multichannel transmission, time-division multiplexing (TDM), and wavelength-division multiplexing (WDM). TDM is commonly performed in the electrical domain to obtain digital hierarchies in telecommunication systems. To overcome the speed limitation of electronic and optoelectronic devices, one can use an optical TDM (OTDM) scheme, in which several optical signals (in number N) modulated at the bit rate B are multiplexed optically to form a composite optical signal at the bit rate NB, using the same optical carrier frequency. Although attractive, this scheme is limited in practice by fibre dispersion.

The other scheme of multichannel transmission, WDM, is now widely used to increase the capacity of fibre links. In WDM, multiple optical carriers with different wavelengths are modulated by using independent electric bit streams, and then combined and transmitted over the same fibre. After transmission, the received optical signal is demultiplexed into separate channels, usually by optical techniques. With WDM, multi-Tbit/s transmission is possible in a single fibre, and several such systems have been demonstrated (Vareille *et al.* 2001), thus making the best use of the optical fibre bandwidth.

Optical amplifiers

As already mentioned, in long-haul transmission systems, fibre loss limitations have to be overcome by periodic signal regeneration. In multichannel systems, signal regeneration by optoelectronic repeaters becomes quite complex and expensive, especially when a hundred or more channels are multiplexed together, as separate optoelectronic repeaters must be used for each channel. By contrast, optical amplification leads to much simpler solutions, as a single device can amplify directly the multichannel optical signal. Several kinds of optical amplifiers can be used, such as the semiconductor laser amplifiers, fibre Raman and Brillouin amplifiers, and rare-earth-doped fibre amplifiers. In this last category, the Erbium-Doped Fibre Amplifier (EDFA) has excellent amplification properties, that make it an ideal component for 1.55μm systems (Desurvire 1994).

In long-haul transmission systems using optical amplifiers, the basic phenomena that limit the capacity-distance product are essentially the signal to noise degradation due to the amplifier noise, the fibre dispersion, and various nonlinear optical phenomena occurring during the signal propagation in the fibre. With optical amplifiers the main source of noise comes from the spontaneous emission that is always present along with the stimulated emission responsible for the amplifier gain. As the spurious light produced by spontaneous emission gets amplified in the (usually high gain) optical amplifiers, this optical noise, named amplified spontaneous emission (ASE), usually becomes larger than the intrinsic photodetector noise. As a result the noise properties of the system are determined essentially by the optical noise. The optical signal to noise ratio (OSNR) is degraded by at least a factor of 2 (3dB) for each amplifier stage, in the case of an ideal amplifier with complete population inversion. For most practical amplifiers the OSNR degradation exceeds 3dB and is typically 6–8dB.

Quality of transmission

Let us now introduce the basic concepts used to characterise the quality of transmission. There are basically two choices for the modulation format of digital optical signals, known as the return-to-zero (RZ) format and the non-return-to-zero (NRZ) format. In the RZ format the optical pulse representing bit "1" is shorter than the bit slot, so that its amplitude returns to zero before the end of the bit duration. By contrast, in the NRZ format, the optical pulse duration is equal to that of the bit slot (the bit slot is equal to clock period). The NRZ format occupies less bandwidth, but is less robust with respect to pulse broadening due to chromatic dispersion. The RZ format is more widely used in high performance systems such as long-haul transmission systems.

The quality of transmission is characterised by the bit error rate (BER). Although this term may suggest a number of errors per second, it is customary to define the BER as the average probability of making an error in the bit identification at the receiver end. Signal transmission with a BER of less than 10^{-9} is typically considered as an error-free transmission. The recent development of high speed error correcting circuits, such as those implementing forward error-correcting codes (FEC) allows to obtain error-free transmission with a BER of typically 10^{-4} to 10^{-3} before the error-correcting scheme. The BER is measured by sending a repetitive pseudo-random bit sequence (PRBS) at the transmitter end, and counting the errors at the receiver end. The first step to qualify at the laboratory level the performance of a long-haul transmission system is to implement a circulating loop configuration (Bergano and Davidson, 1995), and examples of such experiments will be given in the next section.

Another way to estimate the quality of transmission is to observe the so-called "eye-diagram" of the received signal. This diagram is formed by the superposition of the PRBS symbols on the screen of a digital sampling oscilloscope for a given duration. The statistical distributions of amplitudes for the "1" and the "0" symbols give an indication of the quality of transmission. It is in the intersection zone of these distributions that errors are produced. It is possible to characterise the signal degradation linked to the signal-to-noise ratio via the amplitude quality factor Q_a. Let i_1 and i_0 be the received photocurrents corresponding to the 1 and 0 symbols respectively, and σ_1^2 and σ_0^2 their variance. The Q_a factor is given by

$$Q_a = \frac{\langle i_1 \rangle - \langle i_0 \rangle}{\sigma_1 + \sigma_0}. \tag{35}$$

Under the assumption of gaussian statistics for the photocurrents, it is possible to relate the BER to the Q_a factor through the relation

$$BER = \frac{1}{\sqrt{2\pi}} \int_{Q_a}^{\infty} \exp(-x^2)dx = \frac{1}{2}\,\mathrm{erfc}\left(\frac{Q_a}{\sqrt{2}}\right) \approx \frac{1}{Q_a\sqrt{2\pi}} \exp\left(-\frac{Q_a^2}{2}\right), \tag{36}$$

where the approximate form of BER is obtained by using the asymptotic expansion of $\mathrm{erfc}(Q_a/\sqrt{2})$ and is reasonably accurate for $Q_a > 3$. For a BER of 10^{-9} the corresponding value for Q_a is 6. We note that for certain results the Q_a factor is quoted in dB. In this case Q_a (in dB) is related to the linear Q_a factor by the relation

$$Q_a(dB) = 20\log_{10}(Q_a)\,. \tag{37}$$

A linear Q_a factor of 6 corresponds to a Q_a factor of 15.5 dB.

Nonlinear transmission

A great majority of nonlinear effects in the optical fibre comes from the optical Kerr effect. In the simplest approach, it can be expressed as a variation of the optical fibre refractive index $n(\omega, P)$ with the optical power P of the propagating guided wave signal, that is

$$n(\omega, P) = n_0(\omega) + n_2 \frac{P}{A_{\text{eff}}} \tag{38}$$

where n_2 is the nonlinear refractive index ($\sim 2.6 \times 10^{-20} \text{m}^2\text{W}^{-1}$ for silica) and A_{eff} is the guided mode effective area ($\sim 80 \mu\text{m}^2$ for a typical single mode fibre).

The Kerr effect is responsible for a number of nonlinear phenomena in the fibre (Agrawal 1992) such as self-phase modulation (SPM), cross-phase modulation (XPM), and four-wave mixing (FWM). Self-phase modulation concerns the propagation of single short pulses in the fibre and leads to time-dependent nonlinear phase shifts in the complex pulse amplitude. Under certain conditions the linear phase shifts due to chromatic dispersion may be exactly compensated by the nonlinear phase shifts due to SPM, and lead to soliton pulses that propagate without deformation (Hasegawa 1989). The two other effects (XPM and FWM) need to be considered in multichannel WDM systems. Basically XPM refers to the time-dependent nonlinear phase shifts induced on a given pulse by any other pulse overlapping it, such as pulses from the other channels. Four-wave mixing comes from the interaction of two signal waves at frequencies ω_1 and ω_2, and leads to the generation of new spectral components at frequencies $2\omega_1 - \omega_2$ and $2\omega_2 - \omega_1$. For obvious reasons, XPM and FWM constitute a major limitation to the performance of WDM systems, and in recent years a new concept has emerged, *dispersion management*, which provides a powerful way to overcome many of these limitations.

Dispersion management

As introduced earlier in this section, the use of dispersion-shifted fibres provides a practical way to control the chromatic dispersion of a transmission line. It is also possible to combine several fibres with different chromatic dispersion properties in different portions of the transmission line, so as to obtain a desired cumulated dispersion at the end of the line. Thus it has been proposed (Lin and Kogelnik, 1980) to use a highly dispersive fibre associated with a dispersion-compensating fibre (DCF), in order to reduce the interaction of the different WDM channels (by increasing the group velocity mismatch between them), and limit the intensity of XPM and FWM interactions.

On the other hand, the use of soliton propagation as such has not led to practical applications in commercial long-haul transmission lines, due to several limitations, such as the Gordon-Haus jitter (Gordon and Haus, 1986), the interaction between adjacent solitons, or the generation of a continuum wave due to the unavoidable signal attenuation along the fibre (Georges and Charbonnier, 1996). In order to overcome these limitations, it has been proposed to include a dispersion compensation scheme into soliton propagation lines, and very quickly, a distinct improvement of the propagation quality was observed (Knox *et al.* 1995, Nakasawa and Kubota 1995, Suzuki *et al.* 1995). This has led to the development of the dispersion-managed (DM) nonlinear transmission. The structure of a DM transmission line basically consists of alternate portions of fibres, with positive and negative dispersion parameters, so as to obtain a relatively weak *average* dispersion, while keeping a strong *local* dispersion. With this scheme one obtains simultaneously a good resistance to XPM and FWM (strong local dispersion) and a limited jitter (weak average dispersion). It has been shown that this scheme allows for a stable soliton-like propagation (Nijhof *et al.* 1997), sometimes called dispersion-managed soliton. Such DM solitons have some specific and advantageous properties, such as an enhanced power (Smith *et al.* 1996a), a reduction of the Gordon-Haus jitter (Smith *et al.* 1996b), and a reduction of the impact of soliton collisions (Sugahara *et al.* 1997, Georges 1998).

4.2 All-optical regeneration with ultrafast microcavity saturable absorber devices

In the following, we review recent systems experiments demonstrating the use of passive 2R regenerators based on the MQW micro-cavity saturable absorber devices described in Section 3, operating in association with both optical filter and dispersion- managed nonlinear propagation. In a first series of experiments, the saturable absorber device exhibited an average increase of the on-off contrast of 2.5dB over the entire C-band and a recovery time (1/e) of 10ps. According to numerical predictions (Audouin *et al.* 1998) a device with such characteristics can be useful for passive 2R regeneration at bit rates up to 20Gbit/s. It was tested in single-channel 10 and 20Gbit/s transmission loop experiments (Leclerc *et al.* 2000).

Figure 25 shows the experimental set-up of the 20Gbit/s circulating loop experiment, as described in Leclerc *et al.* 2000. Only when operating with the saturable absorber device, an additional Erbium-Doped Fibre Amplifier (EDFA) is placed in the loop before the device to provide 7.5dBm incident power on the saturable absorber device and also to compensate for its 7dB insertion loss.

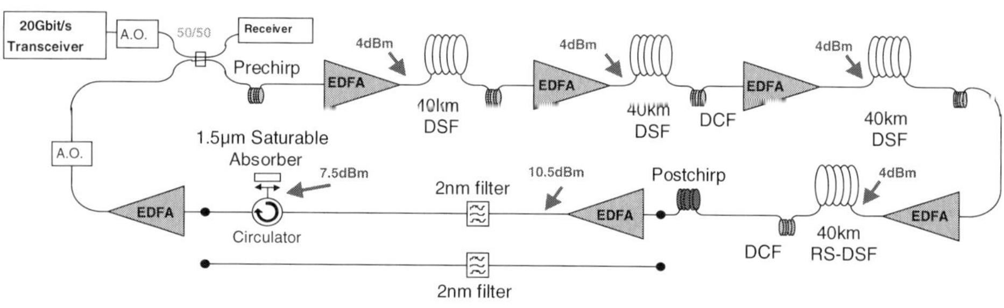

Figure 25. *Loop configuration saturable absorber device and filter configuration in 20Gbit/s experiment. A.O. stands for acousto-optic modulator, EDFA for erbium doped fibre amplifier, RS DSF for red-shift dispersion-shifted fibre, and DCF for dispersion-compensating fibre.*

Figure 26 shows the experimental BER versus propagation distance for two different ASE noise levels in loop. Without saturable absorber device and associated EDFA, the maximum error-free distance (BER $< 10^{-9}$) is measured to be 6,200km at 20Gbit/s line-rate. When inserting the saturable absorber device with the loss-compensating EDFA, a clear improvement in the system performance is observed, yielding an increased error-free distance of 7,800km. In this case, the main system limitation is observed to stem from timing jitter accumulation rather than amplitude fluctuations. In order to further investigate saturable absorber device discrimination efficiency, the ASE noise was purposefully increased by introducing 3dB attenuation before two line EDFA. As expected, the system

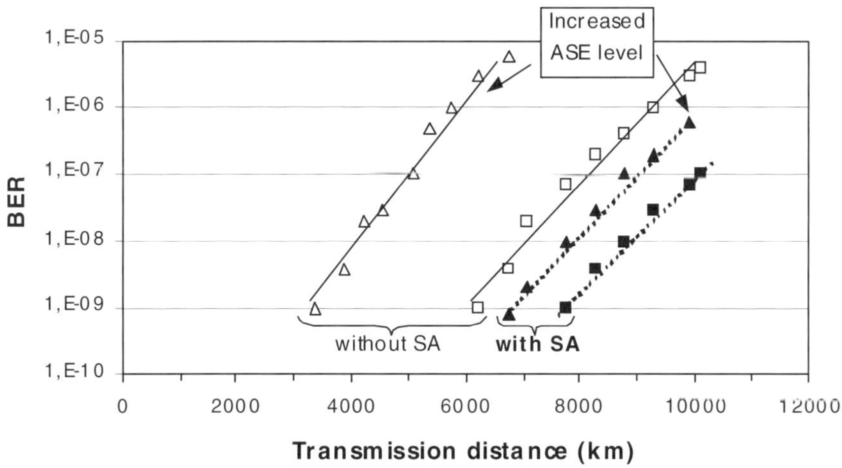

Figure 26. *Experimental bit error rates (BER) versus propagation distance with saturable absorber (SA) device (full symbols) and without it (open symbols) for two different amplified spontaneous emission (ASE) levels in the 20Gbit/s loop.*

performance is significantly degraded and the error-free distance is reduced to 3,400km without saturable absorber device. When implementing the saturable absorber device (with additional EDFA), the error-free distance increases to 6,800km, leading to more than a twofold improvement in transmission distance. Comparison of this result with the previous one obtained without any degradation of the line EDFA noise figure (error-free distance of 7,800km instead of 6,200km at 20Gbit/s) further illustrates the saturable absorber device capability to reduce ASE at 20Gbit/s line-rate. It also demonstrates its regenerative capability in OSNR-limited transmission, when associated with nonlinear transmission and optical filtering.

With a view to extending the properties of passive 2R regeneration schemes based on saturable absorber devices to 40Gbit/s dispersion-managed transmissions, a recent work proposed and numerically demonstrated an improvement of such a regenerator by the insertion of a compression fibre between the saturable absorber device and the optical bandpass filter, as illustrated in Figure 27 (Matsumoto and Leclerc 2002).

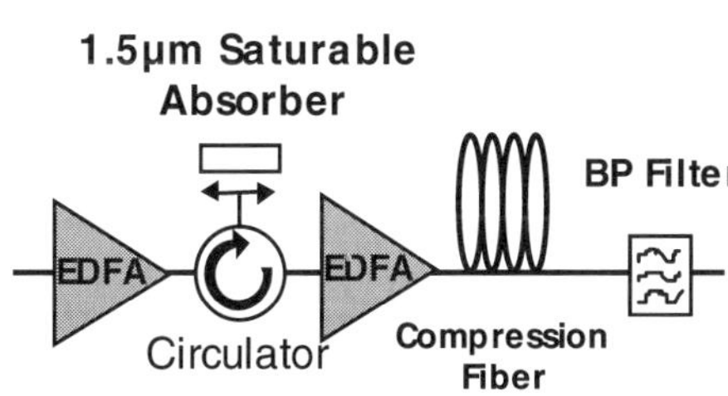

Figure 27. *40Gbit/s optical 2R regenerator based on a saturable absorber device, followed by a nonlinear compression stage made of an optical amplifier (EDFA), a compression fibre and a band-pass (BP) filter.*

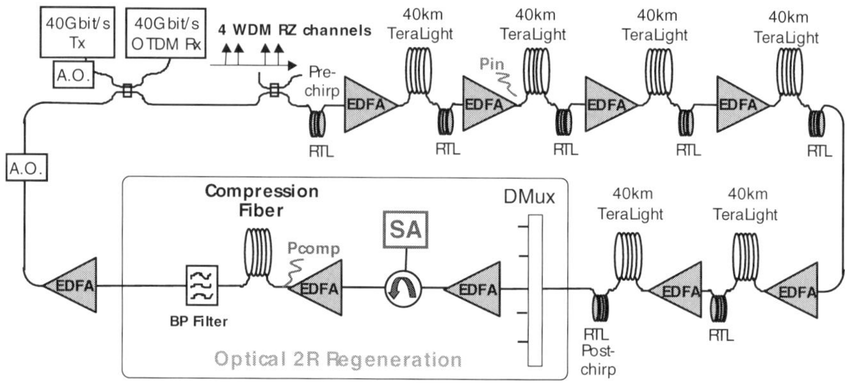

Figure 28. *Experimental loop set-up for 40Gbit/s long-haul transmissions. SA stands for saturable absorber device and RTL for Reverse TeraLight (TM). Other acronyms are described in the legend of Figure 25.*

Such a configuration for the saturable absorber device-based 2R regenerator was experimentally studied and demonstrated in a 40Gbit/s WDM-emulated loop transmission environment (Rouvillain *et al.* 2002). As shown in Figure 28, the optical 2R regenerator is inserted after a demultiplexer in a 240km long dispersion-managed circulating loop.

Figure 29 shows the experimental Q-factor versus propagation distance. Without the 2R regenerator scheme and associated EDFAs, Q-factors of 13dB (BER $< 5\times10^{-6}$) and 11.3dB (BER $< 10^{-4}$) are measured after 1300km and 1650km transmission, respectively. When inserting the optical 2R regenerator, a clear improvement in the system performance is observed, yielding a Q-factor of 13dB at 7600km distance. It should be noted that the main system limitation at such distances stems from timing jitter accumulation and not

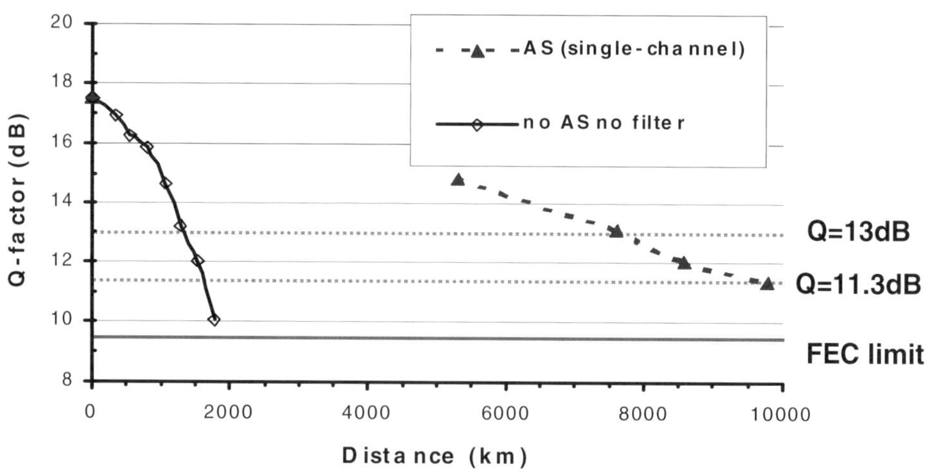

Figure 29. *Experimental Q-factor evolution with transmission distance. AS stands for saturable absorber device, FEC for forward error-correcting code.*

amplitude fluctuations, since passive regeneration by a saturable absorber device controls signal-to-noise ratios but not the time position of the pulses. In order to investigate the regenerator capabilities in a WDM environment, four 200GHz-spaced additional channels were launched into the loop on both spectral sides of the regenerated channel (see Figure 28) so as to emulate a realistic WDM transmission. Indeed, these extra-channels play a disturbing role during the 240km propagation but are extracted from the loop by the de-multiplexing stage in front of the saturable absorber device. When implementing the saturable-absorber-based 2R regenerator, performance of the regenerated centred channel reaches Q=13dB at 2200km distance and 11.3dB after 7600km propagation. Comparison of this result with the previous one obtained for a single channel illustrates the saturable absorber device capability to improve the system performance in a realistic WDM environment and demonstrates its regenerative effectiveness on 40Gbit/s RZ signals. One notes that such system performance, combined with FEC capabilities, can yield an error-free system operation at 40Gbit/s on distances over 7600km.

5 Conclusion

In this chapter, we have reviewed recent advances in the development of a promising type of all-optical processing devices based on ultrafast saturable absorber microcavities. The emphasis has been put on the achievement of useful device parameters for a practical optical switch, in terms of speed, operating power, switching contrast and insertion loss.

From the materials point of view, it has been shown that excitonic absorption saturation in ion-irradiated semiconductor quantum wells provides an adequate basis for the realisation of efficient all-optical switches, particularly when operation at high repetition rates is required. This is due to the low saturation intensity of room-temperature excitons in quantum wells, and to the fact that ion irradiation allows one to reduce the carrier lifetime in a controllable way, through the post-growth introduction of trapping and recombination centres in the quantum well material. The device design was oriented towards a specific goal, the regeneration of optical telecommunication signals. Vertical access microcavity devices were shown to be advantageous in many respects, allowing one to achieve a high switching contrast with a moderate switching power, an intrinsically low polarisation sensitivity and an easy coupling to optical fibres. Finally the ability of our saturable absorber devices to enhance the performances of long distance high bit rate transmission links was demonstrated. Such system demonstrations confirm the regenerative properties of optical regenerators based on semiconductor saturable absorber devices. It is hoped that these demonstrations will be extended in the near future towards multichannel WDM operation, paving the way to the development of passive, low-cost optical 2R regenerator modules, with terabit/s processing capabilities.

There is still room for significant improvement of the device characteristics, through the reduction of saturation energy and insertion losses, along with an increase in the dynamic contrast. Further research in this area should provide valuable information on possible ways to obtain such improved device performances.

Acknowledgments

This chapter includes work carried out by a number of collaborating researchers, as well as former and current students, and has involved collaborations with several other laboratories. We especially thank Guy Aubin, Harry Bernas, Patrick Brindel, Houtaï Choumane, Jean Decobert, Fabrice Devaux, Jean-Christophe Harmand, Olivier Leclerc, Jean-Michel Lourtioz, Emmanuel Lugagne Delpon, Juliette Mangeney, Cristelle Mériadec, Constantin Nelep, Gilles Patriarche, Delphine Rouvillain, Frédéric Seguineau, Alexandre Shen and Nikolaï Stelmakh. We gratefully acknowledge the support of the french National Network on Telecommunication Research (RNRT), through the ASTRE and ASTERIX projects, and the support of Alcatel Research & Innovation, the Région Ile de France, SESAME project n°1377, and the Conseil Général de l'Essonne.

References

Abram I, Robert I, and Kuszelewicz R, 1998, "Spontaneous emission control in semiconductor microcavities with metallic or Bragg mirrors", *IEEE Journal of Quantum Electronics* **34** 71.

Agrawal Govind P, 1992, *Fiber-optic communication systems* (John Wiley & Sons, New York).

Audoin O, Pallise E, Desurvire E, and Maunand E, 1998, "Use of in-line saturable absorbers in WDM soliton systems", *IEEE Photonics Technology Letters* **10** 828.

Avrutin E, and Portnoi M, 1989, "Estimates of the lifetimes of nonequilibrium carriers in a semiconductors irradiated with heavy ions", *Soviet Physics Semiconductors* **22** 968.

Babic D I and Corzine S W, 1992, "Analytic expressions for the reflection delay, penetration depth, and absorptance of quarter-wave dielectric mirrors", *IEEE Journal of Quantum Electronics* **28** 514.

Benjamin S D, Loka H S, Othonos A, and Smith P W E, 1996, "Ultrafast dynamics of nonlinear absorption in low-temperature-grown GaAs", *Applied Physics Letters* **68** 2544.

Bergano N S and Davidson C R, 1995, "Circulating loop transmission experiments for the study of long-haul transmission systems using Erbium-doped fiber amplifiers", *IEEE Journal of Lightwave Technology*, **13**, 879.

Bernstein L, 1966, "Semiconductor joining by the solid-liquid interdiffusion (SLID) process", *Journal of the Electrochemical Society* **12** 1282.

Born M and Wolf E, 1986, in *Principles of Optics*, 6th edition p. 69 (Pergamon, Oxford).

Chemla D S and Miller D A B, 1985, "Room-temperature excitonic nonlinear-optical effects in semiconductor quantum-well structures", *Journal of the Optical Society of America B* **2** 1155.

Desurvire E, 1994, *Erbium-Doped Fiber Amplifiers* John Wiley & Sons, New York.

Garmire E, 1989, "Criteria for optical bistability in a lossy saturating Fabry-Perot", *IEEE Journal of Quantum Electronics* **QE-25** 289.

Georges T and Charbonnier B, 1996, "Continuum generated by chromatic dispersion and power variations in periodically amplified soliton links", *Optics Letters*, **21** 1232.

Georges T, 1998, "Soliton interaction in dispersion-managed links", *Journal of the Optical Society of America B* **15** 1553.

Gibbs H M, Gossard A C, McCall S L, Passner A, Wiegmann W, and Venkatesan T N C, 1979, "Saturation of the free exciton resonance in GaAs", *Solid State Communications* **30** 271.

Gordon J P and Haus H A, 1986, "Random walk of coherently amplified solitons in optical fiber transmission", *Optics Letters* **11** 665.

Hasegawa A, 1989, *Solitons in Optical Fibers* (Springer-Verlag, Berlin)

Hirano A, Kobayashi H, Tsuda H, Takahashi R, Asobe M, Sato K, and Hajimoto K, 1998, "10 Gbit/s RZ discrimination using refined saturable absorber optical gate", *Electronics Letters*, **34** 198.

Højfeldt S, Bischoff S, and Mørk J, 2000, "All-optical wavelength conversion and signal regeneration using electroabsorption modulator", *IEEE Journal of Lightwave Technology* **18** 1121.

Jewell J L, Lee Y H, McCall S L, Harbison J P, and Florez L T, 1988, "High-finesse (Al,Ga)As interference filters grown by molecular beam epitaxy", *Applied Physics Letters* **53** 640.

Knox F M, Forisyak W, and Doran N J, 1995, "10 Gbit/s soliton communication systems over standard fiber at 1.55 μm and the use of dispersion compensation", *IEEE Journal of Lightwave Technology* **13** 1955.

Künzel H, Böttcher J, Gibis R, and Urmann G, 1992, "Materials properties of GaInAs grown on InP by low-temperature molecular beam epitaxy", *Applied Physics Letters* **61**, 1347.

Leclerc O, Aubin G, Brindel P, Mangeney J, Choumane H, Barré S, and Oudar J-L, 2000, "Demonstration of high robustness to SNR impairment in 20 Gbit/s long-haul transmission using 1.5 μm saturable absorber", *Electronics Letters*, **36** 1944.

Lee C C, Wang C Y, and Matijasevic G, 1993, "Au-In bonding below the eutectic temperature", *IEEE Transaction on Components, Hybrids and Manufacturing Technology* **16** 311.

Lin C and Kogelnik H, 1980, "Optical pulse equalization of low-dispersion transmission in single-mode fibers in the 1.3-1.7 μm range", *Optics Letters* **5** 476.

Lugagne Delpon E, Oudar J-L, Bouché N, Raj R, Shen A, Stelmakh N, and Lourtioz J-M, 1998, "Ultrafast excitonic saturable absorption in ion-implanted InGaAs/InAlAs multiple quantum wells", *Applied Physics Letters* **72** 759.

MacLeod H A, 1986, *Thin film optical filters* (Hilger, Bristol)

Mangeney J, Oudar J-L, Harmand J-C, Mériadec C, Patriarche G, Aubin G, Stelmakh N, and Lourtioz, J-M, 2000a, "Ultrafast saturable absorption in heavy-ion irradiated quantum well vertical cavity", *Applied Physics Letters* **76** 1371.

Mangeney J, Lopez J, Stelmakh N, Lourtioz J-M, Oudar J-L, and Bernas H, 2000b, "Subgap optical absorption and recombination center efficiency in bulk GaAs irradiated by light or heavy ions", *Applied Physics Letters* **76** 40.

Mangeney J, Aubin G, Oudar J L, Harmand J C, Patriarche G, Choumane H, Stelmakh N, and Lourtioz J M, 2000c, "All-optical discrimination at 1.5 μm using ultrafast saturable absorber vertical cavity device", *Electronics Letters* **36** 1486.

Mangeney J, Choumane H, Patriarche G, Leroux G, Aubin G, Harmand J-C, and Oudar J-L, 2001, "Comparison of light- and heavy-ion-irradiated quantum-wells for use as ultrafast saturable absorbers", *Applied Physics Letters* **79** 2722.

Marceaux A, Loualiche S, Dehaese O, and Lambert B, 2001, "High-speed 1.55 μm Fe-doped multiple-quantum-well saturable absorber", *Applied Physics Letters* **78** 4065.

Matsumoto M, and Leclerc O, 2002, "Analysis of a 2R optical regenerator utilising self-phase modulation in a highly nonlinear fibre", *Electronics Letters*, **38** 576.

Miller D A B, Chemla D S, Eilenberger D J, Smith P W, Gossard A C, and Wiegmann W, 1982, "Large room-temperature optical nonlinearity in GaAs/GaAlAs multiple quantum well structures", *Applied Physics Letters* **41** 679.

Nakasawa M and Kubota H, 1995, "Optical soliton communication in a positively and negatively dispersion-allocated optical fibre transmission line", *Electronics Letters* **31** 216.

Nijhof J H B, Doran N J, Forisyak W, and Knox F M, 1997, "Stable soliton-like propagation in dispersion managed systems with net anomalous, zero and normal dispersion", *Electronics Letters*, **33** 1726.

Oudar J-L, Kuszelewicz K, Sfez B, Michel J-C, and Planel R, 1992, "Prospects for further threshold reduction in bistable microresonators", *Optical and Quantum Electronics* **24** S193.

Oudar J-L, Rivera T, Kuszelewicz R, and Ladan F, 1994, "Etched arrays of quantum well optical bistable microresonators", *Journal of Physics III, France* **4** 2361.

Rashba E I and Sturge M D eds., 1982, *Excitons*, vol. 2 of Modern Problems in Condensed Matter Science (North-Holland, Amsterdam).

Rouvillain D, Brindel P, Seguineau F, Pierre L, Leclerc O, Choumane H, Aubin G, and Oudar J-L, 2002, "Optical 2R regenerator based on passive saturable absorber for 40 Gbit/s WDM long-haul transmissions", *Electronics Letters*, **38** 1113.

Schmitt-Rink S, Chemla D S, and Miller D A B, 1985, "Theory of transient excitonic optical nonlinearities in semiconductor quantum-well structures" *Physical Review B* **32**, 6601.

Silberberg Y, Smith P W, Miller D A B, Tell B, Gossard A C, and Wiegmann W, 1985, "Fast nonlinear optical response from proton-bombarded multiple quantum well structures", *Applied Physics Letters* **46** 701.

Sosnowski T S, Norris T B, Wang H H, Grenier P, Whitaker J F, and Sung C Y, 1997, "High-carrier-density electron dynamics in low-temperature-grown GaAs", *Applied Physics Letters* **70** 3245.

Smith N J, Knox F M, Doran N J, Blow K J, and Bennion I, 1996a, "Enhanced power solitons in optical fibres with periodic dispersion management", *Electronics Letters*, **32** 54.

Smith N J, Forisyak W, and Doran N J, 1996b, "Reduced Gordon-Haus jitter due to enhanced power solitons in dispersion-managed systems", *Electronics Letters*, **32** 2085.

Smith P W, Benjamin S D, and Loka H S, 1997, "Tailoring of trap-related carrier dynamics in low-temperature-grown GaAs", *Applied Physics Letters* **71** 1156.

Stelmakh N, Mangeney J, Alexandrou A, and Portnoi E L, 1998, "Intensity-invariant subpicosecond absorption saturation in heavy-ion irradiated bulk GaAs", *Applied Physics Letters* **73** 3715.

Sugahara H, Kato H, and Kodama Y, 1997, "Maximum reductions of collision induced frequency shift in soliton WDM systems with dispersion compensation", *Electronics Letters*, **33** 1065.

Suzuki N, Morita I, Edagawa N, Yamamoto S, Taga H, and Akiba S, 1995, "Reduction of Gordon-Haus timing jitter by periodic dispersion compensation in soliton transmission", *Electronics Letters*, **31** 2027.

Takahashi R, Kawamura Y, Kagawa T, and Iwamura H, 1994, "Ultrafast 1.55-μm photoresponses in low-temperature-grown InGaAs/InAlAs quantum wells", *Applied Physics Letters* **65** 1790.

Takahashi R, 2001, "Low-temperature-grown surface-reflection all-optical switch", *Optical and Quantum Electronics* **33** 999.

Vareille G, Julien B, Pitel F, and Marceron J-F, 2001, "3.65 Tbit/s (365x11.6 Gbit/s) transmission experiment over 6850 km using 22.2 GHz channel spacing in NRZ format", *ECOC'01*, paper PD.M.1.7, Amsterdam, Netherlands.

Weiner J S, Chemla D S, Miller D A B, Wood T H, Sivco D, and Cho A Y, 1985, "Room-temperature excitons in 1.6-μm band-gap GaInAs/AlInAs quantum wells", *Applied Physics Letters* **46** 619.

Weiner J S, Pearson D B, Miller D A B, Chemla D S, Sivco D and Cho A Y, 1986, "Nonlinear spectroscopy of InGaAs/InAlAs multiple quantum well structures", *Applied Physics Letters* **49**, 531.

Wherett B S, 1984, "Fabry-Perot bistable cavity optimization on reflection", *IEEE Journal of Quantum Electronics* **QE-20** 646.

Whitehead M and Parry G, 1989, "High-contrast reflection modulation at normal incidence in asymmetric multiple quantum well Fabry-Perot structure" *Electronics Letters* **25** 566.

Yeh Pochi, 1988, *Optical waves in layered media* (Wiley & Sons, New York).

Ultrafast processes in semiconductor optical amplifiers and modulators

Jørn M Hvam

Research Center COM, Technical University of Denmark

1 Introduction

The development of the information society is characterised by a strong growth in the demand for bandwidth to the end users. The evolution is driven by the new possibilities being offered for people to communicate and access information. The explosive growth of the internet is a measure of this development.

The medium of choice for transporting large amounts of data is the optical single-mode fibre, since it combines nearly loss-less propagation with a very large bandwidth. A large installed fibre base already exists and increases in size every year. Present-day techniques, however, do not really utilise the available optical bandwidth, since the electronics needed for data processing at the fibre inputs and outputs limit the data rates. Systems at 10Gbit/s are now standard and 40Gbit/s systems are becoming commercially available in the near future. At higher bit rates, however, the electronics become increasingly complex. Optically transparent solutions that utilise the large fibre bandwidth are therefore explored.

Two major alternatives appear today. The WDM (Wavelength Division Multiplexing) technique utilises the optical bandwidth by combining a number of low bit-rate signals at different wavelengths, which can be discriminated using optical filters. The OTDM (Optical Time Division Multiplexing) technique, on the other hand, employs low bit-rate signals consisting of short optical pulses, which are temporally interleaved to form a single pulse train before being launched on the fibre. In both cases, active optical techniques are required to multiplex and de-multiplex the composite signal, i.e. to combine and select the individual channels. In optical networks consisting of several nodes, additional functionalities such as wavelength conversion and signal regeneration are needed to create an optically transparent network. Particularly for an OTDM system, the quest for devices enabling ultrafast signal processing is imminent. Semiconductor optical amplifiers (SOA) and electro-absorbers (EA) are such devices.

Beyond mere amplification of optical signals to compensate for unavoidable losses associated with other functionalities in the nodes of an optical network, the SOA can also provide functionalities like wavelength conversion and pulse (re)generation via the gain saturation, or gain compression, occurring when the optical input power exceeds a certain saturation threshold. Associated with the gain saturation is (by Kramer-Kronig relations) a possible change of the refractive index of the active region of the SOA. In some cases the index change is exploited as in Mach-Zehnder or Michelson interferometric devices. In other cases the index variations cause an unwanted variation of the instantaneous frequency (chirp). In all cases knowledge of the gain and index dynamics in response to fast electrical and optical signals is vital for fully understanding the potential and the limitations of SOAs for ultrafast signal processing.

A semiconductor waveguide electro-absorber is in principle a reverse biased SOA with two different modes of operation, namely as an EA modulator (EAM) or as a (tunable) saturable absorber (SA). EAMs are used in transmitters to encode an electrical signal onto the optical carrier beam, and also for electrically controlled de-multiplexing of an OTDM signal in the receiver end. Operating as a SA, the EA can be used for optically controlled de-multiplexing as well as wavelength conversion. In combination with a SOA and appropriate feed-back from an optical cavity, the SA makes up a mode-locked semiconductor laser which is an ideal compact source for the generation of ultrashort (tunable) optical pulses.

Both SOA and EA devices are PIN structures grown by metal-organic vapour phase epitaxy (MOVPE) or molecular beam epitaxy (MBE) on n-type substrates of InP or GaAS, depending on the desired wavelength of operation. The intrinsic (I) region is the active region providing the planar wave guiding as well as the electro-optic interaction. It can be composed of bulk semiconductor material or low-dimensional hetero-structures in the form of quantum wells (QW), quantum wires (QWR) or quantum dots (QD) each providing its own special features of light-matter interactions, non-linear optical effects and dynamical properties. The lateral wave guiding is usually provided by conventional ridge wave-guides etched in the upper p-type layer followed by standard planarisation and contacting. The cross-section of a typical EA (or SOA) is shown in Figure 1.

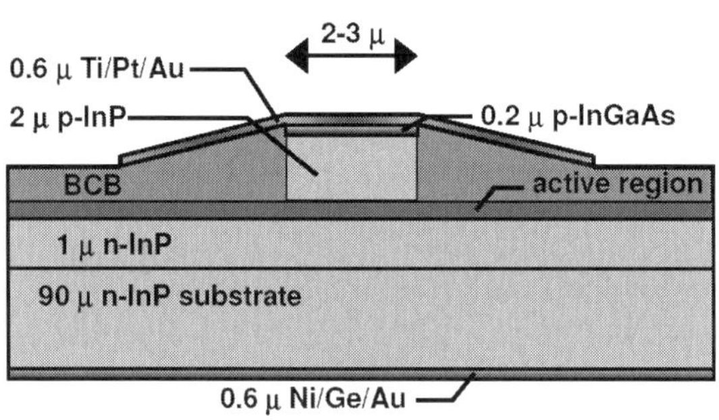

Figure 1. *Cross-section of an InP-based ridge waveguide electro absorber.*

In the present chapter, I shall describe and discuss in some detail ultrafast processes in semiconductor optical amplifiers and electro-absorbers of relevance for the functionality and speed of SOA and EA devices. In Section 2, I shall discuss some basic concepts of light-matter interaction, introduce the formalisms of non-linear optics focusing on third-order nonlinearities, and discuss the relation to carrier dynamics in semiconducting materials. In Section 3, I shall describe some short-pulse propagation and transmission experiments in bulk SOAs with full characterisation of amplitude and phase of the transmitted pulse. In Section 4, the gain and index dynamics of bulk SOAs will be investigated by means of femtosecond pump-probe experiments making use of a novel heterodyne detection scheme. Similar experiments on QD SOAs are discussed in Section 5, including also four-wave mixing experiments determining the homogeneous broadening, or dephasing rate, in QD devices. The heterodyne detection techniques are applied also to the investigations in Section 6 of transfer functions and carrier dynamics of EAs with QW layers in the active region. Conclusions are made in Section 7.

2 Light-matter interactions

Coherent interactions between light and matter are basic ingredients in the optical properties of matter, particularly when it comes to nonlinear optical processes (Bloembergen 1965). For non-resonant interactions the (nonlinear) optical response is instantaneous and follows adiabatically even the fastest laser pulses as long as they can be kept completely non-resonant. This is the preferred regime of nonlinear optical devices for loss-less and fast operation. But the small non-resonant values of the nonlinear coefficients then require very high light intensities (power) and/or long interaction lengths, possibly combined with delicate phase-matching conditions. For resonant excitations, the nonlinear coefficients are strongly enhanced, but the coherence and dephasing of the optical transitions (polarization) and the excited state lifetimes become crucial for the (nonlinear) optical response times.

The simplest systems discussed are those of non-interacting two-level atoms excited resonantly by an intense laser, describing well the optical properties of dilute gases. The dynamics of this system are solved by the optical Bloch equations, forming the basis for the theoretical description of light-matter interactions and introducing the lifetime (T_1) of the excited state as well as the coherence time (dephasing time T_2) of the optical transition (Boyd 1992). In dilute gasses, characteristic life times and dephasing times are in the microsecond or even millisecond range, so that coherent nonlinear optical experiments can be performed alternatively in the time or in the frequency domain (Haroche 1976).

In solids, for example semiconductors, inter-atomic interactions and scattering processes are so fast that truly coherent light-matter interactions occur on a femtosecond time scale. For resonant interactions, however, the optical excitation of real carriers may cause incoherent changes of the optical properties on a time scale of the carrier lifetime typically in the nanosecond regime (Shah 1996).

2.1 Two-level systems

For a system of non-interacting two-level atoms described in a density-matrix formalism, the optical Bloch equations take the form (Boyd 1992)

$$\frac{\partial \rho_D}{\partial t} = -\frac{2i\mu E}{\hbar}(\rho_{10} - \rho_{01}) - \frac{\rho_D - \rho_{D0}}{T_1}, \tag{1}$$

$$\frac{\partial \rho_{10}}{\partial t} = \frac{\partial \rho_{01}^*}{\partial t} = \frac{i\mu E}{\hbar}\rho_D - \left(\frac{1}{T_2} + i\omega_0\right)\rho_{10}, \tag{2}$$

where E is the light electric field and μ is the dipole transition matrix element. The diagonal element $\rho_D = \rho_{00} - \rho_{11}$ is the population difference between the ground state and the excited state The off-diagonal elements $\rho_{10} = \rho_{01}^*$ are the transition amplitudes, and the total damping is given by

$$\gamma = \frac{1}{T_2} + \frac{1}{2T_1}.$$

For a system of N atoms, the macroscopic polarization field P is found to be

$$P = N\mu\left(\rho_{10} + \rho_{01}\right) = N\mu\left(\rho_{10} + c.c.\right), \tag{3}$$

where $c.c.$ denotes the complex conjugate. If the system, after a short-pulse excitation, is left in a coherently excited state at $t = 0$, the polarization field will, according to Equations (2) and (3), decay as

$$\mathbf{P}(t) = \mathbf{P}(0)e^{-\gamma t} \approx \mathbf{P}(0)e^{-t/T_2}. \tag{4}$$

There are two equivalent ways of measuring the damping, or dephasing, of such a two-level system: either by directly measuring the polarization decay with high temporal resolution after a short-pulse resonant excitation, or by measuring the homogeneous line width $2\hbar\gamma$ (FWHM) in a c.w. absorption experiment with high spectral resolution.

In the above, it was assumed that all the atoms had perfectly identical transitions, that is, the system is only homogeneously broadened. In many systems, however, this is not the case. There will be randomly fluctuating environmental conditions for the atomic or electronic transitions that will influence the transition energies in a random way. This gives rise to an inhomogeneous broadening of the observed spectral lines that may completely mask the intrinsic homogeneous broadening or dephasing of the system. In that case, the spectral line-width as well as the polarization decay

$$\mathbf{P}_{\mathrm{inh}}(t) = \mathbf{P}(0)\exp\left(-\gamma t - \frac{\Gamma^2 t^2}{16\ln 2}\right) \tag{5}$$

is governed by the inhomogeneous broadening Γ of the upper atomic level (see Figure 2). In linear spectroscopy there is no simple way of getting around this problem if one wants to measure the homogeneous line width. One must therefore resort to non-linear spectroscopies such as spectral hole burning or photon echo (Boyd 1992). The latter is the result of transient nonlinear four-wave mixing in an inhomogeneous medium discussed further in the next section.

Obviously semiconductors are not simple two-level systems, but rather complicated two-band systems with strong Coulomb, exchange and correlation effects influencing the light-matter interaction (Shah 1996). To fully account for that, the optical Bloch equations above have to be replaced by the semiconductor Bloch equations the solutions of which can only be obtained numerically and in a much less transparent form. In certain cases, however, some insight can be gained by comparing semiconductors with homogeneously or inhomogeneously broadened two-level systems (Figure 2).

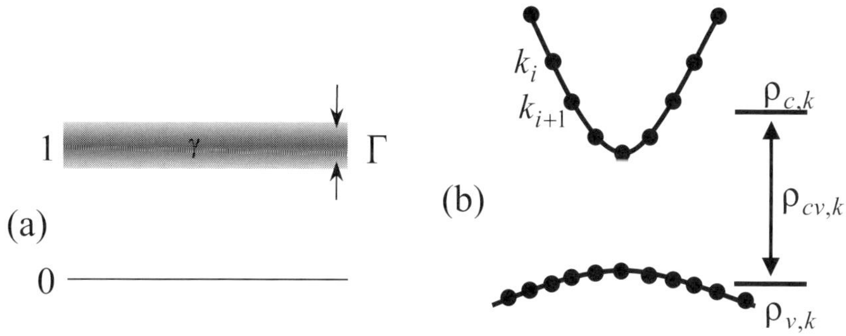

Figure 2. *Level diagram for an inhomogeneously broadened* (Γ) *ensemble of two-level atoms (a) and energy band diagram for direct-gap semiconductor with wave vector states* $\mathbf{k_i}$ *density matrix elements* $\rho_{n,k}$ *(*$n = v,c$*) and interband transition amplitudes* $\rho_{cv,k}$.

2.2 Third-order nonlinear optics

In classical non-linear optics (Bloembergen 1965), the polarization $\mathbf{P}$ of the medium is usually expanded in powers of the electric field vector $\mathbf{E}$:

$$\begin{aligned} \mathbf{P}(\mathbf{r},t) &= \chi^{(1)}\cdot\mathbf{E} + \chi^{(2)}\cdot\mathbf{EE} + \chi^{(3)}\cdot\mathbf{EEE} + \cdots \\ &= \mathbf{P}^{(1)}(\mathbf{r},t) + \mathbf{P}^{(2)}(\mathbf{r},t) + \mathbf{P}^{(3)}(\mathbf{r},t) + \cdots \end{aligned} \tag{6}$$

where $\chi^{(n)}$ is the n-th order nonlinear optical susceptibility. Linear optics is governed by the first term. In systems with inversion symmetry the second order term is identically zero, and thus the lowest order non-linearity is given by the third term.

If the total electric field is a superposition of plane waves, the polarization field will also appear as

$$\mathbf{E}(\mathbf{r},t) = \sum_i \big(\mathbf{E}_i(\mathbf{k}_i,\omega_i)\exp i(\mathbf{k}_i.\mathbf{r}_i - \omega_i t) + c.c.\big), \tag{7}$$

$$\mathbf{P}(\mathbf{r},t) = \sum_j \big(\mathbf{P}_j(\mathbf{k}_j,\omega_j)\exp i(\mathbf{k}_j.\mathbf{r}_j - \omega_j t) + c.c.\big), \tag{8}$$

and the third order polarization can, in the standard non-linear optics notation introduced by Bloembergen (1965), be described as

$$\mathbf{P}^{(3)}(\mathbf{k},\omega) = \chi^{(3)}(\omega;\pm\omega_1,\pm\omega_2,\pm\omega_3)\cdot\mathbf{E}_1(\mathbf{k}_1,\omega_1)\cdot\mathbf{E}_2(\mathbf{k}_2,\omega_2)\cdot\mathbf{E}_3(\mathbf{k}_3,\omega_3). \tag{9}$$

In order to have any noticeable energy transfer to the third order polarization wave, energy and wave-vector conservations have to be met in the above equation, that is,

$$\omega = \sum_{i=1}^{3} \pm\omega_i\,, \qquad \mathbf{k} = \sum_{i=1}^{3} \pm\mathbf{k}_i, \tag{10}$$

where the plus (minus) sign corresponds to an absorbed (emitted) wave. Equations (10) pose, together with the dispersion relations for all participating waves $\omega_i = \omega_i(\mathbf{k}_i)$, very severe limitations to an efficient generation of non-linear signals. These are the phase-matching conditions (Bloembergen 1965, Boyd 1992).

Processes described by Equation (9) are generally called four-wave mixing because the three waves on the right-hand side are mixing in the nonlinear medium to form a third order polarization wave, that serves as an antenna for the out-going fourth wave, the signal wave. In degenerate four-wave mixing, the frequencies of all the waves are the same, that is, $\omega = \omega_1 = \omega_2 = \omega_3 (= 2\omega - \omega)$. If all the wave-vectors are also identical (one beam), phase-matching is automatically fulfilled, and

$$\mathbf{P}^{(3)}(\mathbf{k},\omega) = \chi^{(3)}(\omega;\omega,\omega,-\omega)|\mathbf{E}(\mathbf{k},\omega)|^2\mathbf{E}(\mathbf{k},\omega) \tag{11}$$

The non-linear refractive index $n(I) = n_0 + n_2 I$ and the nonlinear absorption coefficient $\alpha(I) = \alpha_0 + \alpha_2 I$ are related to the third-order susceptibility in Equation (9) by

$$n_2 = \frac{\mathrm{Re}\,\chi^{(3)}}{\epsilon_0 n_0^2 c}, \tag{12}$$

$$\alpha_2 = \frac{\omega\,\mathrm{Im}\,\chi^{(3)}}{\epsilon_0 n_0^2 c^2}. \tag{13}$$

These effects are extremely important for the propagation of intense lasers through media with even small or moderate nonlinear coefficients. Self-focusing effects due to a nonlinear refractive index can destroy optical components, but can also be used constructively for the mode-locking of lasers as e.g. in self-mode-locked Ti:Sapphire femtosecond lasers (Ferguson 1994). Self-phase modulations in optical fibres and the formation of soliton waves are other examples (Mollenauer and Gordon, 1994). Nonlinear absorption (or gain) processes are particularly important in semiconductor lasers and amplifiers for fast optical signal processing in for example, optical communication systems.

Degenerate four-wave mixing can also occur between different (up to three) beams, and in that case the phase matching conditions become more restrictive (Hvam 1994). In the case of two beams, the situation is sketched in Figure 3. In this case, exact phase matching is obtained in the directions $\mathbf{k}_1 = \mathbf{k}_1 + \mathbf{k}_2 - \mathbf{k}_2$ and $\mathbf{k}_2 = \mathbf{k}_1 - \mathbf{k}_1 + \mathbf{k}_2$ Near phase-match is, on the other hand, obtained in the directions $\mathbf{k}' = 2\mathbf{k}_1 - \mathbf{k}_2 - \Delta\mathbf{k}$ and $\mathbf{k} = 2\mathbf{k}_2 - \mathbf{k}_1 - \Delta\mathbf{k}$, where $\Delta\mathbf{k}$ is a measure of the wave vector mismatch in the forward direction. In the back scattering direction the wave-vector mismatch is measured by $\Delta\mathbf{k}_b$.

The FWM signal propagating in the direction $\mathbf{k} \approx 2\mathbf{k}_2 - \mathbf{k}_1$ can be considered as the diffraction of beam #2 ($\mathbf{k}_2$) in the polarization grating formed in the nonlinear medium by interference between the two beams. It has the grating vector $\mathbf{k}_2 - \mathbf{k}_1$ with the corresponding line spacing $\Lambda = 2\pi/|\mathbf{k}_2 - \mathbf{k}_1|$. If the sample has a slab-geometry with a thickness d, the nonlinear diffracted signal propagating in one of the near phase-match directions $\mathbf{k}$ or $\mathbf{k}'$ will have the intensity (Hvam 1994)

$$I_s \propto \frac{d^2 \sin^2\left(\frac{\Delta k \cdot d}{2}\right)}{\left(\frac{\Delta k \cdot d}{2}\right)^2}. \tag{14}$$

For large interaction lengths ($\Delta k \cdot d \gg 1$), there will be negligible intensity in the signal beam, but for $\Delta k \cdot d \ll 1$ (thin sample) the phase mismatch does not lead to severe

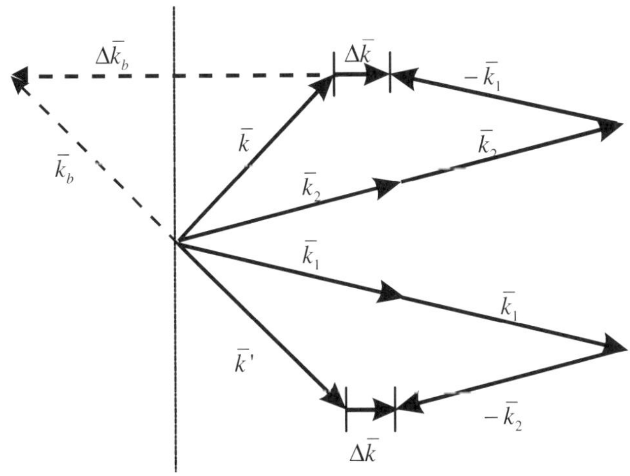

Figure 3. *Four-wave mixing between two beams with directions* $\mathbf{k_1}$ *and* $\mathbf{k_2}$ *yielding near phase-matched FWM signals in the directions* $\mathbf{k}$ *and* $\mathbf{k}'$. *The vertical line is the sample surface and the dashed lines indicate back scattering in the direction* $\mathbf{k_b}$.

reduction of the nonlinear signal. The big advantage of the thin-sample geometry for the purpose of nonlinear optical spectroscopy is, that nonlinear signals are propagating as collimated beams in the background free directions $2\mathbf{k} - \mathbf{k}_2$ and $2\mathbf{k}_2 - \mathbf{k}_1$. By simple spatial filtering one can therefore detect extremely weak nonlinear signals, the detection limit essentially being set by the scattered light intensity. Samples of good optical quality (surfaces) are therefore essential for such experiments.

When the sample is extremely thin (as in a quantum well) or the interaction length for some other reason (linear absorption) is very short, near phase matching can even be achieved in the back scattering direction with intensity comparable to the one in the forward direction, although the signal in both directions will diminish as d^2.

2.3 Four-wave mixing and pump-probe experiments

In a two-beam transient four-wave mixing experiment, the two incoming beams ($\mathbf{k}_1$, $\mathbf{k}_2$) each consists of a train of short, for example, femtosecond, laser pulses from the same coherent laser source. In that case the timing of the two incoming pulse trains is important, since a nonlinear signal in say, the direction $2\mathbf{k}_2 - \mathbf{k}_1$ is only produced if there is a coherent temporal overlap between the polarization from the first pulse from beam #1 and the second pulse from beam #2. Hence, information about the coherence and dephasing of optical transitions in a medium can be obtained by recording the nonlinear signal intensity as a function of the delay τ between the incoming pulses. These are the so-called correlation traces. The situation is sketched in Figure 4a. There is an ambiguity with respect to the sign of τ. In the present work, as in most literature on transient FWM, the delay τ is counted positive when the pulse in beam #1 arrives first, and the signal is detected in the direction $2\mathbf{k}_2 - \mathbf{k}_1$.

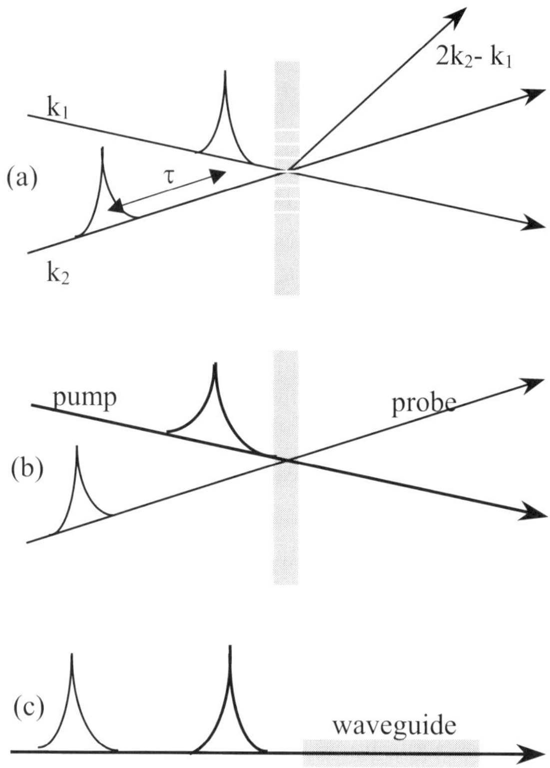

Figure 4. *Transient two-beam four-wave mixing (a) and pump-probe (b) experiments in slab geometry and in waveguide geometry (c).*

Let us for illustrative purposes look at how an ideal homogeneously broadened two-level system, with the complex transition frequency $\Omega = \omega_0 - i\gamma$, will respond to such a transient FWM experiment with delta-like input pulses arriving at $t = -\tau$ and $t = 0$, respectively. The third-order non-linear signal can be found by solving the optical Bloch equations (1) and (2) in a perturbational way by expanding the density matrix in powers of E and solving to third order. We then obtain the third order nonlinear polarization (Hvam 1998)

$$P^{(3)}(\mathbf{r}, t, \tau > 0) \propto \Theta(t)\mu^4 \exp\left[i(2\mathbf{k}_2 - \mathbf{k}_1)\cdot\mathbf{r} - \Omega t + \Omega^*\tau\right]. \tag{15}$$

This means that the intensity of the transient FWM signal propagating in the $2\mathbf{k}_2 - \mathbf{k}_1$ direction is given by

$$I_{TFWM} \propto \left|P^{(3)}(t, \tau > 0)\right|^2 \propto \Theta(t)\mu^8 e^{-2\gamma(t+\tau)}, \tag{16}$$

with the signal decaying in real time as well as a function of the delay with the homogeneous damping rate. In the frequency domain, the signal intensity is expressed by

$$I_{TFWM} \propto \left|P^{(3)}(\omega, \tau > 0)\right|^2 \propto \frac{\mu^8}{|\Omega - \omega|^2} e^{-2\gamma\tau}, \tag{17}$$

from which the homogeneous damping can be determined both from the line-width of the signal and from the decay of the time-integrated signal as a function of the delay between the incident pulses.

In an inhomogeneously broadened system (Figure 2), as for example, represented by a Gaussian distribution of the resonance frequency

$$g(\omega_0') \propto \exp\left(\frac{-4\ln 2(\omega_0' - \omega_0)^2}{\Gamma^2}\right), \tag{18}$$

the macroscopic polarization dies out very rapidly by destructive interference, but the two-beam TFWM experiment then produces a photon echo (PE) (Hvam 1998)

$$I_{PE} \propto \left|P^{(3)}_{\text{inh}}(t, \tau > 0)\right|^2 \propto \mu^8 \exp\left(\frac{\Gamma^2(t-\tau)^2}{16\ln 2}\right) \exp(2\gamma(t+\tau)) \tag{19}$$

and the spectrally resolved time integrated signal as a function of the delay will still reveal the homogeneous line width (γ) through the relation

$$I_{PE} \propto \left|P^{(3)}_{\text{inh}}(\omega, \tau > 0)\right|^2 \propto \mu^8 \exp\left(-8\ln 2\frac{(\omega_0 - \omega)^2}{\Gamma^2}\right) \exp(-4\gamma\tau). \tag{20}$$

The power of the photon echo experiment is that it can measure the homogeneous line width, or damping, even in the case where, in a linear experiment, it is buried in the inhomogeneous broadening. Alternatively, one can perform a cw spectral-hole burning experiment where a spectrally narrow intense pump saturates the absorption (or gain) at the pump wavelength only, and leaves a spectral hole in the inhomogeneously broadened line to be measured by a spectrally broad probe beam. The width of the spectral hole is then a measure of the homogeneous linewidth (Macfarlane 1994).

Figure 4b also illustrates a pump-probe, or differential transmission, experiment where beam #1 is the pump and beam #2 is a (weak) probe beam monitoring the change in transmission induced by the pump. In this case, there is both an initial (T_2) coherent (FWM) contribution and a longer lasting (T_1) incoherent contribution to the differential transmission. It is an advantage of the experiments sketched in Figures 4a,b that the nonlinear (probe) signal is detected in a background free direction, and hence is easily spatially separated from the intense pump beam. If pump-probe experiments are performed in a wave-guide device this cannot be done because the pump and the probe beams have to co-propagate in the waveguide (see Figure 4c). Then other means to separate the pump and the probe beams have to be implemented, as we shall discuss in detail later.

In semiconductors, where the spectral features of absorption and gain are very broad (often 40–50meV), pump-probe experiments are often of a mixed type called dynamic spectral hole-burning, where the development of the spectral hole is followed in real time revealing information about the carrier dynamics.

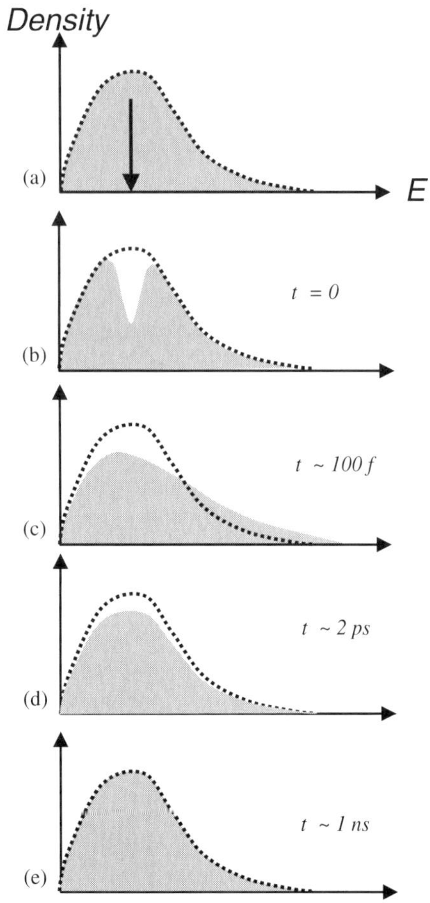

Figure 5. *Electron population (a) in the conduction band of a semiconductor laser after a short spectrally narrow light pulse is incident at $t = 0$. A spectral hole is immediately formed (b) and washed out by carrier-carrier scattering (c). Carrier cooling occurs via carrier-phonon scattering (d) and finally the carrier distribution is recovered by carrier injection (e).*

2.4 Carrier dynamics in semiconductors

Consider a semiconductor biased to gain with a certain density of electrons in the conduction band and holes in the valence band (see Figure 5). If at $t = 0$ an intense, spectrally narrow (within the gain band) light pulse falls on the semiconductor, carriers will, by stimulated emission, be depleted in a narrow spectral range. A spectral hole is formed, instantaneously reducing the gain at the corresponding wavelength, as shown in Figure 5b. In the high density of carriers in the bands, however, carrier-carrier scattering will, within 30–150fs, wash out the spectral hole and create a quasi-Fermi distribution at an elevated temperature (Figure 5c). The cooling of this Fermi distribution to the lattice temperature will occur by carrier-phonon scattering within a few picoseconds (Figure 5d), whereas the full recovery of the gain profile will require injection of carriers (Figure 5e) to replace those lost by stimulated emission. This may take up to several nanoseconds. A full theoretical description of these carrier dynamics requires the solution of the semiconductor Bloch equations with all the relevant scattering terms included (Shah 1996).

3 Short-pulse propagation in bulk SOAs

Ultra short pulses propagating through an SOA will undergo a temporal broadening alone due to the wide spectral content of the pulse. More importantly, however, the high peak power in ultra short pulses tends to dynamically induce nonlinear optical processes in the active material of the SOA causing severe distortions of both amplitude and phase of the transmitted pulse (Mecozzi and Mørk 1997). An experimental set-up for measuring short-pulse transmission in waveguide devices is shown in Figure 6. The input pulses are generated at a repetition rate of 300kHz by an optical parametric amplifier (OPA) pumped by 800nm, 200fs pulses from a Ti:Sapphire regenerative amplifier. The idler from the OPA is tuneable through the entire communication wavelength range and the pulses are sent through a pulse shaper allowing the pulse length to be varied between 200fs and 10ps keeping the pulse near transform limited. Part of the laser beam is coupled through

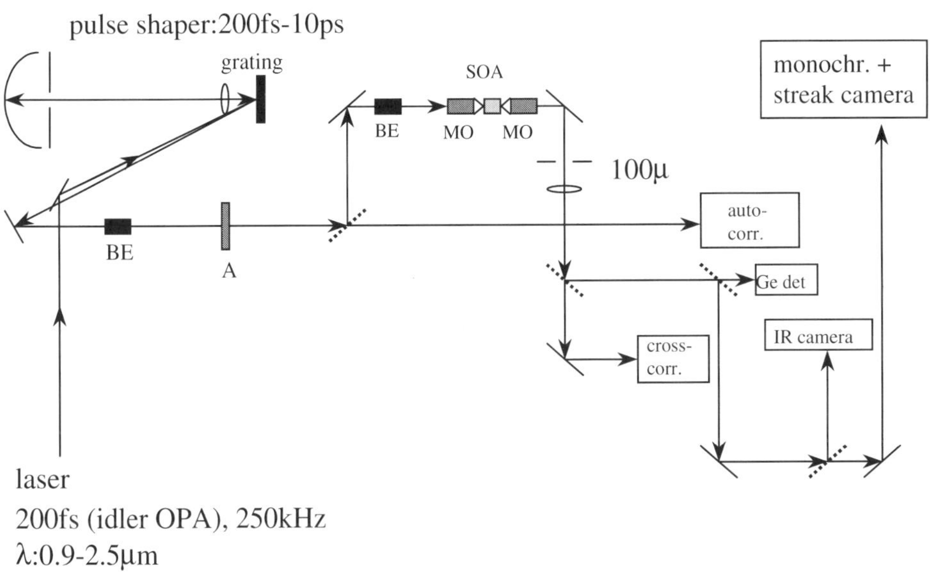

Figure 6. *Experimental setup for short-pulse waveguide (SOA) transmission experiments with beam expanders (BE), attenuator (A) and microscope objectives (MO). An optical parametric amplifier (OPA) delivers the light pulses.*

the device via large numerical-aperture lenses. The rest of the laser beam is used for pulse characterisations via autocorrelation or cross-correlation. The transmitted pulse is characterised with respect to pulse energy, spectral content and/or temporal shape.

3.1 Frequency-resolved optical gating

In order to understand the pulse distortions caused by induced variations in both gain and refractive index of the device it is necessary to be able to fully characterise both input and output pulses, i.e. to measure both amplitude and phase of optical pulses. One of the non-linear optical techniques that have been developed recently for this purpose is frequency-resolved optical gating (FROG) (Trebino *et al.* 1997). The laser pulse train to be characterized is sent to an optical autocorrelator (Michelson interferometer) based on second-harmonic generation. The second harmonic signal is now spectrally analysed as a function of the delay in one arm of the interferometer to form the so-called FROG-trace. From this trace, the temporal variation of both amplitude and phase (or instantaneous frequency) of the original pulse can be uniquely determined applying a rapidly converging computer algorithm to a test pulse shape. By proper pulse shaping, we can control the input laser pulses to be chirp free and transform limited.

Once the input laser pulses are fully characterised, the transmitted pulses can be characterized applying a cross-FROG (XFROG) technique (Linden *et al.* 1998), where the known laser pulse is cross-correlated with the unknown pulse and analysed as in FROG, now using the known pulse as input to the computer algorithm.

We have applied the XFROG technique to measure the instantaneous amplitude and phase (frequency) of femtosecond pulses of different pulse energies transmitted through a bulk InGaAsP SOA in the gain regime (Figure 7a) and at transparency (Figure 7c). In order to understand the non-linear processes involved, we have made model calculations (Figures 7b,d) to compare with the experiments. The model is based on the semiconductor Bloch equations in a density matrix formalism neglecting many-body effects and applying the relaxation time approximation. In an adiabatic approximation, a set of rate equations are set up for the local carrier densities $n(E)$, the total carrier density N and the effective carrier temperature T_n. These rate equations are then supplemented by a propagation equation for a slowly varying field envelope.

In the gain regime (Figure 7b), a temporal broadening of the pulse is observed for increasing pulse energy, which is due to a strong gain compression caused by two-photon absorption. The strong self-phase modulation (chirp) is caused both by coherent (Kerr) effects as well as incoherent (carrier-induced) index changes.

In the transparency regime (Figure 7d), the main features are for increasing input pulse energy a temporal narrowing of the pulse and an increasing (negative) chirp. The chirp is caused by coherent non-linear (Kerr) effects, and the temporal narrowing can only be explained including a spectral artefact caused by the combination of chirp and strong dispersion of the gain curve around transparency (Hughes *et al.* 2001).

The above temporal behaviour is followed by spectral broadenings (self-phase modulation), shifts and break-up (gain compression) as observed in a time-integrated spectral analysis of the transmitted pulses. A more detailed analysis of the gain and index changes caused by intense optical pulses can be obtained from dynamic pump-probe experiments.

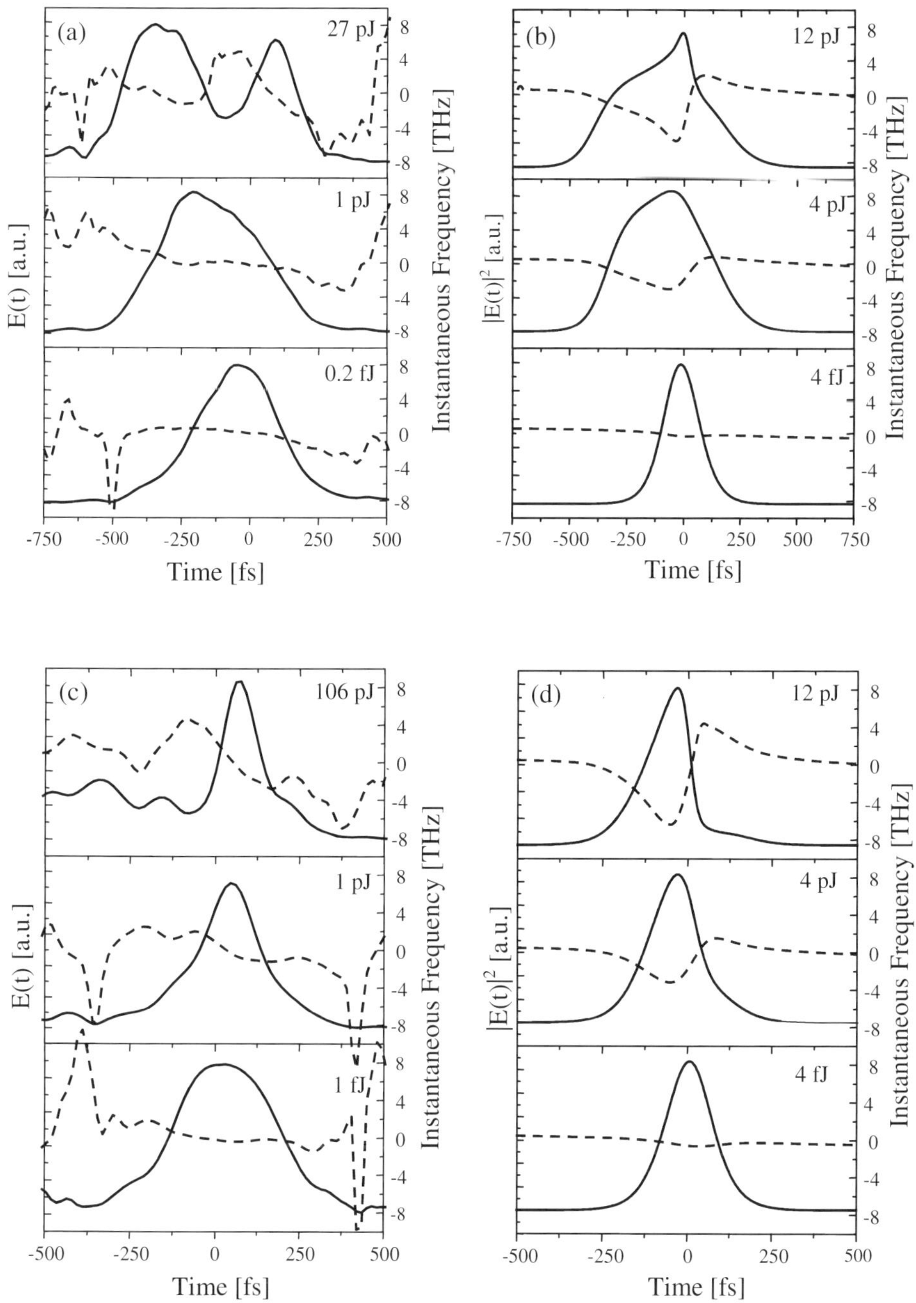

Figure 7. *Instantaneous amplitude (solid curves) and frequency (dashed curves) of 200fs-pulses of different pulse energies transmitted through a bulk InGaAsP SOA in the gain regime (a), (b) and in the transparency regime (c), (d). Experimental results in (a) and (c), and model calculations in (b) and (d).*

4 Gain and index dynamics in bulk SOAs

In order to perform pump-probe experiments, including transient four-wave mixing, in waveguide devices one has to be able to separate the probe (or signal) beams from the pump beam. If both beams are in the same wavelength range, one has to resort to other than spectral filtering. This can be by polarization or by applying a special heterodyne detection scheme (Borri *et al.* 1999a).

4.1 Heterodyne detection

The experimental set-up for pump-probe (and FWM) experiments with heterodyne detection is sketched in Figure 8. The pump and probe beams are frequency shifted from the incident laser by acousto-optic modulators (AOM) applying two different frequencies. Both beams (one via a variable optical pump-probe delay stage) are directed through the sample and the transmitted beams are overlaid the (unshifted) reference laser beam (via a variable real-time delay stage) on a beam splitter, where the two amplitudes are coherently added to form a heterodyne beat signal with a frequency equal to the (probe) AOM radio frequency (RF). In the RF domain both amplitude and phase of the heterodyne

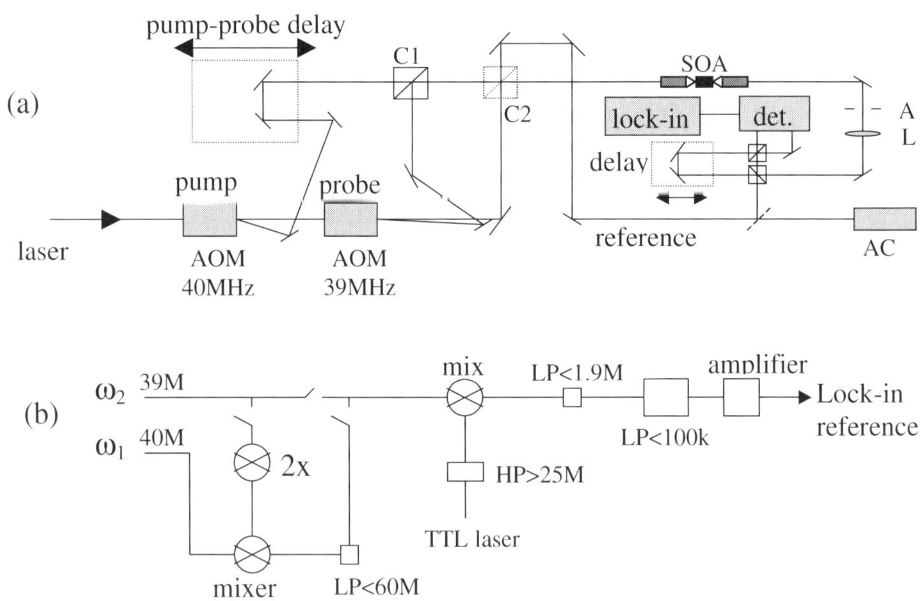

Figure 8. *Experimental set-up with heterodyne detection (a). AOM: Acousto-optic modulator; C1 and C2: broadband non-polarizing cube beam-splitters; A: 200 μm aperture; L: lens, AC: autocorrelator; SOA: semiconductor optical amplifier. (b) Scheme of the electrical mixing to generate the reference signal for lock-in. TTL from regenerative laser amplifier; LP: low-pass filter; HP: high-pass filter.*

signal can easily by measured simultaneously, for example, by a lock-in detector. Since the heterodyne amplitude is proportional to the amplitude of the optical probe and the heterodyne phase is proportional to the phase of the optical probe, we thus have a measure of both amplitude and phase of the optical signal. By varying the optical delay between the probe and the reference signal, we can fully characterize the temporal variation of the amplitude and phase of the probe pulses as in a FROG experiment.

For a fixed (zero) delay between the probe and the reference beams, the full transfer function (real and imaginary part) of a device can be determined as a function of an external parameter, for example, an electrical bias (Romstad *et al.* 2002). By detecting amplitude and phase of the heterodyne signal as a function of pump-probe delay in a differential transmission experiment the dynamic transfer function of a device, gain and index dynamics of an SOA, can be determined.

In degenerate FWM experiments the signal is again degenerate with the input beams. However, frequency shifting the two input beams by different RF frequencies ω_1 and ω_2, the signal is frequency shifted by $2\omega_1 - \omega_2$ (or $2\omega_2 - \omega_1$). Detecting the heterodyne signal at this frequency determines the FWM signal (amplitude and phase) either in real time (varying the real time delay) or time-integrated (varying the pump-probe delay). Dynamic FWM experiments are used to measure the coherence and dephasing (homogeneous line width) of relevant optical transitions in a semiconductor device (see later).

4.2 Pump-probe experiments in bulk SOA

Mørk and Mark (1995) performed pump-probe experiments in bulk InGaAsP SOA forward biased at 10mA for different (pump/probe) wavelengths through the gain (1530nm), transparency (1500nm) and absorption (1480nm) regions. In the gain region there is, following an intense femtosecond pump pulse, an immediate reduction of the gain with a fast recovery over several picoseconds followed by a slow final recovery over about 1ns. In the absorption region there is similarly an absorption saturation followed by a multi component recovery due to coherent and incoherent processes. When the centre frequency of the pump is at transparency the ultrafast coherent processes dominate with only a small contribution from incoherent (carrier-induced) processes.

For comparison with the experiment, the gain dynamics was modelled, using the same rate equations as discussed in Section 3.1. In the calculations, the contributions from spectral hole burning (SHB), two-photon absorption (TPA), carrier heating (CH) and population (N) can be isolated and weighted to fit the experiment. TPA follows the pulse (autocorrelation), the SHB recovers fast (100fs) due to carrier-carrier scattering, the CH (or cooling) is due to carrier-phonon scattering (1ps) and N recovers very slowly (1ns). It is concluded that TPA contributes significantly to the non-linear response on a femtosecond timescale (Mørk and Mark 1995).

With the heterodyne detection technique described above also the phase changes have been measured (Borri 1999a), as shown in Figure 9, revealing the phase dynamics after a short pump pulse in an InGaAsP SOA biased to (a) gain, (b) transparency and (c) absorption respectively. Again, there are fast coherent contributions to the phase change as well as slower incoherent (carrier-induced) contributions. Notice that the carrier-induced (N) index change has different sign in the gain region as in the absorption region. At trans-

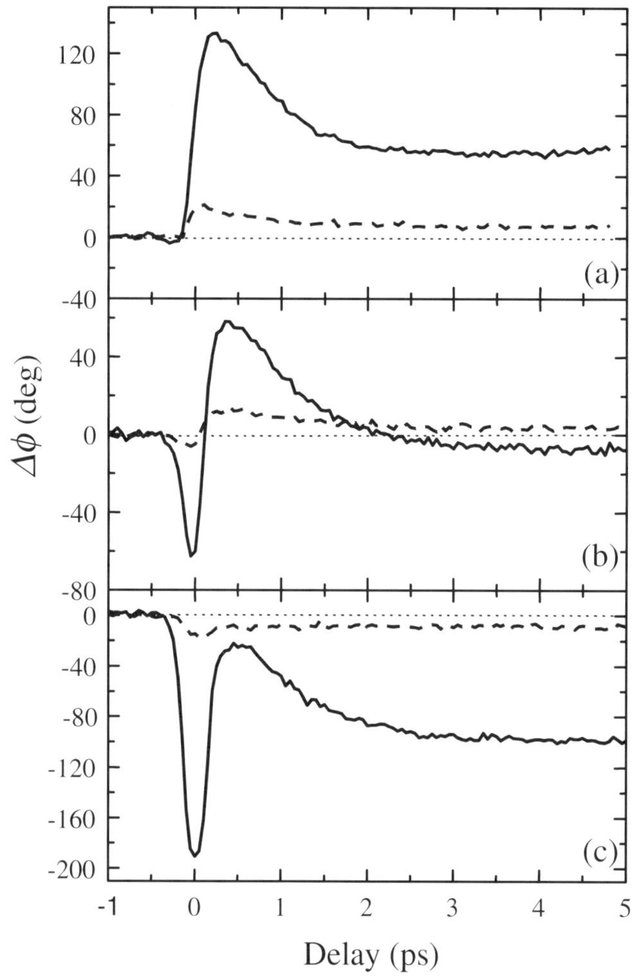

Figure 9. *Pump-induced transient phase changes in InGaAsP SOA for low (11pJ) pump pulse energy (dashed curves) and for high (460pJ) pulse energy (full curves), when the SOA is biased to (a) gain, (b) transparency and (c)absorption.*

parency, model calculations, shown in Figure 10, containing contributions from TPA, SHB, CH and N add up (TOT) to fit the experiments well. At high pump intensity and long devices the experiments can only be modelled satisfactorily taking into account the spectral artefact caused by the index (instantaneous frequency) change in combination with the strong gain dispersion near transparency (Mørk and Mecozzi 1996).

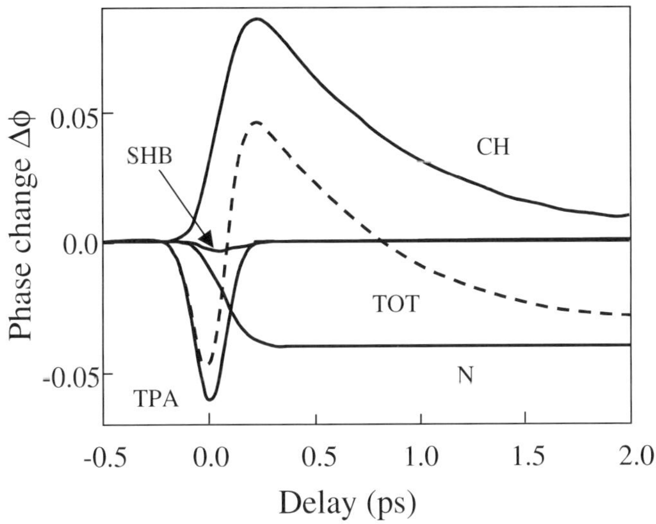

Figure 10. *Pump-induced dynamical phase changes near transparency, but on the absorption side. The calculated (Mørk and Mecozzi 1996) contributions from carrier density (N), carrier heating (CH), spectral hole burning (SHB), and two-photon absorption (TPA) are shown by solid curves, and the total contribution is shown by the dashed curve.*

5 Quantum-dot SOAs

Recently, self-assembled semiconductor quantum dots have caught a lot of attention as the gain medium in the next generation of semiconductor lasers and amplifiers (Bimberg *et al.* 1999). Anticipated advantages of quantum-dot based lasers and amplifiers over conventional bulk and quantum well devices are better energy selectivity and temperature stability, ultrahigh material gain and better slope efficiency, low threshold current density and chirp-free operation. Modulation bandwidths and switching speeds are key issues to be investigated in connection with these novel quantum-dot based devices.

We have measured transmission properties as well as gain and index dynamics of quantum dot amplifiers using femtosecond pulses in transmission experiments and in pump-probe experiments (Borri *et al.* 2000a) applying the heterodyne detection technique as described in Section 4. The homogeneous linewidth of the quantum dot transitions is measured in four-wave mixing (FWM) experiments (Borri *et al.* 2000b).

5.1 Transmission and gain

The transmission coefficient, or device gain G, defined as $G = E_{\mathrm{out}}/E_{\mathrm{in}} = e^{g\,L}$, where E_{in} (E_{out}) is the pulse energy at the input (output) of the device and g is the modal gain coefficient, is shown in Figure 11 as a function of the input pulse energy for 150fs pulses centred at 1.08μm and for different device bias currents. For small input energies,

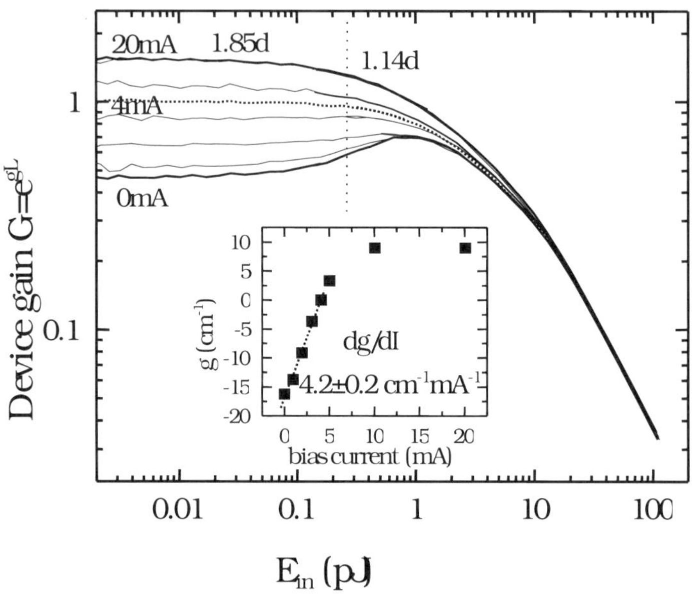

Figure 11. *Device gain of 150fs-pulses transmitted through a quantum-dot amplifier as a function of the input pulse energy for different bias currents as indicated. The waveguide ridge is 8μm wide and 475μm long. Inset shows the modal gain as a function of the bias current with a linear fit before saturation (Borri et al. 2000a).*

an absorption of −3.35dB is found at zero bias, transparency occurs at 4mA, and a maximum gain of 1.85dB is reached at 10mA. From the corresponding modal gain g (see inset of Figure 11), a differential gain of $4.2\text{cm}^{-1}\text{mA}^{-1}$ is deduced before saturation. As anticipated, this is significantly higher than for a similar quantum-well device. With increasing input energy, a bleaching of the absorption and a depletion of the gain occurs due to stimulated transitions induced by the pulse. For high input energies (> 1pJ), an additional reduction due to two-photon absorption (TPA) is present at all bias currents.

5.2 Gain and index dynamics

We have investigated the gain and index dynamics in pump-probe experiments for pump input energies still in the small signal regime, where the contribution of TPA is of minor importance. In Figure 12, the gain change in dB, as deduced from the probe transmission change, is plotted as a function of the pump-probe delay for bias currents of 0mA (absorption), 4mA (transparency) and 20mA (gain). A striking difference between the absorption and the gain recovery dynamics appears. At zero bias, pump induced bleaching of the absorption occurs, reaching a maximum value of 2.8dB and recovering over several picoseconds. At 4mA forward bias, corresponding to (average) transparency, the transmission changes are small, and at 20mA forward bias, a gain compression of up to −1.3dB is recovering in less than 0.3ps. The data show that the pump-induced SHB in the absorption case builds up with the time-integral of the pulse, according to a recovery

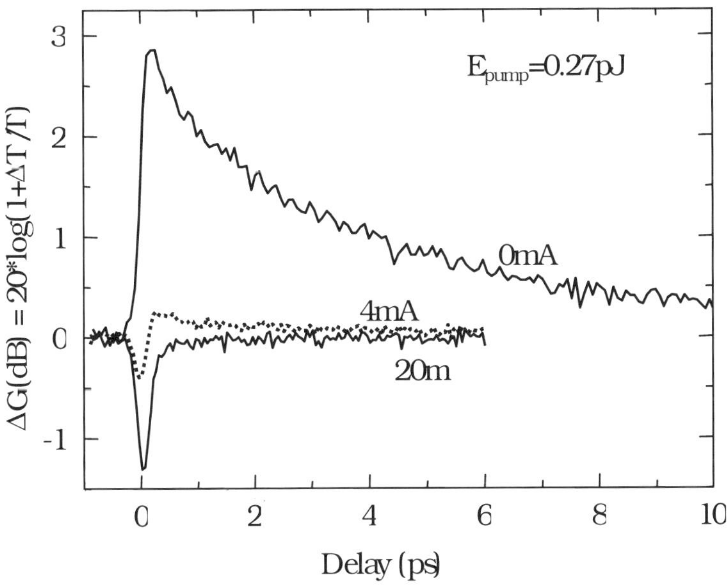

Figure 12. *Pump-induced (0.27pJ) dynamic gain changes measured for different bias currents yielding absorption (0mA), transparency (4mA), and maximum gain (20mA) in the device of Figure 11 (Borri et al. 2000a).*

time much longer than the pulse duration, while in the gain case it recovers on a time comparable to, or shorter, than the pulse duration. This is also reflected in the higher maximum absorption bleaching compared to the gain compression observed in Figure 12.

In the absorption case the dots are initially empty. When the pump arrives, bound electron-hole pairs (excitons) are created in the dot ground state, leading to a bleaching of the absorption by spectral hole-burning (SHB). This recovers by spontaneous carrier recombination or by carrier escape from the dot ground state. Typical spontaneous lifetimes in QDs are of several hundreds of picoseconds, as measured by time-resolved photoluminescence, much longer than the measured absorption recovery. The latter should then be attributed to carrier escape by phonon absorption, that is, carrier heating. The absorption recovery can be fitted by a bi-exponential response function with two equally weighted time constants of 1.25ps and 5.9ps attributed to hole and electron escape times (Sosnowski *et al.* 1998). The carrier heating also appears in the phase dynamics.

At transparency, the SHB effects are small and the remaining contribution to the gain changes is mainly due to TPA (around -0.5dB). In the case of gain, the dots are initially occupied and, by stimulated emission, the pump leads to a reduction of the exciton population (gain compression) that recovers by relaxation/capture of carriers into the dot ground state. The measured recovery time is, therefore, a direct measure of the carrier relaxation time from excited states into the dot ground state in a QD amplifier under working conditions (Borri *et al.* 2000a). The used bias current of 20 mA corresponds to a maximum modal gain that is just enough for ground state lasing of the investigated structure, and is not an unrealistic working point for QD lasers. From a fit to the gain

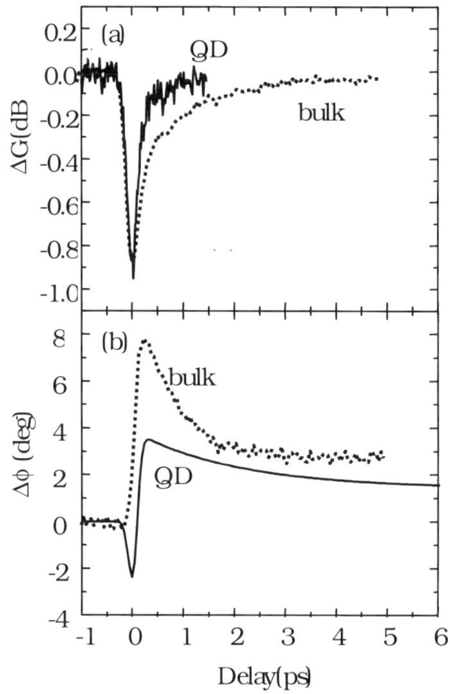

Figure 13. *Comparison of dynamical gain (a) and phase (b) changes in the quantum-dot SOA of Figure 11 (full curves) and a bulk InGaAsP SOA (dotted curves). Reduced carrier heating is observed in both gain and phase dynamics of the quantum-dot device.*

recovery at 20mA, taking into account TPA, we estimate a time constant of 115±10fs, slightly shorter than the pulse duration. The measured ultrafast SHB recovery in the gain case thus indicates ultrafast carrier relaxation in electrically pumped QD amplifiers, which is very promising for high-speed applications.

The carrier relaxation occurs probably via Auger scattering from the excited states, partly occupied at 20mA, into the dot ground state, as recently theoretically proposed (Ferreira and Bastard 1999). Note that the occurrence of such ultrafast scattering processes implies fast dephasing times that is, large homogeneous broadening (10–20meV), which we have confirmed by four-wave mixing experiments, as discussed in the next section. We should comment that the fit of the gain data at large delay (up to 6ps) also shows a minor contribution with a slower exponential decay of 0.7±0.1ps time-constant, possibly due to carrier heating (CH). Such a CH effect also appears in the phase dynamic (see Figure 13) but with a recovery time of 2.0±0.5ps. Still, the carrier heating effects are small compared to bulk SOAs as shown in Figure 13. It should be emphasized, though, that the fast gain (and index) recovery observed here is for a low repetition rate of the pump pulses leaving the carrier reservoirs in the wetting layers unaffected. For higher repetition rates (bit rates) the carrier relaxation and heating effects in the wetting layer (quantum well) may still limit the response time of quantum-dot SOAs.

5.3 Homogeneous linewidth and dephasing

Some of the predicted superior performances of quantum-dot lasers rely on a delta-like density of states, which is modified by the homogeneous and inhomogeneous broadenings. Presently the inhomogeneous broadening due to size, strain and compositional fluctuations in real In(Ga)As quantum-dot structures usually exceeds the homogeneous broadening, but recent improvements in growth technology have reduced the observed linewidth to 21meV in InAs/InGaAs quantum dots (Nishi *et al.* 1999). The question now arises: what is the homogeneous linewidth in a QD laser under operational conditions?

The room-temperature dephasing time T_2 in InAs/InGaAs dots, determining the homogeneous linewidth by $\gamma = 2\hbar/T_2$, was determined by femtosecond time-resolved four-wave mixing (TR–FWM) with heterodyne detection. A homogeneous linewidth γ=5meV, weakly dependent on the optical excitation intensity, was found and attributed to phonon interaction (Borri *et al.* 1999b). We have also determined the dephasing time in an SOA device that is forward-biased to a gain value near threshold for lasing (Borri *et al.* 2000b). In Figure 14, the time-integrated FWM field amplitude is shown as a function of

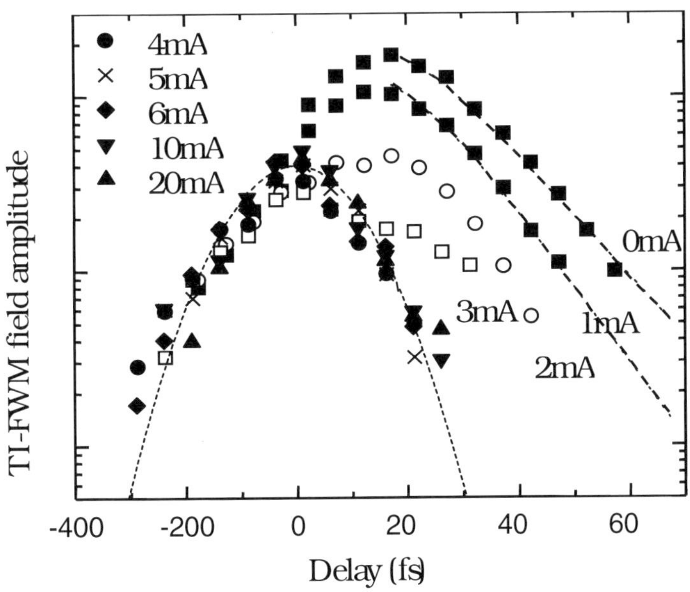

Figure 14. *Time-integrated FWM signal (field amplitude) at different bias currents (0-20mA) as a function of the delay between the two exciting pulses. The thin dashed curve is the calculated instantaneous response.*

the delay between the incident pulses for different bias currents. Up to 2mA bias current the FWM has a delayed maximum as in a photon echo. The dashed lines in Figure 14, at 0 and 1mA, are numerical simulations to the FWM decay at positive delay, obtained by solving the optical Bloch equations of an inhomogeneously broadened system for a 140fs Gaussian pulse and with dephasing times in the range of 200–250fs. The device reaches transparency (zero net number of stimulated transitions) for a bias current close to 3mA, as measured by pump-probe experiments. At this bias the TI–FWM response has, at zero delay, the minimum value compared to all other bias currents, as expected at optical transparency for a nonlinear signal induced by a density grating. However, the delay dependence is quite peculiar, with the appearance of a delayed signal after the initial decay. When further increasing the bias current from 4mA up to 20mA a fast decay of the TI–FWM signal, with no delayed maximum, is observed which follows an instantaneous response (thin dashed curve in Figure 14). Comparisons with the numerical simulations give an upper limit of 50fs for the dephasing time at currents larger than 4mA. The transition from a photon echo to a free polarization decay with increasing bias current is confirmed by real-time resolution of the FWM signal, which can be achieved by varying the delay between the signal and reference beams for each delay between the two input (pump-probe) beams (see Figure 8).

The shortening of the dephasing in the presence of electrically injected carriers indicates an increasing carrier-carrier scattering of the QD ground state. Below 2mA the average occupation number is less than one exciton per dot, and carrier-carrier scattering inside the same dot is unlikely. Thus, scattering with the thermalised carriers occupying the wetting layer (with a high density of states) is responsible for the slight reduction of the dephasing time. At transparency, the average occupation number is one exciton per dot and the optically induced net density change is zero so that there is a minimum in the FWM signal.

For bias currents above transparency, most dots are occupied by two or more excitons, and the FWM signal is dominated by the fast response. Note that $T_2 = 50$fs corresponds to $\gamma = 26$meV, comparable to the pulse spectral width in the experiment and comparable also to the inhomogeneous broadening due to dot size fluctuations in the best quantum dot samples (Nishi *et al.* 1999). The differential transmission experiments performed in the gain region, which probe the population inversion lifetime, showed a 100fs recovery time of the spectral hole burned by the pump pulse. This indicates the occurrence of fast carrier-carrier scattering also in the energy relaxation dynamics, possibly due to intra-dot Auger scattering. However, the population lifetime is still significantly above the measured dephasing time, which is thus dominated by elastic scattering processes.

6 Quantum-well electro absorbers

When the device in Figure 1 is reverse biased it is acting as an electro absorber (EA). With a multiple quantum well configuration as the active layer, the electro absorption effect is strongly enhanced by the quantum-confined Stark effect (Miller *et al.* 1986). In the following, I shall briefly discuss the complex transfer function of such an EA acting as an electro absorption modulator (EAM), and the absorption and index dynamics when the EA is acting as a saturable absorber (SA). In both cases the heterodyne detection scheme is applied to determine amplitude and phase of the transmitted light.

6.1 Transfer function of a multiple-quantum well EAM

To fully specify the optical pulses produced by an EAM one has to know both real and imaginary parts of the transfer function for variable bias voltage. This is achieved by the heterodyne detection technique, sketched in Figure 8, using only the probe and reference

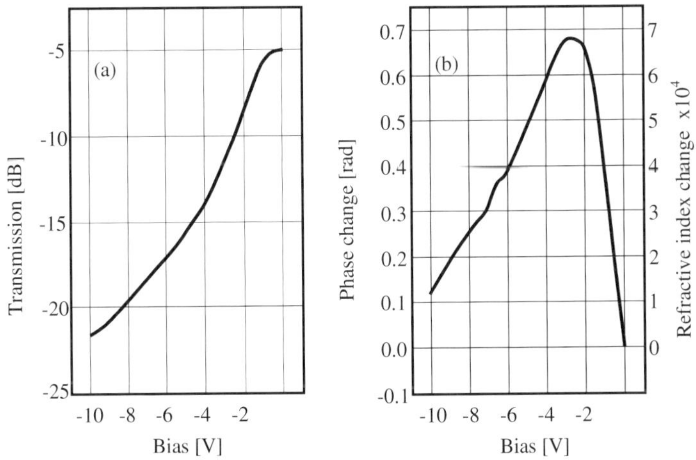

Figure 15. *Intensity (a) and phase (b) transfer curves measured as a function of reverse bias for an EA modulator. From the phase measurement, the refractive index change is calculated and shown on the right axis.*

beams (with zero delay between them) (Romstad *et al.* 2002). The amplitude and phase of the heterodyne (RF) signal reflect the amplitude and phase of the transmitted optical signal, which are then easily monitored as a function of a bias voltage sweep (on a millisecond timescale). An example is shown in Figure 15 for a 250μm long device operating at 1550nm. The relation between the bias-induced changes in the real and imaginary part of the optical susceptibility χ is usually described by the line-width enhancement factor, or α-parameter, as

$$\alpha = \frac{\Delta\chi_R}{\Delta\chi_I} = -\frac{4\pi\nu_0}{c}\frac{\Delta n}{\Delta g}, \tag{21}$$

where Δn is the refractive index change, Δg is the change in (modal) gain coefficient and ν_0 is the centre frequency of the light.

From the experimental results in Figure 15, it is observed that the α-parameter of the EAM investigated is negative for $V_{\rm bias} < -3$V. The variation in the instantaneous frequency ν of the transmitted pulse is related to the variation in phase ϕ (refractive index n) as

$$\frac{d\nu(t)}{dt} = -\frac{1}{2\pi}\frac{d^2\varphi(t)}{dt^2} \propto -\frac{d^2n(t)}{dt^2}. \tag{22}$$

This means that for $\alpha < 0$, the leading edge of the pulse is red shifted and the trailing edge of the pulse is blue shifted with respect to the centre frequency (positive chirp). This is favourable for a pulse to be transmitted through a standard single mode fibre at 1550nm where the dispersion is anomalous implying that higher frequencies travel faster than lower frequencies. A chirped pulse with $\alpha < 0$ will thus be shortened (up to a limit) as it propagates.

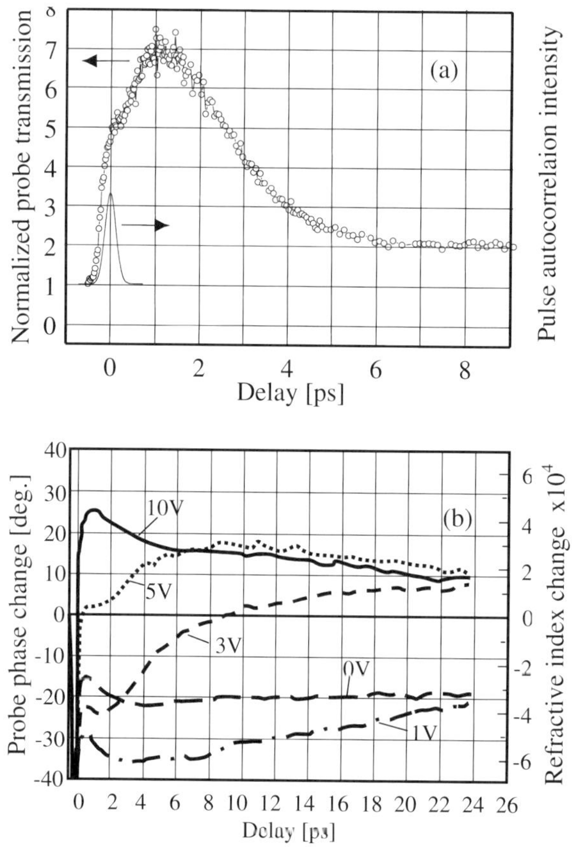

Figure 16. *(a): Pump-induced relative change of probe transmission in EA saturable absorber. (b): Pump-induced phase change and refractive index change for different reverse bias, as indicated. The pump pulse input energy is 3pJ.*

6.2 Absorption and index dynamics of EA saturable absorber

When the EA is reverse biased at DC, it will act as a saturable absorber for intense light pulses transmitted through the device. The absorption and index dynamics following an intense incident pulse can be studied in a pump-probe experiment again applying the heterodyne detection technique. The result is shown in Figure 16 for a 200fs pulse of 3pJ. The normalised probe transmission is shown (a) for a reverse bias of 10V, and the probe phase change is shown (b) for different reverse biases between 0 and 10V, both as a function of pump-probe delay. The pulse autocorrelation function is also shown.

The instantaneous (coherent) changes in the probe pulse are due to spectral hole burning (amplitude) and the optical Stark effect (phase). The following (incoherent) dynamical changes are due to carrier-carrier scattering (on a time-scale of 100fs) and carrier-phonon scattering (on a time-scale of 1ps). During this initial period, absorbed carriers will also escape the quantum wells by tunnelling and thermionic emission and cause a screening of the strong electric field from the reverse bias (Cavailles *et al.* 1992).

This gives rise to a significant bleaching of the absorption (and change of phase) that will only recover after the carriers are fully swept out of the device, typically in 100–200ps.

Note in Figure 16 that the slow component of the phase change changes sign from negative to positive when the reverse bias is increased beyond 3V. This is in agreement with the change in sign of the a-parameter for increasing reverse bias discussed in the previous section.

For ultrafast signal processing the slow component in the recovery should be suppressed as much as possible, which is achieved by applying a strong reverse bias. Thus for a reverse bias of 6V, a 10ps switching window with $>$ 9dB extinction ratio has been realised. Similarly, the long recovery time has been sufficiently suppressed to demonstrate wavelength conversion at 40Gbit/s. Again the chirp introduced by the strong negative bias is favourable for short-pulse propagation in standard single-mode fibre.

7 Conclusions and outlook

The demand for ultrafast optical and electro-optical components in transparent optical networks has made the study of ultrafast processes in semiconductor materials and devices an important one. It is of paramount importance to be able to investigate and understand the physical processes governing the light-matter interaction, the carrier dynamics, and thereby the ultimate limits to the speed of nonlinear optical signal processing. As the data rate goes beyond 40Gbit/s, the limitations by parasitic (device related) effects may be exchanged by fundamental materials related limitations that can only be overcome by proper materials design and engineering. We have seen that in semiconductor devices like optical amplifiers and electro absorbers the active region has gradually changed character from bulk semiconductor to quantum wells and most recently to quantum dots.

By quantum confinement of the carriers in one, two, or three dimensions the light-matter interactions can be significantly modified (enhanced) and the optical properties, including dynamics, can be engineered to match the required functionalities and specifications. In the present article, I have illustrated the current developments by highlighting some progress made in our own laboratories on quantum-well and quantum-dot based waveguide devices like optical amplifiers and electro absorbers. The waveguides have been by conventional index guiding. It can be foreseen that in the nearest future also the optical confinement will undergo dramatic changes by making use of photonic crystal effects as the photonic counterpart of the electronic properties in crystalline semiconductors. This may lead to a revolution in integrated optics and possibly to a much closer integration of optics/photonics and electronics.

Acknowledgments

Progress in research and technology is to a large degree based on a well-functioning teamwork in a strong scientific environment. I would like to thank the whole staff of the Optoelectronics Area at Research Center COM for providing this environment. In particular, I would like to thank Dan Birkedal, John Erland, Jesper Mørk, Paola Borri, Wolfgang Langbein and Francis Romstad for their contributions to this work.

References

Bimberg D, Grundmann M, and Ledentsov N N, 1999, *Quantum Dot Heterostructures* (Wiley, Chichester) .

Bloembergen N, 1965, *Nonlinear Optics* (Benjamin Inc., New York, Amsterdam) .

Borri P, Langbein W, Mork J, Hvam J M, 1999a, *Optics Communications* **169** 317.

Borri P, Langbein W, Mørk J, Hvam J M, Heinrichsdorff F, Mao H M, and Bimberg D, 1999b, *Phys Rev B* **60** 7784.

Borri P, Langbein W, Hvam J M, Heinrichsdorff F, Mao H M, and Bimberg D, 2000a, *IEEE J Sel Top Quantum Electron.* **6** 544.

Borri P, Langbein W, Hvam J M, Heinrichsdorff F, Mao H M, and Bimberg D, 2000b, *Appl Phys Lett* **76** 1380.

Boyd R W, 1992, *Nonlinear Optics* (Academic Press, San Diego).

Cavailles J, Miller D, Cunningham J, Wa P, and Miller A, 1992, *IEEE J Quantum Electron* **28** 2486.

Ferguson A I, 1994, Advances in Solid State Lasers, in *Nonlinear Spectroscopy of Solids: Advances and Applications*, editor Di Bartolo B (Plenum Press, New York, London).

Fereira R and Bastard G, 1999, *Appl Phys Lett* **74** 2818.

Haroche S, 1976, in *High Resolution Laser Spectroscopy*, editor Shimoda K (Springer-Verlag, Berlin Heidelberg).

Hughes S, Borri P, Knorr A, Romstad F, and Hvam J M, 2001, *IEEE J Sel Top Quantum Electron* **7** 694.

Hvam J M, 1994, Nonlinear Spectroscopy of Inorganic Materials: Excitons in Semiconductors, in *Nonlinear Spectroscopy of Solids: Advances and Applications*, editor Di Bartolo B (Plenum Press, New York, London).

Hvam J M, 1998, Coherent dynamics in Semiconductors, in *Ultrafast Dynamics of Quantum Systems: Physical Processes and Spectroscopic Techniques*, editor Di Bartolo B (Plenum Press, New York).

Linden S, Giessen H, and Kuhl J, 1998, *Phys Stat Sol (b)* **200** 119.

Macfarlane R M, 1994, High Resolution Nonlinear Spectroscopy of Rare Earth Ions in Solids, in *Nonlinear Spectroscopy of Solids: Advances and Applications*, editor Di Bartolo B (Plenum Press, New York, London).

Mecozzi A and Mørk J, 1997, *J Opt Soc Am B* **14** 761.

Miller D A B, Chemla D S, Schmitt-Rink S, 1986, *Phys Rev B* **33**, 6976.

Mollenauer L F and Gordon J P, 1994, Solitons in Optical Fibers and their use in Ultra Long Distance, High Bit Rate Transmissions, in *Nonlinear Spectroscopy of Solids: Advances and Applications*, editor Di Bartolo B (Plenum Press, New York, London).

Mørk J and Mark J, 1995, Time-resolved Spectroscopy of Semiconductor Laser Devices: Experiments and Modelling, in *SPIE Proc* **2399** 146.

Mørk J and Mecozzi A, 1996, *J Opt Soc Am B* **13** 1803.

Nishi K, Saito H, Sugoa S, and Lee J S, 1999, *Appl Phys Lett* **74** 1561.

Romstad F, Birkedal D, Mørk J, and Hvam J M, 2002, *IEEE Photonics Technology Letters* **14** 621.

Shah J, 1996, *Ultrafast Spectroscopy of Semiconductors and Semiconductor Nanostructures* (Springer-Verlag, Berlin Heidelberg New York).

Sosnowski T S, Norris T B, Jiang H, Singh J, Kamath K, and Bhattacharya P, 1998, *Phys Rev B* **57** R9423.

Trebino R, DeLong K W, Fittenhoff D N, Sweetser J N, Krumbgel MA, Richman B A, and Kane D J, 1997, *Rev Sci Instrum* **68** 3277.

Towards Polymer Lasers and Amplifiers

Ifor D W Samuel

University of St Andrews, Scotland

1 Introduction

Conventional semiconductors such as silicon and gallium arsenide are expensive, brittle, crystalline materials. The development of techniques to grow these materials in the remarkable purity required for devices took decades, and now enables modern electronics and optoelectronics. Polymers, or plastics are organic materials with entirely different properties. They are long chain-like molecules consisting of many units which repeat along the backbone. Like silicon and gallium arsenide they are found throughout everyday life in applications from packaging to clothing, and from furniture to bulletproof vests. There are three main reasons for their widespread use. First is that there is an enormous range of structures, leading to an enormous range of properties. This structural diversity has been beautifully exploited in nature in polymers such as proteins and DNA. The second main reason for polymers being so widespread arises from the first, and is that their properties can be tuned by changing their structure—for example to give hard polymers (e.g. for cooling fans), or soft polymer (e.g. for packaging or clothing). The third and perhaps most important reason for polymers being so widespread is that they can readily be processed or shaped: they can be moulded into desired shapes, or extruded into films, tubes and fibres. This simple processing or manufacture also leads to the low cost associated with most plastics.

Nearly all polymers are electrical insulators. However, there is an important different class of polymer, namely conjugated polymers. Here "conjugated" means that their structure consists of alternating single and double bonds linking carbon atoms. The result is that electrons are delocalised along the polymer chain, and the materials have semiconducting electronic properties. This is exciting as it provides a new class of semiconductor combining novel semiconducting electronic properties with the simple processing of polymers, and the scope for tuning properties by changing the structure. Semiconducting polymers can be dissolved in common solvents and deposited by very simple processes such as spin-coating or ink jet printing. This opens up the possibility of making a whole range of semiconducting electronic devices in a very simple way (Samuel 2000, Heeger 2001).

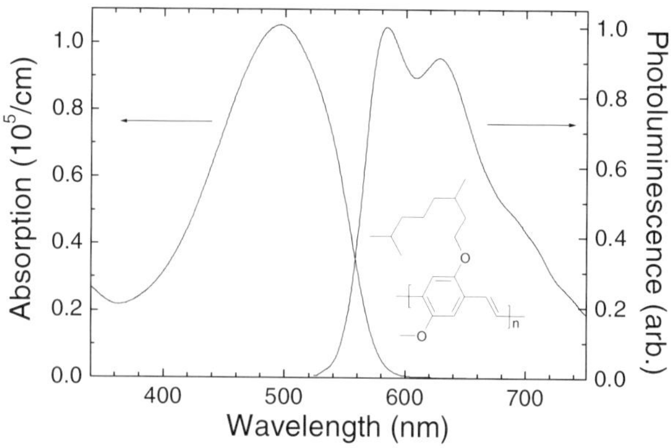

Figure 1. *Absorption and photoluminescence spectra of a thin film of a typical conjugated polymer. Notice the strong absorption and broad spectra.*

So far, polymer light-emitting diodes have been studied in most detail (Friend *et al.* 1999). In the simplest case, these devices consist of a semiconducting polymer film approximately 100nm thick in between two contacts. When a voltage is applied between the contacts, electrons are injected from one contact and holes from the other. They move under the applied field and can meet up to form an excited state (exciton) which can then emit light. This is very promising as a display technology because red, green and blue light-emitting polymers can be made and used to make a light-emitting colour display. Such a display could be flat and even flexible and as a result of being emissive would have excellent viewing angle. The drive to bring these devices to market has led to much-improved materials and tremendous advances in our understanding of these materials and the devices made from them. It means that it is now time to look beyond LEDs to other polymer semiconductor devices including polymer transistors, solar cells, lasers and optical amplifiers. Such devices would benefit from the scope for tuning properties and simple fabrication of the polymers. In this chapter we will examine the potential for polymer lasers and optical amplifiers.

In addition to the general features of semiconducting polymers outlined above, the materials have a number of specific advantages for lasers (Tessler 1999, McGehee and Heeger 2000, Turnbull *et al.* 2001a) and optical amplifiers (Lawrence *et al.* 2002). First, they have broad spectra (see Figure 1), making them attractive for tuneable lasers and capable of broadband optical amplification. The latter point is significant for ultrafast data communications in which the large bandwidth of information will demand broadband optical amplifiers. Another attractive feature is that conjugated polymers have very strong absorption of order 10^5cm^{-1} in the solid state. This means that a film of thickness 100nm can be optically pumped from the side and still absorb nearly all the incident light. It also means that the materials should have similarly strong gain. This could enable very compact lasers and optical amplifiers, in contrast to fibre devices which are up to 100m long. Another feature apparent from Figure 1 is that the absorption and emission spectra are well separated so that self absorption of emission is weak, and the material behaves as a 4 level system with low threshold.

Further advantages of conjugated polymers are that there is little concentration quenching, the possibility of electrical pumping and their compatibility with polymer fibre. Dyes are highly fluorescent in solution, but often non-fluorescent in the solid state due to formation of dimers or aggregates which quench the luminescence. In contrast, conjugated polymers can be strongly luminescent in the solid state, with photoluminescence quantum yields above 50% in some materials. Another advantage of conjugated polymers over dyes is that they are semiconductors, so that in the longer term there is the potential for them to be electrically pumped.

The energy gaps of conjugated polymers cover the visible region of the spectrum. This is attractive for lasers, as inorganic semiconductor sources do not yet satisfactorily cover all of the visible spectrum. The visible bandgap is a disadvantage for telecommunications applications in which sources and amplifiers are at 1.3–1.5μm, due to the low loss of silica fibre at those wavelengths. However, the wavelength for data communications, which is mainly over much shorter distances, is not so rigidly defined and in terms of both processing and wavelength, conjugated polymers could have excellent compatibility with polymer optical fibre (POF). POF is attractive for fibre to the home because it can readily be made in a graded index profile allowing a large core to be used with modest dispersion. In contrast, single mode silica fibres have micron dimension cores which then require very precise (and therefore expensive) connectors. In the very long term we may expect communications to move to shorter wavelength because of the larger bandwidth available. Recent advances in photonic crystal fibre now suggest a possible way of carrying visible light over large distances.

There are two features of conjugated polymers likely to be disadvantages for lasers and amplifiers. The first is that their photostability is inferior to inorganic semiconductors. It would certainly be necessary to exclude oxygen and water rigorously from any device. Photostability is a concern, but the demonstration of LEDs lasting tens of thousands of hours is encouraging. The other feature that is not favourable is that the excited state lifetimes are generally short (a consequence of the very strong absorption) so that the natural radiative lifetime is of order 1ns, and the actual radiative lifetimes are less than this due to non-radiative decay that is, of order a few hundred picoseconds. The short excited state lifetime makes it harder to maintain a population inversion.

2 Microstructured Semiconducting Polymers

In parallel with the great progress in organic semiconductors, the field of photonic crystals has advanced rapidly over the past decade (Yablonovitch 1993, Krauss and De la Rue 1999). Wavelength scale microstructure can strongly modify the emission and propagation of light, opening the way to novel photonic devices. Semiconducting polymers offer simple fabrication of photonic devices, and so are convenient systems for the study of microstructure. In particular they offer the opportunity for simple fabrication of microstructured devices. Microstructure can be used to control both spontaneous and stimulated emission. This is illustrated below in the context of polymer LEDs and lasers.

2.1 Microstructured Polymer LEDs

In light-emitting diodes, the light is generated in a semiconductor layer of much higher refractive index than the surroundings. The result is that only a small fraction of the light

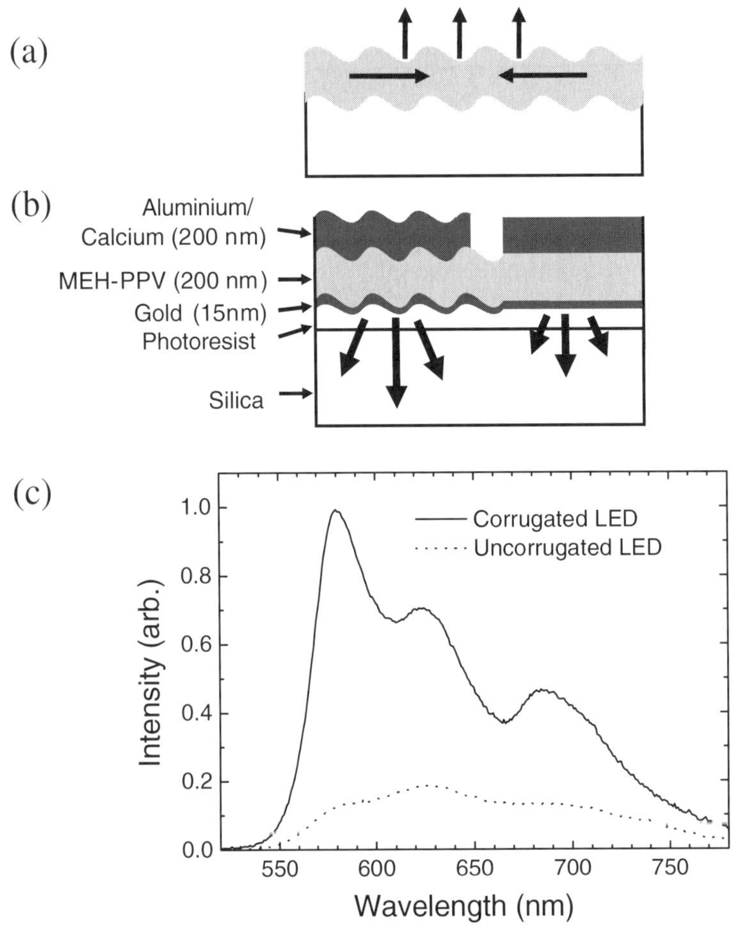

Figure 2. *(a) Schematic showing how wavelength scale corrugation can be used to Bragg scatter waveguided light out of an LED. (b) A corrugated (left) and planar (right) conjugated polymer LED. (c) Comparison of light emission from LEDs in (b).*

generated escapes from the device, whilst most of the light is totally internally reflected and trapped in waveguide modes of the semiconductor. This results in a major loss of efficiency of the light-emitting diodes. For an isotropic emitter, the fraction of light directly emitted from the device is $1/4n^2$ where n is the refractive index of the materials. In the case of organic LEDs, one contact is reflective, so that a factor $1/2n^2$ can escape. Refractive indices of organic semiconductors typically lie in the range 1.5–2.0, and so only a fifth of the light generated in the device escapes. The problem is even more severe in inorganic semiconductors because of their higher refractive index.

In collaboration with Professor Barnes at the University of Exeter, we have explored a way of using wavelength scale microstructure to overcome this problem (Lupton *et al.* 2000, Matterson *et al.* 2001). The idea is shown schematically in Figure 2(a). The light-emitting layer is corrugated like a diffraction grating. The period of the corrugation is chosen so that waveguided light is Bragg scattered out of the polymer layer. Hence light which would otherwise have been trapped in waveguide modes and wasted, can be recovered. In order to test this idea we first investigated the photoluminescence of a corrugated polymer film, and showed that the spectrum and intensity of light-emission was increased. For an orange-red polymer, the required period of the corrugation was 400nm. Implementing this idea in an LED is a challenging problem as a very strong electric field is applied to the polymer film, and any non-uniformities could lead to dielectric breakdown. The structure investigated is shown in Figure 2(b) and consists of a polymer film in between contacts. Microstructured and conventional planar LEDs were made side by side and their light output compared under identical experimental conditions (same drive current, light collection geometry etc.). The result is shown in Figure 2(c) which compares the spectrum measured in the forward direction from both corrugated and uncorrugated LEDs. It can be seen that the light emission is considerably stronger for the corrugated LED, and measurements made over a wide range of angles show that overall the efficiency of the device is doubled. Hence wavelength scale microstructure could prove a powerful way of increasing the efficiency of LEDs.

2.2 Microstructured Polymer Lasers

A variety of experiments including the observation of gain in transient absorption measurements and amplified spontaneous emission have shown that optical excitation leads to gain in conjugated polymers (Tessler 1999, McGehee and Heeger 2000). A laser consists of a gain medium and a resonator. The resonator is usually a pair of mirrors, but Bragg scattering can also be used to apply feedback and so make a laser. This can be achieved using the structure shown in Figure 2(a) and pulsed optical excitation (Kallinger *et al.* 1998, Turnbull *et al.* 2001a). First order Bragg scattering of waveguided light sends it vertically out of the structure, and so leads to output coupling. However, second order Bragg scattering reflects the waveguided light back on itself and so provides feedback. Hence a corrugated layer of a gain medium can be a simple laser structure requiring no alignment. Such devices are called distributed feedback (DFB) lasers.

1-D Polymer DFB Laser

The fabrication of a polymer DFB laser is very simple: a polymer film is spin-coated onto a corrugated silica substrate—a process that takes only a minute. Of course the preparation of the corrugated substrate is more demanding as the required period is typically 400nm. One way of making such structures is to coat a silica substrate with photoresist, expose it to the interference pattern of two laser beams, develop the photoresist and then etch the pattern into the substrate.

The operating characteristics of such a polymer laser are shown in Figure 3. At low excitation densities (Figure 3(a)) a relatively broad spectrum is observed consisting of Bragg-scattered photoluminescence, with a dip at 610nm due to the stop-band of the grating (Turnbull *et al.* 2001). As the intensity is increased, a narrow feature appears just on the long wavelength edge of the stop band, and increases in intensity as the pump power is increased (Figure 3(b)). The output energy of the laser is shown as a function of the pump pulse energy in the inset to Figure 3(b). A clear threshold is observed at 600nJ.

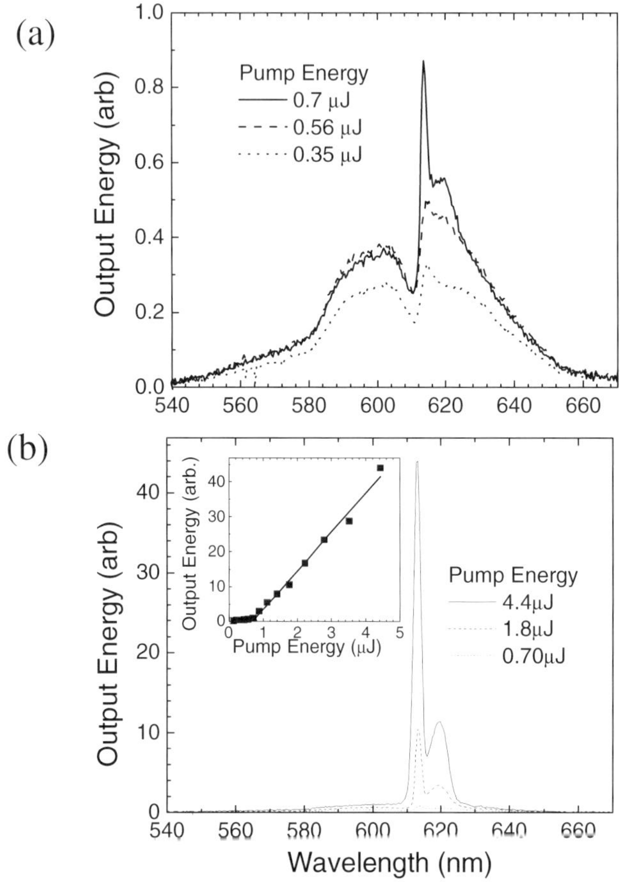

Figure 3. *Spectra of a 1-D polymer DFB laser as a function of pump energy.*

The first reported polymer laser consisted of a polymer microcavity approximately 100nm thick. As a result of the very short interaction length, it required an extremely powerful pump source—a regenerative amplifier. The DFB geometry is much more attractive because it takes advantage of the strong absorption of the polymer to absorb the pump light, but gain occurs in the plane of the film giving a much longer interaction length and reduced threshold.

We have been able to make an important contribution to understanding the detailed operation of polymer DFB lasers. This has come from comparing detailed measurements of the laser output with the photonic mode structure. In this way we find how the microstructure determines the laser emission. Figure 4(a) shows the geometry of the experiment. A corrugated semiconducting polymer film is excited optically and the emission spectrum as a function of observation angle (θ) is measured. The result of this measurement is shown in Figure 4(b). Close to zero degrees and 614nm, very strong emission is observed. This is the lasing mode. At 610nm there is a minimum in the emission, corresponding to the stop-band of the grating. The figure also has an "X" shape which is characteristic of the process. Consider a measurement made at some finite angle θ in

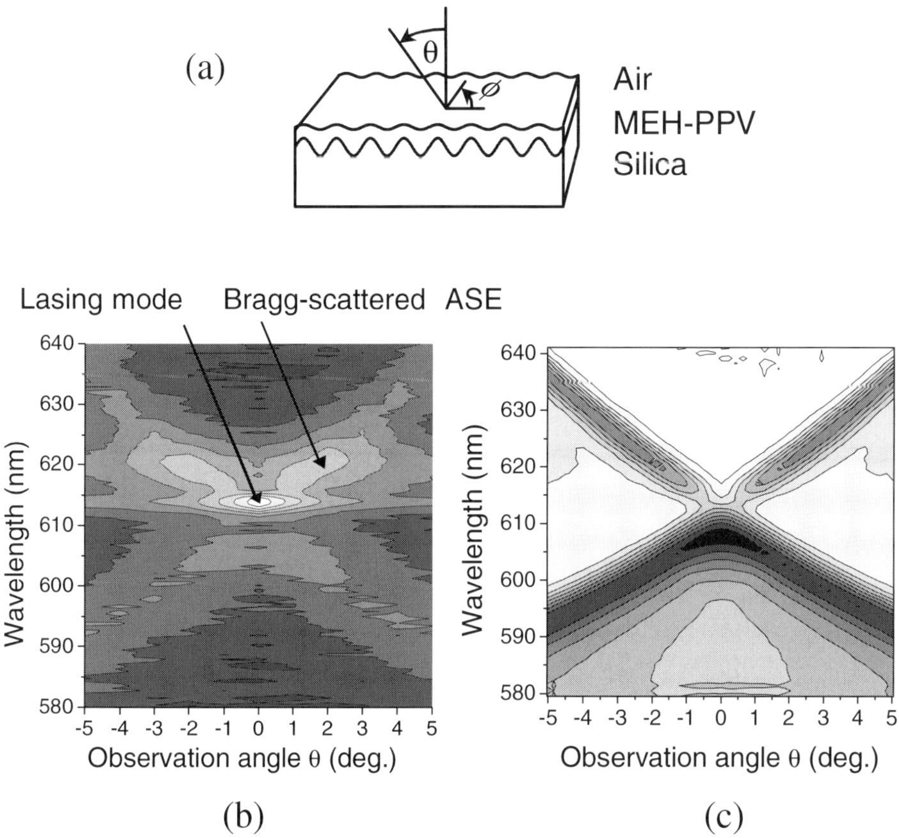

Figure 4. *(a) Geometry of angle dependent transmission measurements (b) Gray scale image showing angle dependence of emission from a 1-D polymer DFB laser. Light regions represent the strongest emission. (c) Gray scale image showing angle dependence of transmission through a 1-D polymer DFB laser. Dark regions represent low transmission i.e. strong coupling to waveguide modes.*

Figure 4(a). For light travelling to the right, one particular wavelength will be Bragg scattered (through an angle less than 90 degrees) into the direction of observation. Light travelling to the left is also Bragg scattered, but to be observed must be scattered through an angle of more than 90 degrees. This process will occur for a different wavelength of light, and so for any finite value of θ, two wavelengths will be observed: one from light travelling to the right, and the other from light travelling to the left. As the angle is increased, the wavelength difference increases, giving the "X" shape observed.

The above explanation is confirmed by measurements of the photonic mode structure. This is done by angle-dependent transmission measurements. When light which would normally be transmitted is Bragg scattered into a waveguide mode, transmission will be reduced. Hence dips in the transmission signal indicate coupling to waveguide modes and allow the effect of the corrugation on the propagation of light to be observed. The experiment measures coupling of light from outside the laser structure into it, but of course the inverse process of coupling from inside the laser to outside also occurs and is highly relevant to the laser operation. The results of these measurements on the laser structure of Figure 4(b) are shown in Figure 4(c). Dark regions correspond to reduced transmission, that is, to coupling to waveguide modes. It is immediately apparent that there are remarkable similarities between the laser operation in Figure 4(b) and the photonic mode structure in Figure 4(c). For example Figure 4(c) also shows the "X" shape due to Bragg scattering, and the stop-band at 610nm can be seen as a minimum of transmission. When we look at zero degrees, we see that coupling is much stronger on the short wavelength side of the stop-band (around 606nm) than on the long wavelength side (around 614nm). This means that output coupling from the laser is weaker at 614nm, so the quality factor of the resonator is higher, and the laser operates at this wavelength. Further details of this work are given in Turnbull *et al.* (2001a).

A Novel Tuneable Polymer Laser

The wavelength at which feedback occurs, and hence the lasing wavelength are determined by the period of the grating. As semiconducting polymers have a substantial gain bandwidth, the emission could be tuned to shorter wavelength using a shorter grating period, and to longer wavelength by a longer grating period. An elegant way of doing this is to use a chirped grating, that is one is which the period varies as a function of position, as shown in Figure 5(a). A polymer film was spun on top of this grating, and excited

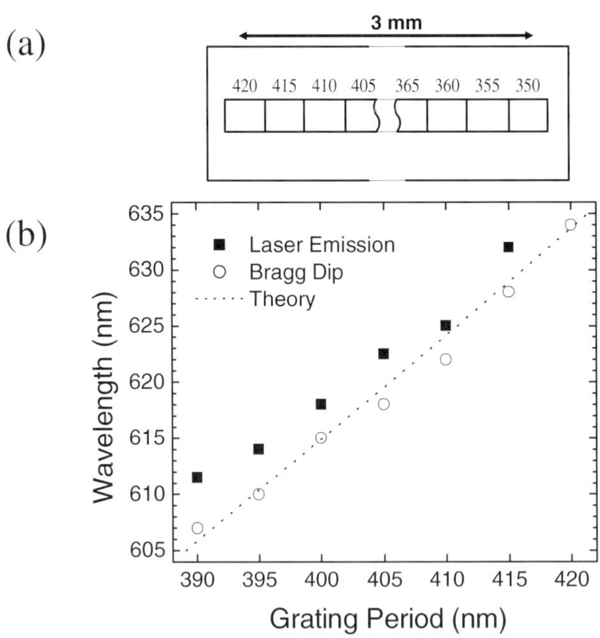

Figure 5. *A novel tuneable polymer laser. (a) Chirped grating structure used. The numbers (from 420 to 350) indicate the period of the grating in nm. (b) Wavelength of laser emission as a function of grating period for structure in (a).*

optically. The laser could be tuned in a very simple and convenient way by translating the substrate so that the pump beam was incident on different parts of the grating. The emission wavelength as a function of grating period is shown in Figure 5(b). Also shown are the calculated lasing wavelength using the refractive index of the polymer (theory) and the position of the Bragg dip corresponding to the stop band of the grating. All are found to be in good agreement with one another.

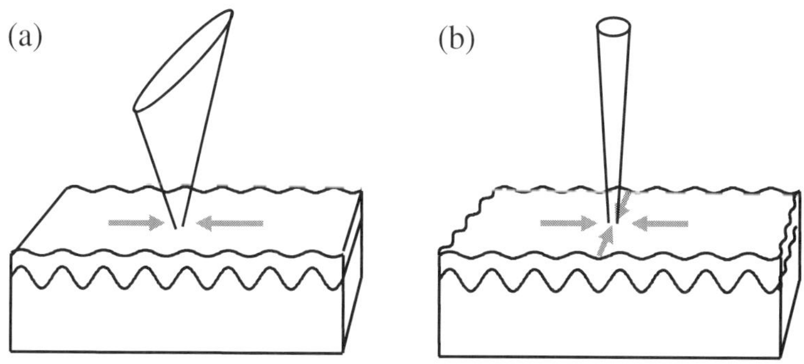

Figure 6. *(a) Diagram showing how Bragg scattering by one-dimensional corrugation leads to fan-shaped emission. (b) Diagram showing how Bragg scattering by two-dimensional corrugation gives a well-defined mode perpendicular to the substrate.*

2-D Polymer DFB Lasers

So far we have discussed lasers consisting of a simple corrugated substrate. In such one-dimensional DFB lasers, feedback only occurs in one direction. The grating out-couples light at a well defined angle θ (in Figure 4(a)), but gives no control of direction in the plane so that the emission is fan-shaped, as shown in Figure 6(a). An improved laser can be made using two perpendicular gratings to give feedback in two perpendicular directions. This approach has two important advantages. First it reduces the effective mode volume and so should reduce the threshold for lasing. The second advantage is that emission will be perpendicular to both gratings, and is therefore expected to be in a well-defined beam perpendicular to the substrate, as illustrated in Figure 6(b).

We, and others, have made such two-dimensional lasers (Riechel *et al.* 2000, Turnbull *et al.* 2003a). Like their one-dimensional counterparts the spectrum narrows above a threshold; the graph of output power versus input power shows a clear threshold. However, the threshold is reduced to be of order 10nJ, and a near to diffraction limited output beam is obtained. The laser operation can also be fully understood in terms of the photonic mode structure measured by angle-dependent transmission (Turnbull *et al.* 2003a).

Using this approach we have been able to make a major advance in reducing the size of the pump laser. As mentioned earlier, the first polymer laser was pumped by a regenerative amplifier giving enormous peak powers, and subsequent polymer lasers have generally been pumped by large Q-switched Nd:YAG or nitrogen lasers. As a result of our work towards reducing threshold, we have recently been able to demonstrate a polymer laser pumped by a microchip laser the size of a (large) matchbox (Turnbull *et al.* 2003b).

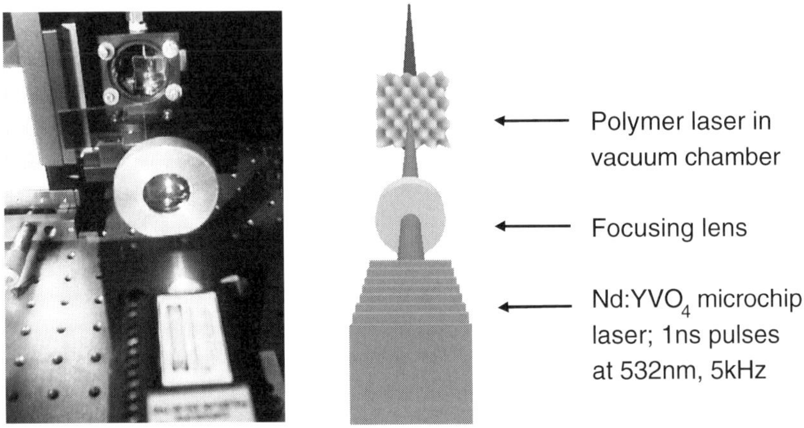

Figure 7. *Photograph and diagram of a compact polymer laser consisting of a 2D corrugated polymer film pumped by a matchbox-sized microchip laser.*

The configuration used is shown in Figure 7, and is a significant step towards practical polymer lasers. Because it is a DFB laser, needing no alignment, it even works when handheld in the pump laser beam, and the polymer laser beam is clearly visible. It has proved a useful tool for proving to colleagues that (optically pumped) polymer lasers do work!

The work described above has involved a progression from feedback in one direction to feedback in two directions. The next stage of this work is to explore feedback in all directions by making circular grating structures of the type shown in Figure 8. Early results are encouraging and with optimisation this approach should lead to further reductions in threshold.

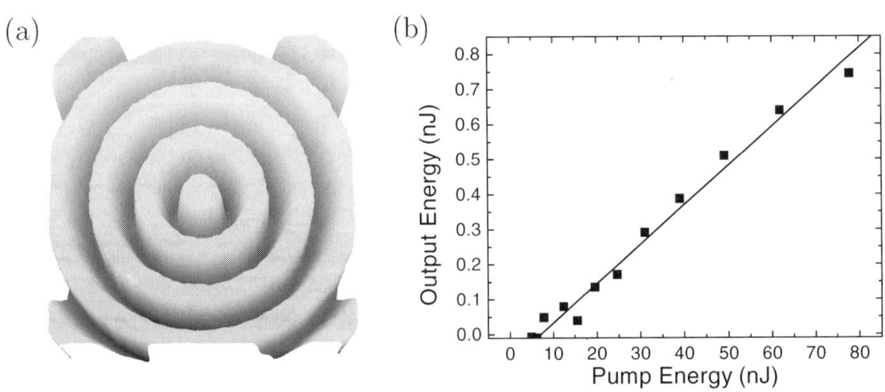

Figure 8. *(a) A circular grating structure of diameter 2μm with grooves of depth 0.1μm. (b) Performance characteristics of a circular grating polymer laser.*

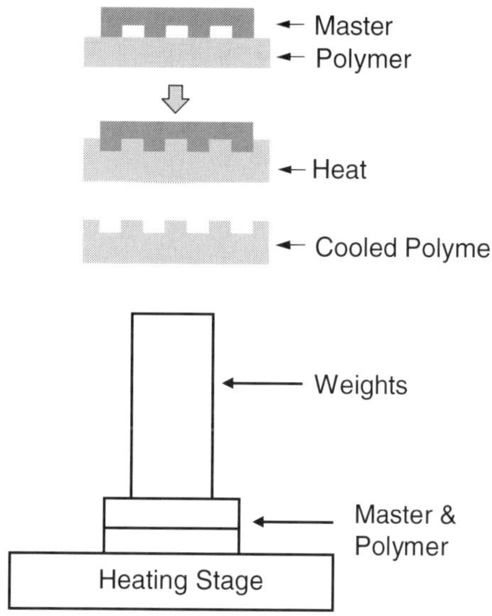

Figure 9. *Processing sequence (top) and experimental configuration (bottom) for hot embossing lithography.*

Fabrication of Microstructure

The successful polymer lasers described here all use wavelength scale microstructure to provide feedback. In fact the microstructure typically has a period of 400nm, which is below what can be achieved by conventional photolithography. One way of making such microstructure is by electron beam lithography. Although very accurate it is a slow process using very sophisticated equipment. An alternative is to expose photoresist to the interference pattern of two laser beams, but this also requires considerable care. These complicated techniques would reduce the advantage of simple fabrication stated at the beginning as one of the motivations for organic semiconductors. We have therefore explored the scope for simple fabrication using a family of techniques called soft lithographies. A master structure that has been made by conventional techniques is taken and pressed into the semiconducting polymer which is simultaneously heated above its glass transition temperature as shown in Figure 9. Above this temperature, the polymer softens, and the pattern is transferred to the polymer, and will remain there permanently once the polymer is cooled below the glass transition temperature. The apparatus involved is very simple—a hotplate is used to heat the polymer, and pressure is applied using weights. Remarkably this simple method successfully transfers both one and two dimensional corrugation of 400nm period into the polymer (Lawrence *et al.* 2002a). It provides a very simple way of making the required microstructure, and provides a vivid demonstration of the simple processing of organic semiconductors. Very recently we have reported an even simpler approach using solvent instead of heat to soften the polymer and allowing a DFB laser to be patterned in only two minutes (Lawrence *et al.* 2003).

Prospects for Electrical Pumping

All polymer lasers are currently optically pumped. The most important reasons for this are that the current densities required to achieve lasing are higher than most polymers can tolerate. Our strategy has been to reduce this gap by improving the optical design of polymer lasers to reduce threshold. Electrical pumping also introduces additional problems of losses associated with the contacts and absorption of light by the injected charges. As a step from the polymer films described above towards a real device with metal contacts, we have shown that it is possible to build an (optically pumped) polymer laser on a corrugated metal grating (Andrew *et al.* 2002). However, making an electrically pumped polymer laser remains a very challenging problem.

3 Towards Polymer Optical Amplifiers

An optical amplifier is a device which takes a weak light pulse and amplifies it into a strong one. Like a laser, it works by stimulated emission, but unlike a laser there is no cavity. Optical amplifiers are commonly used in telecommunications to counteract the effects of losses in fibres. They are also needed when a weak signal must be divided between several receivers. High speed communications needs amplifiers which give gain over a wide range of wavelengths, and polymers are worth exploring in this context because of their broad spectra and other advantages mentioned in the introduction. A range of experiments on semiconducting polymers and lasers made from them has shown that the materials have gain, but there were no reports of polymer optical amplifiers, that is, devices that used a polymer to amplify a weak light pulse into a much stronger one.

We therefore investigated the feasibility of making a polymer optical amplifier. We chose to work with a polymer solution in a 1cm cuvette. This was excited by the output of a nitrogen laser, and the presence of gain detected by a weak probe beam passing through the polymer cuvette. The probe beam was generated by a dye laser pumped by the same nitrogen laser, and could be tuned to allow the gain to be measured over a range of wavelengths. The results are shown in Figure 10. There are three important features in the data. First gain is observed. Second, the gain is very large—over 35dB (corresponding to amplification by a factor of 3000) in only 1cm of gain material. The third point is that high gain is maintained over a range of at least 50nm, which corresponds to a bandwidth of 50THz. Hence we have demonstrated compact broadband optical amplification that we suggested might be one of the advantages of polymers. By optimising the concentration and going to small input energies, even higher gains of up to 44dB can be obtained. The effect of input pulse energy on gain is shown in Figure 10(b), and the solid line shows the fit to the well known expression for a homogeneously saturated pulsed amplifier (Koch *et al.* 1982):

$$E_{\text{out}} = E_{\text{sat}} \ln\left\{1 + \left[\exp(E_{\text{in}}/E_{\text{sat}}) - 1\right] \exp(G)\right\}$$

where E_{in} is the input pulse energy, E_{out} is the output pulse energy, E_{sat} is the saturation energy and G is the small signal gain coefficient. Further details of these promising results are given in Lawrence *et al.* (2002b), and we are now working towards solid state versions of this device.

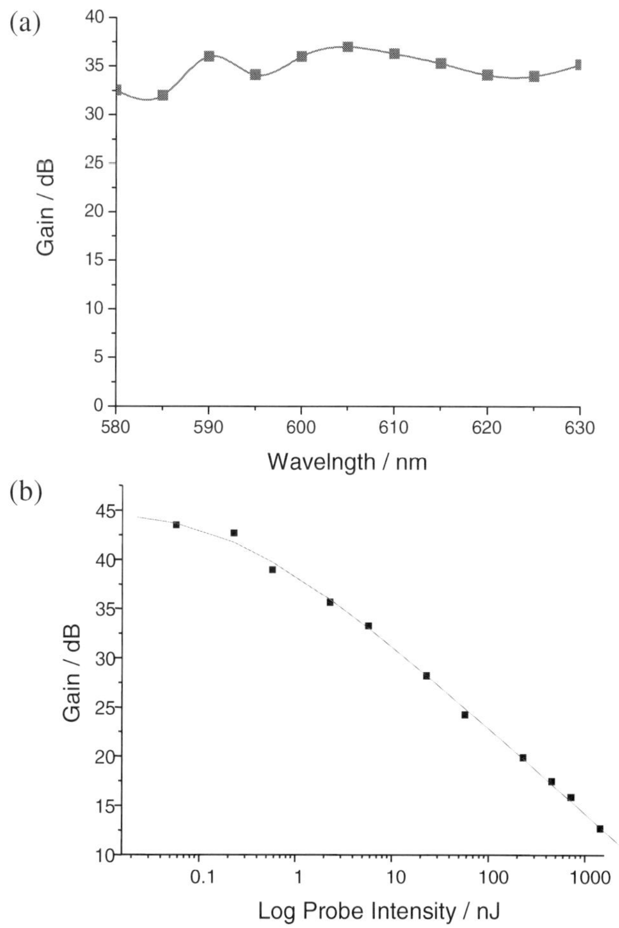

Figure 10. *(a) Example of gain measurements of a polymer amplifier as a function of wavelength. (b) Effect of probe pulse energy on gain of a polymer amplifier.*

4 Conclusion

Conjugated polymers are easily processed semiconductors that can be used to make a range of optoelectronic devices such as LEDs, lasers and very recently optical amplifiers. Much further work is needed—for example the lasers and amplifiers are pulsed at present. However, recent progress has been rapid allowing a compact optically pumped polymer laser to be made, and also demonstrating a broadband optical amplifier. There is considerable scope to control the emission, propagation and amplification of light by microstructure, which can now be made in a very simple way using an embossing process. These factors should combine to give further advances in polymer photonics.

Acknowledgements

It is a great pleasure to acknowledge the work of Dr G.A. Turnbull and Mr J.R. Lawrence who performed the experiments reported here. I should also like to acknowledge collaborations with the groups of Prof W.L. Barnes in Exeter, and Prof. A. Shore in Bangor. Finally I should like to acknowledge the Royal Society for a University Research Fellowship, and the SHEFC and EPSRC for financial support.

References

Andrew P, Turnbull G A, Barnes W L and Samuel I D W, 2002, *Appl. Phys. Lett.* **81** 954.

Friend R H, Gymer R W, Holmes A B, Burroughes J H, Marks R N, Talianai C, Bradley D D C, Dos Santos D A, Bredas J L, Loglund M and Salaneck W R, 1999, *Nature* **397** 121.

Heeger A J, 2001, *Rev. Mod. Phys.* **73** 681.

Kallinger C, Hilmer M, Haugeneder A, Perner M, Spirkl W , Lemmer U, Feldmann J, Scherf U, Mullen K, Gombert A and Wittwer W, 1998, *Adv. Mater.* **10** 920.

Koch T L, Chiu L C and Yariv A, 1982, *Opt. Commun.* **40** 364.

Krauss T F and De la Rue R M, 1999, *Proc. Quant. Elec.* **23** 51.

Lawrence J R, Andrew P, Barnes W L, Buck M, Turnbull G A and Samuel I D W, 2002a, *Appl. Phys. Lett.* **81** 1955.

Lawrence J R, Turnbull G A and Samuel I D W, 2002b, *Appl. Phys. Lett.* **80** 3036.

Lawrence J R, Turnbull G A and Samuel I D W, 2003, *Appl. Phys. Lett.* **82** 4023.

Lupton J M, Matterson B J, Samuel I D W, Jory M J and Barnes W L, 2000, *Appl. Phys. Lett.* **77** 3340.

Matterson B J, Lupton J M, Safonov A F, Salt M G, Barnes W L and Samuel I D W, 2001, *Adv. Mater.* **13** 123.

McGehee M D and Heeger A J, 2000, *Adv. Mater.* **12** 1655

Riechel S, Kallinger C, Lemmer U, Feldmann J, Gombert A, Wittwer V and Scherf U, 2000, *Appl. Phys. Lett.* **77** 2310.

Samuel I D W, 2000, *Phil. Trans. R. Soc. Lond.* **A 358**, 193.

Tessler N, 1999, *Adv. Mater.* **11** 363.

Turnbull G A, Andrew P, Jory M J, Barnes W L and Samuel I D W, 2001a, *Phys. Rev.* **B 64** 125122.

Turnbull G A, Krauss T F, Barnes W L and Samuel I D W, 2001b, *Synth. Met.* **121** 1757.

Turnbull G A, Andrew P, Barnes W L and Samuel I D W, 2003a, *Phys. Rev.* **B 67** 165107.

Turnbull G A, Andrew P, Barnes W L and Samuel I D W, 2003b, *Appl. Phys. Lett.* **82** 313.

Yablonovitch E, 2001, *Sci. Am.* **285** 46.

Optical spintronics for switching

Alan Miller

University of St Andrews, Scotland

1 Introduction

The ability to control and manipulate the spin of electrons and holes in semiconductors has become a very active area of research, driven by the prospect of developing fundamentally new types of electronic and optoelectronic devices (Awschalom and Kikkawa 1999). Spin injection from ferromagnetic metals into semiconductors is one of the routes being investigated for the realisation of spin transistors, giving rise to the term *spintronics*. Spin dynamics may also have a role in quantum computing elements, for instance the spin states of single electron quantum dots can act as qubits (Loss and DiVincenzo 1998). The physics of spin relaxation and spin transport in semiconductors has therefore become crucial in the development of 'spin electronics'. In this chapter, we consider the consequences of spin polarisation on the optical properties of semiconductors. In particular, we emphasise spin-dependent nonlinear optical interactions associated with excitons in low dimensional semiconductors at room temperature. Time resolved experiments that access nonlinear optical interactions on picosecond and femtosecond timescales provide insights into electron spin relaxation and spin transport in semiconductors, as well as offering prospects for new all-optical switching devices.

2 Selection rules in quantum wells

Optical transitions between energy bands in low dimensional semiconductors offer a means of producing 100% spin-polarised electrons and holes. Figure 1 compares typical band structures in bulk and quantum well, direct gap, III–V semiconductors with cubic zinc blende crystal structures. Bulk semiconductors have band gaps characterised by a low effective mass conduction band and two valence bands with different effective masses. These valence bands are degenerate at $k = 0$, and split into 'heavy' and 'light' hole bands at finite k. In fact, each of the bands shown is doubly degenerate, corresponding to the two possibilities for electron spin, $m_j = \pm 3/2$(heavy hole) and $m_j = \pm 1/2$ (light hole).

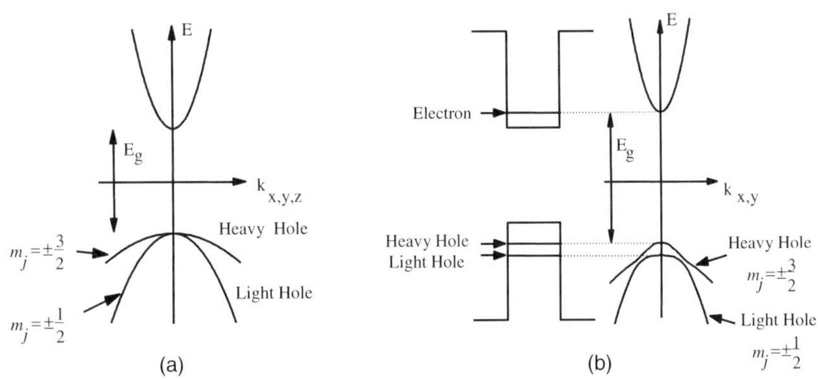

Figure 1. *Band structures of (a) bulk and (b) quantum well III–V semiconductors*

Carrier confinement in quantum wells leads to lifting of the degeneracy of the heavy and light hole valence bands at the Brillouin zone centre ($k = 0$). Figure 1(b) illustrates a Type I quantum well system (lowest conduction band and highest valence bands in the same material). Here the valence bands are distorted such that the heavy hole becomes lighter and the light hole heavier. Strain can further complicate the shape of the bands near $k = 0$; however we will limit our discussion to unstrained quantum wells.

Figure 2 gives wave-functions for the conduction and valence bands at $k = 0$. These functions highlight the form, symmetry and spin states based on the atomic eigenfunctions from which they arise. Note that the symmetries of the wave-functions for the heavy hole and conduction bands have a relatively simple (single spin) form. On the other hand, the light hole bands contain components of both spins (similarly, the spin-orbit split bands, lower in energy and not shown here, also have mixed spin wave-functions). Optical transitions are allowed between states of like spin.

Interband transitions from the $|X \pm iY\rangle$ symmetry heavy hole valence bands to the symmetric $|S\rangle$ states of the conduction band give the result that circularly polarised light (right and left circular polarisation states represented by σ_+ and σ_- respectively) will produce spin polarised electrons and holes. Note that a given circular polarisation produces electrons of opposite spin for transitions originating from the heavy and the light valence bands. The same selection rules apply in bulk semiconductors, but the ability to spectrally isolate transitions from the heavy hole band allows the creation of distributions

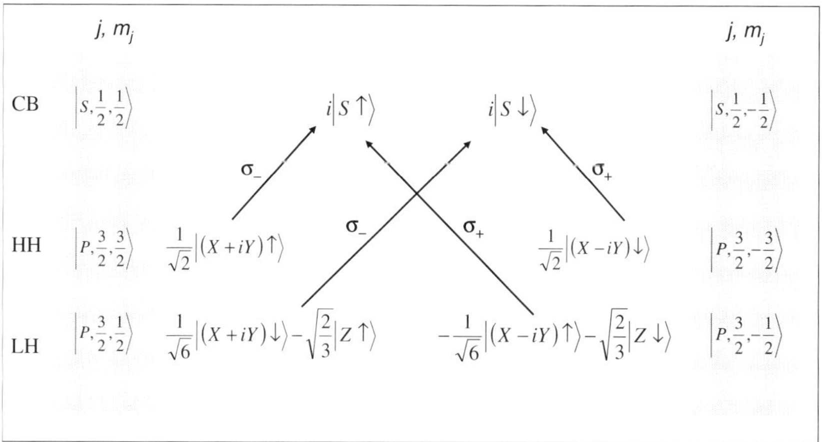

Figure 2. *Wavefunctions and optical selection rules for transitions between the heavy hole (HH) and light hole (LH) valence bands and the conduction band (CB) states.*

of 100% polarised electrons and holes in quantum wells.

To complete our picture of optical transitions from the highest valence band levels, Figure 3 shows the allowed transitions available for linearly polarised light between the light hole and the conduction bands. These transitions are not available in a quantum well under the condition of light travelling perpendicular to the wells. We are now in a position to describe spin dependent optical phenomena in quantum wells, but first we will briefly review optical nonlinearities associated with excitons in semiconductors.

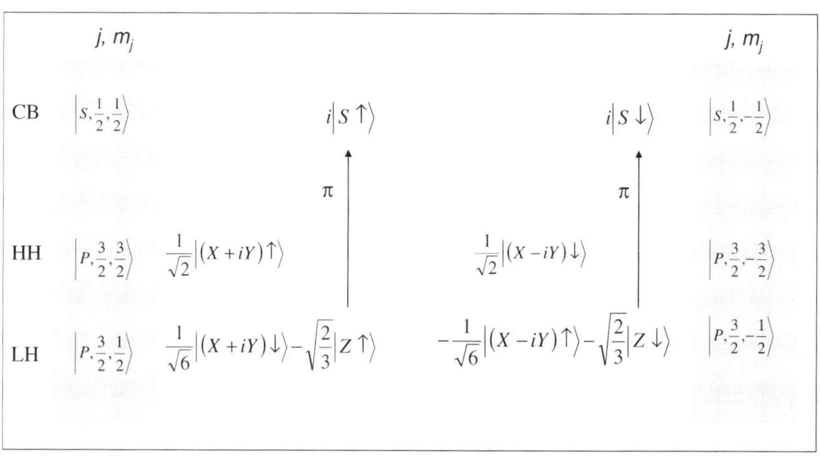

Figure 3. *Allowed transitions between the light hole band and the conduction band for linearly polarised light.*

3 Optical nonlinearities

All-optical switching requires knowledge of optical nonlinearities and their timescales. Optical nonlinearities associated with excitons in quantum wells are very sensitive and offer methods of accessing spin-related phenomena. Figure 4 illustrates the typical spectral dependence of the absorption coefficient of a GaAs quantum well at room temperature.

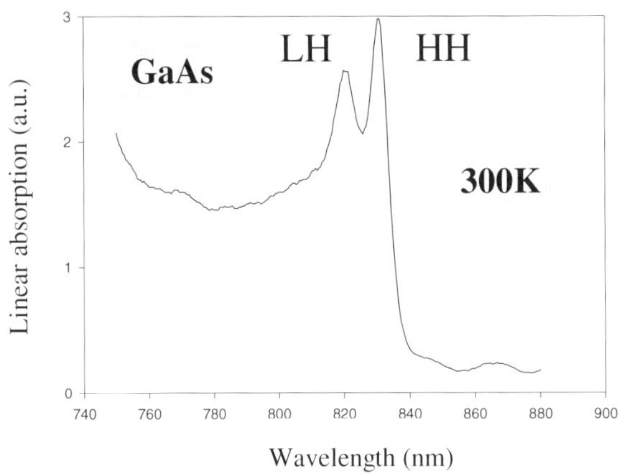

Figure 4. *Absorption coefficient of GaAs quantum wells showing the exciton doublet.*

Excitons associated with transitions from both heavy hole (HH) and light hole (LH) valence bands are clearly resolved compared to bulk semiconductors at room temperature. This results from the stronger electron-hole overlap and larger binding energy for excitons in two dimensions; see Figure 5.

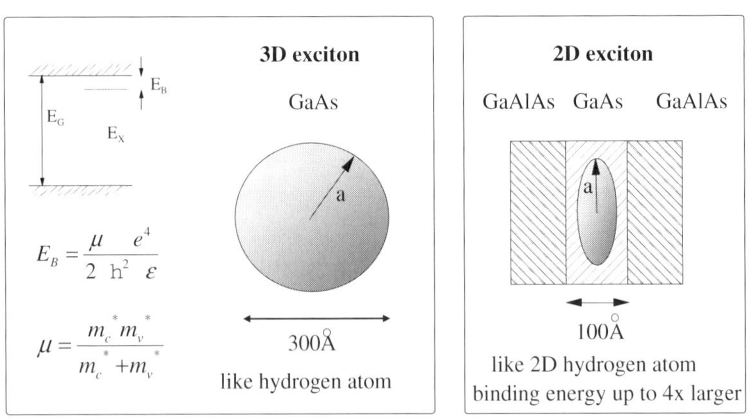

Figure 5. *Excitons in bulk and quantum well semiconductors*

Moderate densities (greater than 10^{10}cm^{-2}) of excess carriers can easily quench excitonic absorption features (Miller *et al.* 1983). At room temperature excitons ionise into free electrons and holes on sub-picosecond timescales and these free carriers inhibit the creation of more excitons. The excitonic saturation can be divided into phase-space filling (PSF) and Coulomb contributions (Schmitt-Rink *et al.* 1985). PSF arises from the Pauli exclusion principle and is therefore spin dependent since each k-state in the band can accommodate only one spin up and one spin down electron. There is therefore a spin dependent reduction of the exciton oscillator strength when k states associated with the exciton are occupied. Coulomb screening results in a decrease in the potential of a charge when other charges are present. The addition of a charge (occupying any k-state) causes a redistribution of the carriers and a reduction in the overall energy of the system, resulting in a smaller oscillator strength. The exciton binding energy is also reduced. To first order, Coulomb screening is independent of spin.

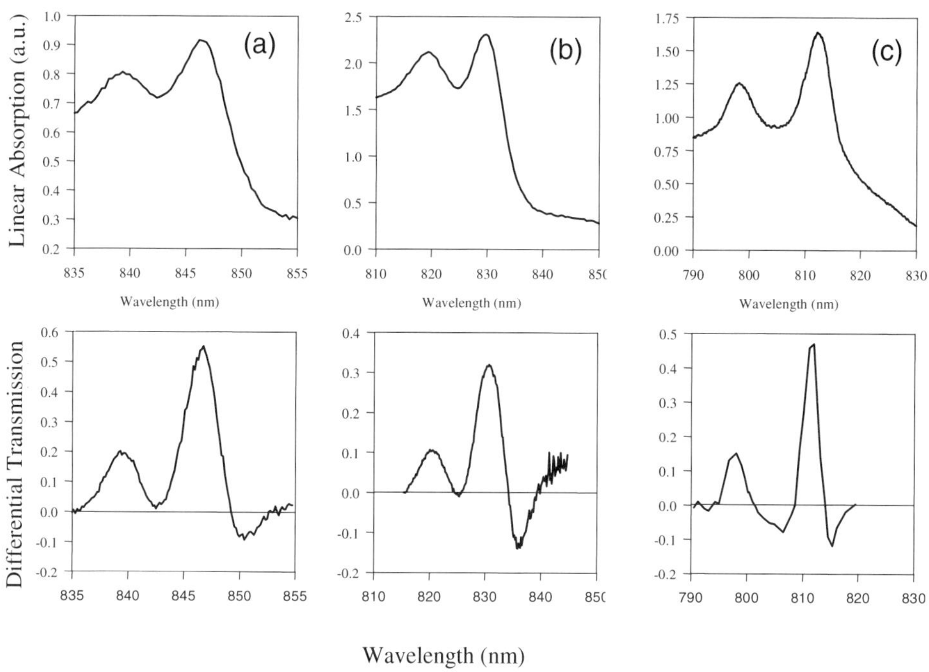

Figure 6. *Comparison of absorption coefficients and transmission changes of heavy and light hole excitons due to excess free carriers in room temperature GaAs quantum wells with (a) 9.0, (b) 6.5, and (c) 4.4nm well widths.*

Figure 6 shows the spectral transmission changes in GaAs quantum wells at room temperature due to exciton saturation in pump-probe measurements for three different quantum well widths. Positive changes are due to a combination of PSF and Coulomb screening. The negative signals on either side of the heavy and light hole excitons arise from broadening of the excitons. Optically created free carriers increase the scattering rate, and hence reduce the exciton lifetime, producing broadening of the exciton spectral width that was otherwise determined by phonon scattering at room temperature. In GaAs quantum wells it was found that carrier induced screening remains an important contribution to excitonic transmission changes, although PSF becomes increasingly dominant in narrower wells, (Miller *et al* 1999). Table 1 shows the relative strengths of mechanisms contributing to exciton saturation from time resolved saturation measurements.

Well width (nm)	9.0	6.5	4.4
Spin relaxation time (ps)	180	90	50
Free carrier screening	1	1	1
Free carrier PSF	1.5	3.2	3.6
Exciton screening	0	0	0
Exciton PSF	11.1	13.2	14.0

Table 1. *Spin relaxation times and relative magnitudes of different components contributing to exciton saturation in GaAs wells of different width (Miller et al. 1999)*

Circularly polarised light can be used to produce circular dichroism in quantum wells. This dichroism can be employed to measure the relative magnitudes of the PSF and Coulomb contributions because of the spin dependence of PSF. This also provides a means of directly measuring spin relaxation times as described in the next section.

4 Spin relaxation

Spin relaxation times can be determined by resonant pump probe techniques. Electron spin decay times are found to be on the order of a few picoseconds (Tackeuchi *et al.* 1996, Britton *et al.* 1998). Because of interband scattering and the mixed spin nature of the light hole band, hole spin decay is much faster, on sub-picosecond times.

Experimental measurements of the spin relaxation rates can make use of the optical nonlinearity associated with exciton saturation. As illustrated in Figure 1, circularly polarised pulses resonant with the heavy hole transition, excite only one electron spin state. If the sample is then probed with a delayed pulse with the same circular polarisation then there will be an enhancement of the transmission due to PSF and Coulomb screening as long as the spin polarisation is maintained. For the case of a probe pulse with the opposite circular polarisation, an empty spin state is examined and then only Coulomb contributions will be observed. As the populations of the two spin states equalise over time, the transmission of the two cases will equalise on the timescale of the spin relaxation. Typical results are shown for an InGaAs quantum well sample in Figure 7.

Three mechanisms are proposed to explain spin relaxation of electrons in semiconductors: D'Yakonov-Perel (DP), Elliot-Yafet (EY) and Bir, Aronov and Pikus (BAP). The DP and EY mechanisms are derived from band structure effects and the BAP mechanism is a result of magnetic interactions.

The DP mechanism is based on the breaking of spin degeneracy ($E(k,\uparrow) \neq E(k,\downarrow)$) as a consequence of spatial inversion (Goldoni and Fasolino 1993, D'Yakonov and Perel

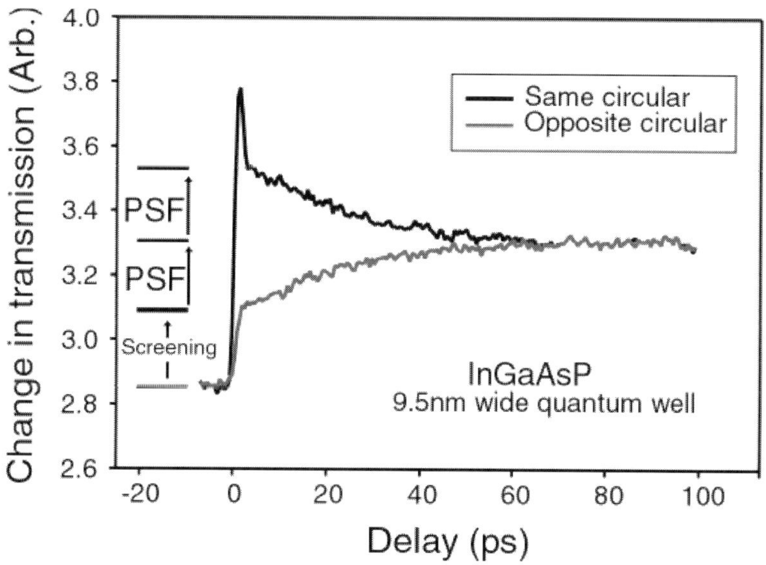

Figure 7. *Change in transmission as a function of delay for same (SCP) and opposite (OCP) circular polarisations of the pump and probe at 1550nm in InGaAs quantum wells at room temperature.*

1971). The spin-orbit interaction results in a small splitting of the conduction band. This is equivalent to the existence of a pseudo-magnetic field, defined by the momentum direction, which causes a precession of the electron spins. Scattering of the electrons changes the momentum, which leads to a rotation of the precession axis and thus allows the spins to flip. If the time between collisions is less than the precession period the electron spins are unable to follow the frequent changes of the precession axis. Therefore, as the collision rate increases this mechanism becomes less effective.

The EY mechanism considers the effect of spin-orbit coupling on normal electron-phonon and electron-impurity scattering (Elliott 1954, Yafet 1963, Tackeuchi *et al.* 1999a). The mechanism is a consequence of mixing of the valence band wavefunctions with conduction band wavefunctions at finite k values. This mixing allows electrons to spin flip due to momentum scattering from optical and acoustic phonons and/or impurities. The spin relaxation rate increases with an increasing momentum relaxation rate, contrary to the DP mechanism.

The BAP mechanism is based on the magnetic interaction between an electron and hole where the spins are flipped in the event of a collision (Bir 1975). The probability of this is determined by the exchange interaction between electrons and holes. This mechanism is important when excited electrons are surrounded by a high concentration of holes so is dominant in heavily p-doped semiconductors and at low temperatures.

Measurements in GaAs MQWs concluded that the DP mechanism dominated at room temperature (Britton *et al.* 1998). The DP effect is enhanced due to the increased spin-orbit splitting of the conduction band. This gives greater local electric fields hence accelerating the precessions that cause the spins to flip. A theoretical comparison of the spin relaxation rates for bulk ($\tau_{s,3D}$) and quantum well semiconductors ($\tau_{s,2D}$) by D'Yakonov and Kachorovskii 1986) (DK theorem) predicted an enhancement due to the DP mechanism given by the equation,

$$\frac{1}{\tau_{s,2D}} = \frac{4}{(k_f L)^4} \frac{1}{\tau_{s,3D}}, \tag{1}$$

where L is the well width and k_f the Fermi wavevector for the 3D density.

It has been useful to determine the spin relaxation rates measured in this way, as a function of well width, L in different semiconductors. Strictly, the DP theory depends on the first electron confinement energy, E_{1e}, in the wells rather than the well width. The confinement energy depends on the well width, depth, alloy concentration and effective mass. The spin relaxation rates are plotted against E_{1e} in Figure 8. It is immediately evident that there is an increase in relaxation rate for increasing confinement energy.

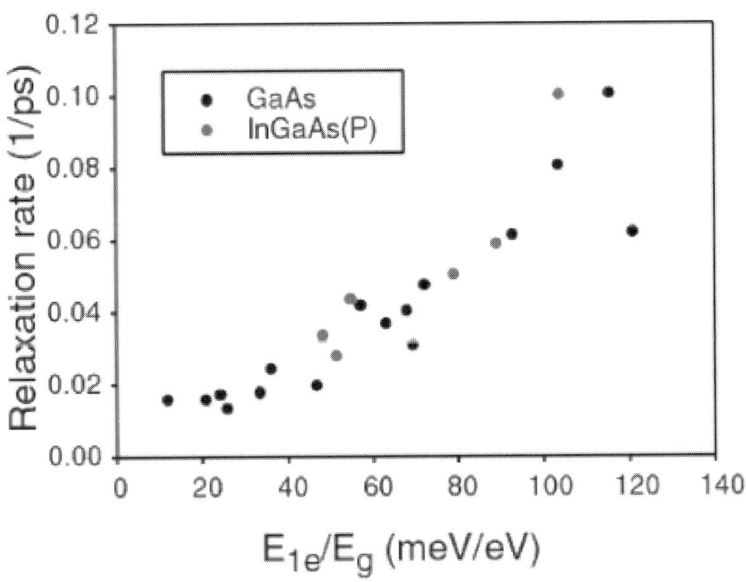

Figure 8. *Spin relaxation rates for InGaAsP and GaAs quantum wells as a function of the first electron confinement energy divided by the bandgap. (Britton et al. 1998, Tackeuchi et al. 1999b and Marshall et al. 2002).*

It can be observed that the InGaAsP points fit along the same line as the GaAs points. From this we can draw the conclusion that the dominant spin relaxation mechanism is the same for both systems. It has been found that the DK theorem differs by a factor of ten from experimental measurements (Lau *et al.* 2000). This is because of the assumption that the in-plane momentum of the quantum well, k, is much less than π/L at an energy $k_B T$ above the band edge. This is not satisfied for common well widths at room temperature. Lau *et al.* (2000) considered both the inversion symmetry and the spin-orbit coupling in order to calculate the spin splitting. This modified DP model agrees well with experimental results for a number of multiple quantum well materials.

Summarising, the results suggest that the dominant mechanism for spin relaxation arises from the spin-splitting of the conduction band (the D'Yakonov-Perel mechanism). However, it is likely that there could be contributions from the Elliott-Yafet process as both mechanisms are drawn from band-structure effects. The rapid spin relaxation times in InGaAsP at 1.55μm show good potential for incorporation in fast optoelectronic switching devices.

5 Optical switching

The induced circular dichroism demonstrated in Section 5 can be exploited in polarisation switching devices. In the previous section only absorptive saturation was discussed. Here we extend this to nonlinear refractive circular dichroism.

An AlGaAs multiple quantum well etalon structure was developed as a polarisation switch by Nishikawa *et al.* in 1996 utilising the effects of spin polarisation on linearly polarised light at 800nm. They demonstrated a room temperature all-optical spin polarisation switch exhibiting a 7ps time constant using a 4ps switching pulse and an on/off ratio of 4:1. Of interest for telecommunications applications are the results on InGaAs quantum wells around 1500nm. J. Hyland *et al.* (1998) demonstrated room temperature switching times of 5 to 20ps at 1.52 microns in these materials. In these examples, spin polarised electrons were created using circularly polarised light resonant at the exciton resonance as described previously. We have seen that this produces circular dichroism in the transmission through changes in the absorption coefficient, but a causal consequence of these modifications are corresponding changes in refractive index for the two senses of circular polarisation. When time delayed pulses that are linearly polarised are incident in the semiconductor conditioned in this way, we need to consider the effects on the two components of circular polarisation. Different refractive indices for the two circular components give rise to a phase difference and hence a rotation of the major axis of the polarisation. The net effect of absorptive and refractive changes on the probe will be a rotated elliptical polarisation. If the electric field of the two circular polarisation components are represented by:

$$E_+ = \exp(i\omega t)\exp(-\alpha_l z)\exp(in_l kz), \tag{2a}$$

$$E_- = \exp(-i\omega t)\exp(-\alpha_r z)\exp(-in_r kz), \tag{2b}$$

where α_l and α_r are the absorption coefficients and n_l and n_r the refractive indices experienced by each of the polarisations. The propagation distance through the sample, light frequency and wave-vector are denoted by z, ω and k respectively.

Therefore, a circularly polarised pump will cause a transient rotation of the polarisation and induced ellipticity of the linear probe as well as a net transmission change. This can be utilised to create an optical polarisation switch. This consists of a conventional pump-probe set-up with a circularly polarised pump and a linearly polarised probe. An analyser cross-polarised with the initial probe beam is positioned after the sample. An

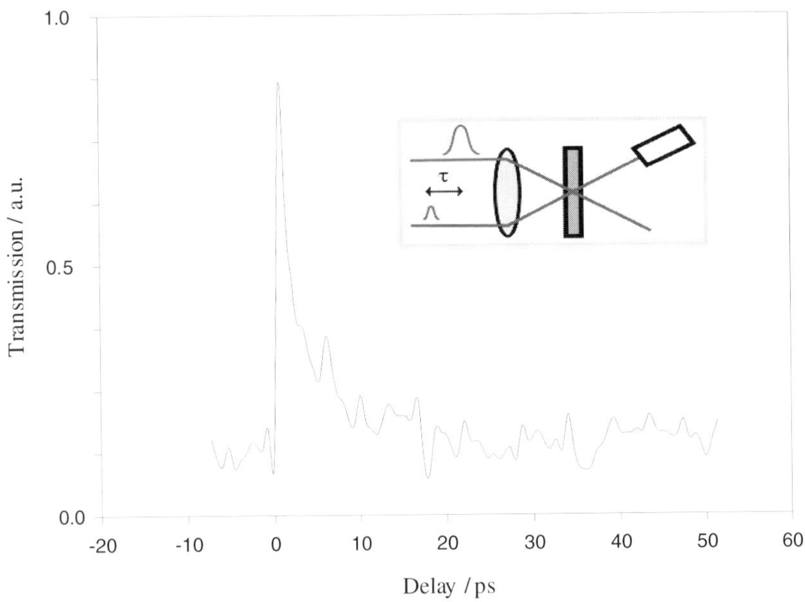

Figure 9. *All optical polarisation switching and recovery in an InGaAs quantum well at room temperature*

example of such an experiment shows the resulting polarisation rotation and recovery in Figure 9 for InGaAs quantum wells (Hyland *et al* 1999). The pump beam is chopped and the modulation is transferred to the probe via both the absorptive and refractive changes. Only when there is a net spin polarisation will there be transmission through the analyser and a signal detected. This creates a polarisation switch that decays with the spin relaxation time. Ideally, we would like to have minimum ellipticity and a rotation of 90 degrees.

Conclusions

We have illustrated here that spin dependent optical nonlinearities in semiconductors prove useful for a number of reasons. Spin relaxation times can be conveniently measured by accessing circular dichroism in pump-probe measurements; fundamental optical interactions with excitons can be exposed using spin as a method of distinguishing different optical saturation mechanisms; and spin dependent optical nonlinearities offer opportunities for all-optical switching devices. These techniques can be extended to other low dimensional systems such as quantum dots. In addition, transport properties can be monitored by employing spin dependent optical nonlinearities (Cameron *et al.* 1996). Electron spin therefore offers a variety of opportunities for exploitation that overall gives us a deeper understanding of optical-semiconductor interactions.

References

Awschalom D D and Kikkawa J M, 1999, *Physics Today* **52(6)** 33.

Bir G L, Aronov A G, and Pikus G E, 1975, *Sov. Phys. JETP* **42** 1976.

Britton R S, Grevatt T, Malinowski A, Harley R T, Perozzo P, Cameron A R and Miller A, 1998, *Appl. Phys. Lett.* **73** 2140.

Cameron A R, Riblet P, Miller A 1996, *Phys. Rev. Lett.* **76** 4793

D'Yakonov M I and Kachorovskii V Y, 1986, *Sov. Phys. Semicond.* **20** 110.

D'Yakonov M I and Perel V I, 1971, *Sov. Phys. JETP* **33** 1053.

Elliott R J, 1954, *Phys. Rev.* **96** 260.

Flatte M, 2000, *'Spin relaxation in InGaAs(P).'*

Goldoni G and Fasolino A, 1993, *Phys. Rev. Lett.* **70** 2051.

Hyland J T, Kennedy G T, Miller A, and Button C C, 1998, *IEEE Photonics Technol. Lett.* **10** 1419.

Lau W H, Olesberg J T, and Flatte M E, 2000, *Phys.Rev* **B64** 162301(R)

Loss D and DiVincenzo D P, 1998, *Phys. Rev.* **A 57** 120.

Marshall D, Mazilu M, Miller A, Button C C, 2002, *J. Appl. Phys.* **91** 4090

Miller A, Manning R J, Milsom P K, Hutchings D C, Crust D W, and Woodbridge K, 1989, *J. Opt. Soc. Am. B-Optical Physics* **6** 567.

Miller A, Riblet P, Mazilu M, White S, Holden T M, Cameron A R and Perozzo P, 1999, *J. Appl. Phys.* **86** 3734.

Miller D A B, Chemla D S, Eilenberger D J, Smith P W, Gossard A C, and Wiegmann W, 1983, *App. Phys. Lett.* **42** 925.

Nishikawa Y, Tackeuchi A, Yamaguchi M, Muto S, and Wada O, 1996, *IEEE J. of Selected Topics in Quant. Elect.* **2** 661.

Schmitt-Rink S , Chemla D S, and Miller D A B, 1985, *Phys. Rev. B* **32** 6601.

Tackeuchi A, Nishikawa Y, and Wada O, 1996, *Appl. Phys. Lett.* **68** 797.

Tackeuchi A, Kuroda T, Muto S, Nishikawa Y, and Wada O, 1999a, *JJAP Part 1-Regular Papers Short Notes & Review Papers* **38** 4680.

Tackeuchi A, Kuroda T, Muto S, and Wada O, 1999b, *Physica B* **272** 318.

Yafet Y, 1963, *Solid State Phys.* **14** 1.

Localised waves in femtosecond optics

Peeter Saari

University of Tartu, Estonia

1 Introduction

How do we reach the right decision with minimal intellectual effort? How do we understand whether a proposition or project is physically correct and sound, or flawed or even a bluff, without carrying out a time-consuming detailed analysis? The history of science and technology has a well-known example on how to deal with such questions: in 1775, referring to the energy conservation law, the French Academy of Sciences ceased to consider *perpetuum mobile* projects which had come flooding in during the previous two centuries. Since then the powerful heuristic and practical value of physical laws, especially of those formulated like taboos, was widely comprehended and such laws began to play a central role in the evolution of physics. One could even speculate that in its phylogenesis any scientific discipline acquires maturity and usefulness when it formulates certain basic taboos as cornerstones of the discipline. At this point it is important to realise that if a law of Nature says: "You cannot . . . ," it may be an instruction to move ahead rather than a restriction killing good ideas. Recall that all technological miracles that surround us have been created on the basis of physical laws, not thanks to fantasies from fairy tales.

We could say that the erudition of a student of physics reaches maturity when he or she has learned how to think effectively and creatively with the help of factual knowledge, rules, laws and methods studied and, most importantly, within the imposed restrictions.

Successful experience of solving physically sophisticated problems develops an ability to find correct answers almost intuitively—quickly and without undergoing any routine analysis. Needless to say such ability, let us call it a correct physical instinct, makes its owner a valuable expert in his subfield.

However, some physical truths, as they gain general acceptance, enter textbooks, become stock rules and loose their exact content for the majority of the physics community. Moreover, in this way, superficially understood rules may turn into superfluous taboos inhibiting the study of new phenomena. For example, conviction that "uniformly moving charge does not radiate" caused a considerable delay in discovering and understanding the Cherenkov effect. Even the refined statement "uniformly moving charge does not radiate in vacuum" is not exact because it excludes the so-called transition radiation known about for only half a century (see, e. g., Jackson 1999) despite being a purely classical effect of macroscopic electrodynamics.

Similarly, well established and normally efficient methods or concepts may become routine, and their applicability cease to be questioned. For example, the special theory of relativity has imprinted the equivalence of inertial reference frames on our minds so that we are eager to jump into so-called co-propagating frames. This usually makes problems of relativistic physics easier to understand. However, this is not always so. One of the greatest physicists of the 20th century, J S Bell tells an illustrative story in his small but instructive book (Bell 1993). One day in the CERN canteen the following problem came up for discussion between Bell and a distinguished experimental physicist. Three small spaceships, A, B, and C (Figure 1) drift freely in a region of space remote from other matter, without rotation and without relative motion. B and C are equidistant from A.

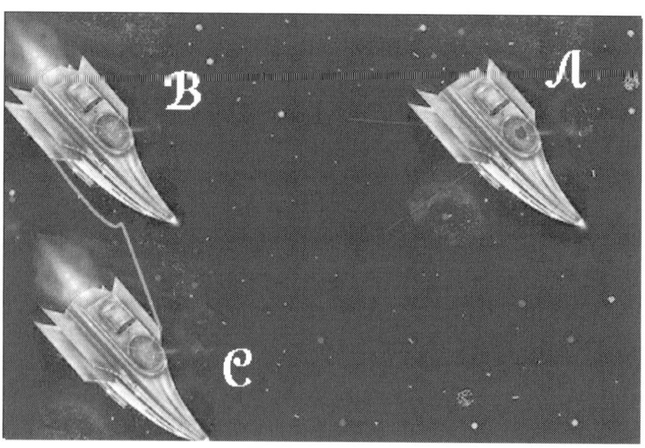

Figure 1. *The spaceships and the thread before starting the acceleration programme.*

On reception of a signal from A the motors of B and C are ignited and the ships accelerate gently. Let ships B and C be identical, and have identical acceleration programmes. Then, as reckoned by an observer in A, they will have at every moment the same velocity, and so remain displaced one from the other by a fixed distance. Suppose that a fragile thread is tied initially between B and C. If it is just long enough to span the required distance initially, then as the rockets speed up and the velocities reach the order of c, it will become too short, because of its need to undergo Lorentz-Fitzgerald contraction, and must finally break.

The thread must break because the natural contraction imposes stress. Although it seems to follow straightforwardly from the well-known longitudinal contraction effect in relativistic kinematics, is this really so? Bell's opponent refused to accept that the thread would break, and regarded Bell's assertion, that indeed it would, as a personal misinterpretation of special relativity. They decided to appeal to the CERN Theory Division for arbitration. There emerged a clear consensus that the thread would not break! Theorists feel obliged to jump into the co-propagating frame, that is, to work out how things look to astronauts in ships B and C. At first glance there is no motion in this reference frame, therefore people give the wrong answer. Of course, theorists get the right answer on further reflection. They find that B, for example, sees C drifting further and further behind, so that a given piece of thread can no longer span the distance. So, it is only after a time-consuming sophisticated analysis, and perhaps only with a residual feeling of unease, that they finally accept a conclusion which is perfectly trivial in terms of the Lorentz-Fitzgerald contraction observed in A's rest frame. Bell concludes that those with a less sophisticated education often have stronger and sounder instincts.

The main objective of this paper is to introduce properties and possible applications of extremely short light pulses localised in space by means of so-called meso-optical elements (Soroko 1996) or by common focusing. These pulses are solutions of linear wave equations and are considered as propagating in vacuum or in a linear homogeneous medium, that is, they have nothing to do with solitons or other localization phenomena known in contemporary nonlinear optics. Nevertheless, their startling features challenge more than one superficially understood "physical truth" or routinely applied concept. In addition to what was already implied above we will demonstrate this superficiality of taboos such as "the group velocity cannot exceed c, the speed of light in vacuum" or "the mass of a photon cannot be other than zero". Also questioned will be other widespread convictions such as "the treatment of fields having complicated spatio-temporal dependence is much easier in the spectral (Fourier) representation" or "as opposed to solving a wave equation exactly, the Huygens principle is a qualitative alternative tool". Hopefully this paper provokes a questioning of certain textbook truths and may refine the reader's physical instincts.

The remaining part of the preamble section serves as an introduction of some basic concepts and terminology. Following the preamble, there are two further sections. Section 2 is devoted to so-called propagation-invariant localised wavepackets which fly with a superluminal velocity $v \geq c$ and without any apparent diffraction or spread over large distances. Section 3 describes why and how subcycle pulses undergo temporal reshaping in passing through the focus.

1.1 Bessel beam

Although wave fields possessing Bessel-function-shaped radial profile had been known for some time, it was the intriguing demonstration of the "diffraction-free" optical Bessel beam (Durnin 1987) that triggered investigation of (pseudo)nondiffracting waves and their generation. This has lead to expanding applications in various fields of science and

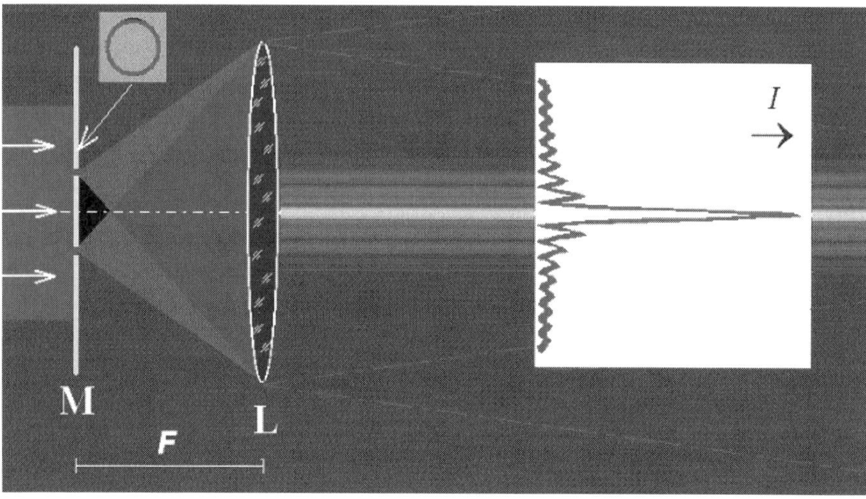

Figure 2. *Generation of the zero-order Bessel beam from a monochromatic laser beam falling from the left to an opaque mask M with an annular slit placed in the focal plane of a collimating lens L. The latter forms a fan of plane waves whose directions of propagation lie on a cone around the axis. The inset shows the radial dependence of the intensity (modulus squared) of the beam.*

technology. The simplest, although rather inefficient, method of generating the Bessel beam from a monochromatic laser beam is depicted in Figure 2.

A monochromatic zero-order Bessel beam can be viewed as an axisymmetric generalization of the interference pattern of a pair of plane waves, which propagate in the zx-plane at angles $\pm\theta$ relative to the z axis. In fact, if one sums up the fields of all such pairs over the **k**-vector cone around the propagation axis, the well-known lateral interference profile factor $\cos(k_\perp x)$ of the pair of waves is replaced by its cylindrical counterpart $J_0(k_\perp \rho)$, where J_0 is the zero-order Bessel function of the first kind, ρ is the transverse distance from the propagation axis, $k_\perp = k \sin\theta$ is the transverse component of the wave vector, and $k = \omega/c = 2\pi/\lambda$ is the wavenumber. Thus, one immediately obtains a physically transparent expression for the field of the beam propagating toward the positive direction of the axis z :

$$\Psi_B(\rho, z, t) \propto J_0(\rho k \sin\theta) \exp\left[i(k_\parallel z - \omega t)\right], \tag{1}$$

where $k_\parallel = k\cos\theta$ is the wave vector projected along the axis of propagation and serves as the propagation constant of the cylindrical wave. The phase velocity $\omega/k_\parallel = kc/k\cos\theta = c/\cos\theta$ is clearly superluminal. The reason is the same as why the crossing point of scissors moves faster than either branch. What about the group velocity of a quasi-monochromatic Bessel beam along the axis of propagation? The answer depends on the functional dependencies between k, $k_\parallel$, and $k_\perp$, in any case all obeying the dispersion equation

$$\frac{\omega^2}{c^2} = k^2 = k_\parallel^2 + k_\perp^2 \; . \tag{2}$$

If $k_\perp$ is held constant then the group velocity $d\omega/dk_\| = c\cos\theta$ is *sub*luminal. This is the case usually encountered in textbooks since it means that tubes formed in space by zeros of the field are the same for any frequency, as it must be in waveguides. However, in the case where $\theta(\omega)$ is held constant we get a *super*luminal group velocity $d\omega/dk_\| = c/\cos\theta$. This result and even more surprising related ones will be considered and interpreted in Section 2.

The field whose amplitude is defined (up to a constant factor) by Equation 1, is propagation-invariant (nondiffractive) as far as the so-called axicon angle θ is constant, because then the z, t-dependence enters into Ψ through the single propagation variable

$$d = z\cos\theta - ct = (z - vt)c/v \tag{3}$$

where $v = c/\cos\theta$ is the superluminal velocity along the z-axis. Of course, only an ideal infinite-energy and infinite-diameter Bessel beam can be propagation-invariant up to an infinite distance. In reality, diameter D of the aperture of a Bessel beam generator is always finite (see Figure 2) and therefore the interference-caused Bessel-function profile of the near-axis field exists only up to a distance of the order $\sim (D/2)\tan^{-1}\theta$.

What is the mass of photons of the Bessel beam field? Before answering this question it should be stressed that while mass m in popular scientific books on the theory of relativity as well as in many general physics textbooks is the quantity that increases with velocity and actually is an equivalent of energy ($E = mc^2$), in contemporary theoretical physics m stands for the rest mass as an invariant quantity (see Adler 1987). If one applies the basic relativistic relation

$$\frac{E^2}{c^2} = p^2 + m^2c^2 \tag{4}$$

between energy, momentum, and mass of a particle to a photon through substitutions $E = \hbar\omega$ and $\mathbf{p} = \hbar\mathbf{k}$ in the case of a plane wave, one obtains the textbook result $m = 0$, since both E/c and p are equal to $\hbar k$. However, the mass of a photon of a field which contains a standing-wave component is not equal to zero (Okun 1989). This is just the case for the Bessel beam and from $p = \hbar k_\| = \hbar k\cos\theta$ we obtain for the mass of a single photon of the cylindrical wave $m = (\hbar\omega/c^2)\sin\theta$. It is not so surprising if we notice that Equation 2 is equivalent to Equation 4 if multiplied by a constant factor $\hbar^2$ which plays here only a role of a translation tool between "wave language" and "particle language".

Finally, note that as it is generally adopted in treatments of light wave propagation problems, with the exception of some extraordinary polarizations and very large angles θ, there is no need to deal with electromagnetic field vectors, that is, a scalar theory is sufficient. That is why we are not going to obscure this paper with vectorial equations. Fields $\mathbf{E}$ and $\mathbf{H}$ can be routinely derived from a scalar wave field, for example by the Hertz potentials.

1.2 Back from the Fourier world

Spectral description is convenient since the Fourier transformation replaces differential equations and/or convolution integrals with simple algebraic operations. Moreover, up to recent decades, optical fields have been traditionally treated as quasi-monochromatic, as the bandwidth of light waves is very narrow in comparison with the carrier frequency.

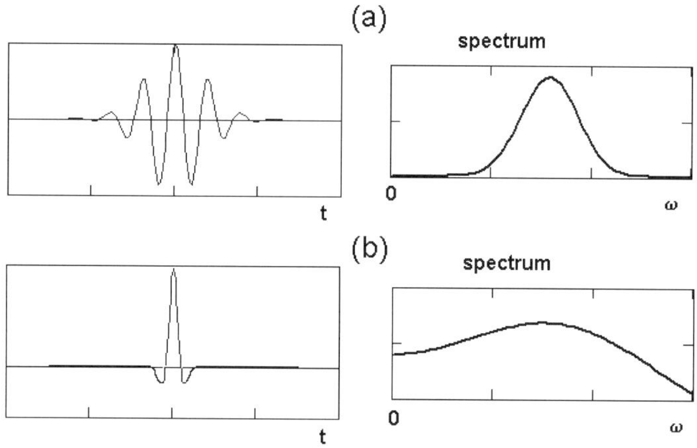

Figure 3. *A few-cycle (a) and near-cycle (b) pulse and their Fourier spectra.*

With the decrease in the duration of optical pulses down to the femto-attosecond region, there is a growing need to go beyond the quasi-monochromatic limit and to go back to "time-domain thinking". There is an ultimate milestone on going away from the narrow-band limit—where the spectrum of the pulse extends to the frequency scale origin and beyond. This causes an overlapping of positive-frequency and negative-frequency Fourier components and, as a result, the concept of an analytic signal with its Hilbert-transform-related real and imaginary parts remain the only alternative for a description of the pulse and its envelope (Figure 3).

Time domain descriptions of optical fields, where instead of the harmonic oscillatory functions the role of basic elementary constituents are played by Dirac delta pulses, is often physically more comprehensible. For example, consider a real measurement of the spectrum of an optical pulse. Of course, there is no such device as time-to-frequency-domain Fourier transformer in reality, since already the infinite time limits of the Fourier integral contradict the causality requirement. Obviously a real spectral instrument carries out the integration over a finite temporal interval and the spectrally resolved output signal depends on time and has a finite spectral resolution $\Delta\omega$. But what defines the temporal resolution $\Delta\tau$ of the spectrometer? This determines how precisely we can record temporal changes in the spectrum provided that the light detecting device is fast enough. Relying on the complementarity between temporal and spectral descriptions one might suppose that $\Delta\tau \approx \Delta\omega^{-1}$. Since the wavelength resolution of a small grating spectrometer is about 1nm which is in the centre of the visible spectrum corresponding to $\Delta\omega = 6.23 \times 10^{12}$rad/sec, the temporal resolution should be about 160fs. In reality it is much longer depending on imperfections of the optics. In order to understand why, one has to study the origin of the response of the spectrometer. Textbook derivations of the spectral response of an ideal diffraction grating are based on interference, which sums certain geometric series and is rather nonintuitive altogether. In contrast, an evaluation of the response of a spectrograph or monochromator, including imperfections, is fairly simple in the time-domain picture (Saari *et al.* 1981). Let us consider here qualitatively, without formulas, formation of the impulse response in a simplified schematic spectrometer.

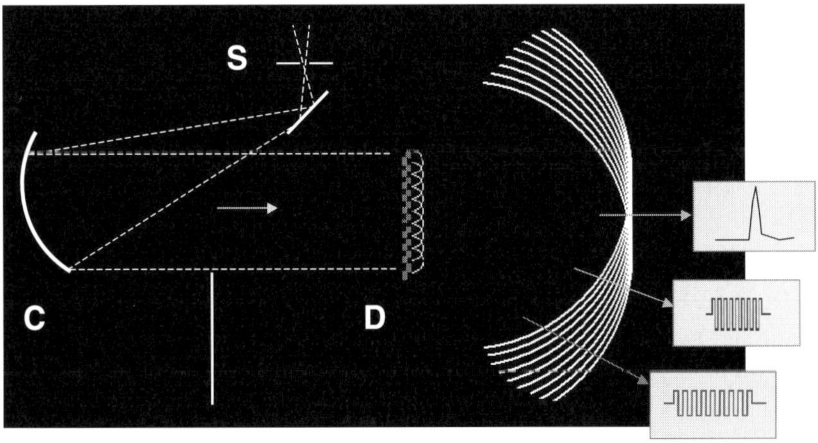

Figure 4. *Formation of the impulse response of a spectrometer. Suppose for clarity that an ultrashort light pulse, shorter than a half-cycle, flies through the entrance slit S and the collimating mirror C directs it as a flat plane wave pulse toward the diffraction grating D. Immediately after hitting the grating, each groove of which acts as a Huygens-type secondary source, a fan of cylindrical waves flies toward output mirror (not shown). Insets depict waveforms seen under different angles of diffraction; the larger the angle the longer the response as well as the carrier wavelength of the output wave (extracted, e. g., by a slit). For clarity the pulse is depicted as falling normally onto the grating, and the number of grooves, and hence the number of cycles in the fan of waves, is shown by three orders of magnitude smaller than it is in reality.*

As obviously follows from Figure 4, the temporal resolution $\Delta\tau$ is determined simply by the width of the illuminated surface of the grating: it is precisely equal to c^{-1} times the projection of this width onto the direction of diffraction for a given wavelength. As the gratings are typically about 10cm wide, $\Delta\tau$ turns out to be of the order of hundreds of picoseconds. Moreover, it follows from Figure 4 that the temporal dependence of the output field (the impulse response to be recorded) mimics the transmission (or reflection) profile of the grating. If the periodic profile of the latter is not exactly sinusoidal, the output wave also contains the higher harmonics of the carrier frequency in full accordance with known existence of higher diffraction orders in the case of such gratings. The number of oscillations in the output waveform is nothing but the number of grooves illuminated on the grating, while the wavelength, of course, depends on the angle of diffraction. The frequency response function of the spectrometer and, hence, the spectral resolution, commonly measured by a source of monochromatic light, are obtained from the Fourier transform of the impulse response. Imperfections of collimating optics and of real gratings, that is, those with non-constancy of the groove spacing, cause slight variations in the periodicity of zero crossings in the output waveform. In other words a phase modulation may increase the width $\Delta\omega$ of the spectral response by several orders as compared to its ideal minimum-uncertainty value $\Delta\tau^{-1}$ (Saari *et al.* 1981). Had we tried to reach these results by the frequency-domain analysis, we would have spent much more effort.

2 Superluminal nonspreading pulses

In this section we give an overview of electromagnetic versions of the so-called X-type waves found as peculiar solutions to the homogeneous wave equation (Lu and Greenleaf 1992) and intensively studied in recent years (see, e. g., reviews by Besieris *et al.* 1998, Recami 1998, Saari 2001, Salo *et al.* 2000). In the spectral language the X-type waves are certain ultrawideband superpositions of Bessel beams. However, the physical nature and the origin of the name given to these pulsed waves are easier to comprehend from a time-domain treatment presented in the next subsection.

The superluminality of the X waves—not only of the phase and group velocities but of the rigid movement of the pulse as a whole —is now an experimentally verified fact (Saari and Reivelt 1997, Mugnai *et al.* 2000, Alexeev *et al.* 2002). It is interesting to point out physically meaningful resemblance between the shape of the X waves and that of the tachyon (Recami 1998). The shape of the tachyon was derived (Recami 1986) in times of great activity in theoretical study of these hypothetical superluminal particles. To these years belongs paper (Faingold 1976) where a double-cone-shaped "electromagnetic tachyon" as a result of light reflection by a conical mirror was considered. As a matter of fact, this quarter-of-century-old paper seems to be the pioneering work on X-type waves. Recognizing that the trailing half of the double-conical shape of X-type waves resembles a three-dimensional wakefield or Mach cone known from acoustic shock wave of supersonic jets or from the Cherenkov effect, we have found a new representation of the localised waves as if generated by certain superluminally moving Huygens-type sources. This representation, which we introduce in Subsections 2.2 and 2.3, in a sense generalises the Huygens principle into a domain of superluminally moving secondary sources. Why the startling superluminality is yet in accordance with the relativistic causality, will be briefly discussed in Subsection 2.4. The last subsection lists some ideas of possible applications of X-type pulsed waves. The limited space does not allow us to consider other, perhaps more sophisticated types of the propagation-invariant localised solutions to the wave equation, for example, so-called focus wave modes (Brittingham 1983, Ziolkowski 1989, Overfelt 1991, Besieris *et al.* 1998) only very recently realised experimentally (Saari and Reivelt 2002).

2.1 Physical nature of X-type waves

In order to make the physical nature of the X-type superluminal localised waves more comprehensible, we first discuss a simple representation of them as a result of interference between plane wave pulses.

The simplest , although rather inefficient, set-up for generating the pulsed wave shown in Figure 5 is the same as depicted in Figure 2. However, the continuous input wave has been replaced by ultra-wideband pulses. The input pulses must be short enough, a few cycles at most, to assure that the output wave possesses the X-shaped profile instead of being just a short truncated Bessel beam. With reference to Figure 5(a) let us consider a pair of plane wave bursts possessing identical temporal dependences and the wave vectors in the plane $y = 0$. Their propagation directions given by unit vectors $\mathbf{n}_{/}=[\sin\theta,\ 0,\ \cos\theta]$ and $\mathbf{n}_{\backslash}=[-\sin\theta,\ 0,\ \cos\theta]$ are tilted under angle θ with respect to the axis z. In spatio-temporal regions where the pulses do not overlap their field is given simply by the burst

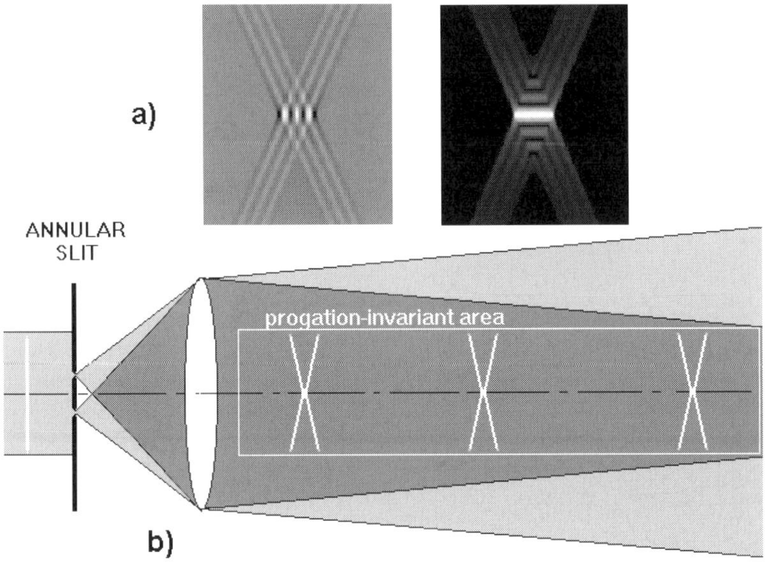

Figure 5. *X-type scalar wave formation from a scalar plane wave pulses containing about three cosinusoidal cycles. As linear gray-scale plots in a plane of the propagation axis and at a fixed instant, shown are (a, left inset) the field of the wave (real part if the plane waves are given as analytic signals) and (a, right inset) its amplitude (modulus). Note that the central bullet-like part of the wave would stand out even more sharply against the sidelobes if one plotted the distribution of the intensity (modulus squared) of the wave; (b), a simple set-up for generation of finite-aperture optical Bessel-X pulses.*

profile as $\Psi_P(\eta - ct)$, where η is the spatial coordinate along the direction $\mathbf{n}_/$ or $\mathbf{n}_\backslash$, respectively. In the overlap region, if we introduce the radius vector of a field point $\mathbf{r} = [x,\ y,\ z]$, the field is given by superposition

$$\Psi_P(\mathbf{r}\cdot\mathbf{n}_/ - ct) + \Psi_P(\mathbf{r}\cdot\mathbf{n}_\backslash - ct) = \Psi_P(x\sin\theta + z\cos\theta - ct) + \Psi_P(-x\sin\theta + z\cos\theta - ct)\ , \quad (5)$$

which is nothing but the well-known two-wave-interference pattern with doubled amplitudes. Altogether, the superposition of the pulse pair—the two branches denoted by \ and / forming the letter X—makes up an X-shaped propagation-invariant interference pattern moving along the axis z with speed $v = c/\cos\theta$ which is both the phase and the group velocity of the wave field in the direction of the propagation axis z. This speed is superluminal in a similar way as one gets a faster-than-light movement of a bright stripe on a screen when a plane wave light pulse is falling at angle θ onto the screen. Let us stress that here we need not deal with the vagueness of the physical meaning inherent in the group velocity in general, rather the whole spatial distribution of the field moves rigidly with v because the time enters into Equation 5 only together with the coordinate z through the propagation variable $z_t = z - vt$.

Let us further superimpose axisymmetrically all such pairs of waves whose propagation directions form a cone around the axis z with the top angle 2θ, in other words, let the pair of unit vectors be

$$\mathbf{n}_/ = [\sin\theta\cos\phi,\ \sin\theta\sin\phi,\ \cos\theta]\,,$$
$$\mathbf{n}_\backslash = [\sin\theta\cos(\phi+\pi)\,,\ \sin\theta\sin(\phi+\pi)\,,\ \cos\theta]\,,$$

where the angle ϕ runs from 0 to 180 degrees. As a result, we get an X-type superluminal localised wave in the following simple representation

$$\Psi_X(\rho, z_t) = \int_0^{\pi} d\phi\,\left[\Psi_P\left(\mathbf{r}_t\cdot\mathbf{n}_/\right) + \Psi_P\left(\mathbf{r}\cdot_t\mathbf{n}_\backslash\right)\right] = \int_0^{2\pi} d\phi\,\Psi_P\left(\mathbf{r}_t\cdot\mathbf{n}_/\right)\,, \tag{6}$$

where $\mathbf{r}_t = [\rho\cos\varphi,\ \rho\sin\varphi,\ z_t]$ is the radius vector of a field point in the co-propagating frame and cylindrical coordinates (ρ, φ, z) have been introduced restricting ourselves in this paper to axisymmetric or so-called zeroth-order X-type waves only. Hence, according to Equation 6, the field is built up from interfering pairs of identical bursts of plane waves. Figure 5 gives an example in which the plane wave profile Ψ_P contains three cycles.

The less extended the profile Ψ_P, the better the separation and resolution of the branches of the X-shaped field. In the superposition, the points of constructive interference lie on the z axis, where the highly localised energy "bullet" arises in the centre, while the intensity falls off as ρ^{-1} along the branches and much faster in all other directions (note that in contrast with Figure 5a in the case of interference of only two waves the on-axis and off-axis maxima must be of equal strength). The optical carrier manifests itself as one or more (depending on the number of cycles in the pulse) halo toroids which are nothing but residues of the concentric cylinders of the intensity characteristic of the Bessel beam. That is why we use the term 'Bessel-X pulse' (or wave) to draw a distinction from carrierless X waves. By making use of an integral representation of the zero-order Bessel function $J_0(v) = \pi^{-1}\int_0^{\pi}\cos[v\cos(\phi-\varphi)]\,d\phi$, where φ is an arbitrary angle, we get (for details, see Saari and Reivelt 1997) the common representation of X-type waves as a superposition of monochromatic cylindrical modes (Bessel beams) of different wavenumbers $k = \omega/c$:

$$\Psi_X(\rho, z, t) = \int_0^{\infty} dk\,S(k)J_0(\rho k\sin\theta)\,\exp\left[i\,(zk\cos\theta - \omega t)\right]\,, \tag{7}$$

where $S(k)$ denotes the Fourier transform of the profile Ψ_P. Equation 7 gives again for both the phase and the group velocities (along the z-axis, the direction of the propagation of the packet of the cylindrical waves) the superluminal value $v = c/\cos\theta$. Using variables $k_{||} = k\cos\theta$ and $z_t = z - vt$ introduced earlier Equation 7 reads

$$\Psi_X(\rho, z, t) = \int dk_{||}F(k_{||})J_0\left(\rho k_{||}\gamma^{-1}\right)\exp\left(ik_{||}z_t\right), \tag{8}$$

where we have defined $\gamma = 1/\sqrt{v^2/c^2 - 1}$. If the spectrum is "white" but decreasing (decay constant Δ) from the strongest dc-component toward higher frequencies exponentially $F(k_{||}) = \Delta\exp(-k_{||}\Delta)$, then the integral is nothing but the Laplace transform of $J_0(\rho k_{||}\gamma^{-1})$ in the variable $k_{||}$, which gives the analytic expression of the simplest carrierless X wave

$$\Psi_X(\rho, z, t) = \frac{\Delta}{\sqrt{(\Delta + i(z - vt))^2 + \rho^2\gamma^{-2}}}\,. \tag{9}$$

The last expression resembles the point-charge potential. That is why we will derive it once more from a completely different starting point.

2.2 Huygens principle revisited

In order to proceed, we show how the Huygens-type fictitious sources, sources commonly considered as lying on a surface and enabling a qualitative description of the wavefield, can give any exact solution of the homogeneous (source-free) wave equation (HWE), if only they are allowed to take a suitable distribution in space and time.

Any spherically symmetrical solution to HWE can be expressed similarly to the one-dimensional HWE solution as a sum of two counter-propagating waves

$$U_{\odot}(r,t) = [f(r-ct) + g(r+ct)]\, r^{-1}.$$

However, in order not to allow unphysical singularity at $r = 0$, the expanding wave must be a replica of the imploding one: $g(x) = -f(-x)$. Without loss of generality one can centre the time argument origin somewhere into the pulse so that $f(x) = 0$ if $|x| > r_c$, where r_c is the radius of a bounding sphere into which the field is completely confined at the moment $t = 0$. Therefore for any distant ($r > r_c$) observation point and for $t > 0$ such that $t - 2r/c < 0$ we get

$$U_{\odot}(r,t) = -U_{\odot}(r, t - 2r/c).$$

A particular (and singular) solution $D(r,t)$ obeying the initial conditions

$$D|_{t=0} = 0, \quad \text{and} \quad \frac{\partial}{\partial t} D|_{t=0} = \delta(\mathbf{r}) = -\frac{1}{2\pi r}\delta'(r)$$

is therefore

$$D(r,t) = \frac{1}{4\pi r}\left[\delta(r-ct) - \delta(r+ct)\right] \equiv G_{+}(r,t) - G_{-}(r,t), \tag{10}$$

where $G_{\pm}$ denote the causal (retarded) and anticausal (advanced) Green functions, respectively. Thus, the function D represents a spherical delta-pulse-shaped wave, first (at negative times t) converging to the origin (the right term) and then (at positive times t) diverging from it. The minus sign between the two terms, which results from the requirement that a source-free field cannot have a singular point, is of crucial importance as it assures vanishing of the function at $t = 0$. This change of the sign, when the wave goes through the collapsed stage at the focus, is also responsible for the 90 degrees phase factor of the Huygens-Fresnel-Kirchhoff principle and for the Gouy phase shift peculiar to all focused waves.

With this function, sometimes called the Riemann or Schwinger function, as an elementary constituent, *any* solution to three-dimensional HWE can be expressed as convolution integral

$$U(\mathbf{r},t) = \int \left[D(R,t)\dot{\rho}_g(\mathbf{r}') + \frac{\partial}{\partial(ct)} D(R,t)\rho_u(\mathbf{r}') \right] d\mathbf{r}'$$

where $R = |\mathbf{r} - \mathbf{r}'|$ and the source density-type functions are determined by the field "snapshot" at the time origin moment $\dot{\rho}_g(\mathbf{r}) = \partial/\partial(ct)\; U(\mathbf{r},0)$ and $\rho_u(\mathbf{r}) = U(\mathbf{r},0)$. However, unlike solution of a radiation problem, since D contains not only the retarded Green function but the advanced one as well, $\dot{\rho}_g$ and ρ_u describe distribution of fictitious sources, i.e. sources coupled with sinks of the same strength. Generally, there is a certain freedom in how these sinks-and-sources should behave in time and space in order to result in the given free wavefield.

2.3 X-waves as wakewaves

From Equation 10, using the common procedure (see, e. g. Jackson 1999) one can calculate the Liénard-Wiechert potentials for a moving point charge q flying, for example, with a constant velocity v along axis z. However, as D includes not only the retarded Green function but also the advanced one and therefore what is moving has to be considered as a source coupled with a sink at the same point. For such a Huygens-type source there is no restriction $v \leq c$ and for superluminal velocity $v/c > 1$ we obtain axisymmetric scalar and vector potentials $\Phi(\rho, z, t)$ and $\mathbf{A}(\rho, z, t) = \Phi(\rho, z, t)\mathbf{v}/c$, where ρ is the radial distance of the field point from the axis z, the velocity vector is $\mathbf{v} = [0, 0, v]$, and (in cgs units, Lorentz gauge)

$$\Phi(\rho, z, t) = \frac{2q}{\sqrt{(z-vt)^2 + \rho^2\gamma^{-2}}} \left\{ \begin{array}{c} \Theta\left[-(z-vt) - \rho\gamma^{-1}\right] \\ -\Theta\left[(z-vt) - \rho\gamma^{-1}\right] \end{array} \right\}, \tag{11}$$

where $\Theta(x)$ denotes the Heaviside unit step function. As argued already, for the sake of simplicity we do not calculate the electromagnetic field vectors $\mathbf{E}$ and $\mathbf{H}$ (or $\mathbf{B}$), neither will we consider dipole sources and sinks required for obtaining non-axisymmetric fields. We will restrict ourselves to scalar fields obtained as superpositions of the potential given by Equation 11. The first of the two terms in Equation 11 gives an electromagnetic Mach cone of the superluminally flying charge q. In other words it represents nothing but a shock wave emitted by a superluminal electron in vacuum. The mathematical expression for such a wave was found by Sommerfeld in 1904, three decades earlier than Tamm and Frank worked out the theory of the Cherenkov effect, however Sommerfeld's result was dismissed as an unphysical result after the special theory of relativity appeared. The second term in the Equation 11 describes a leading and reversed Mach cone collapsing into the superluminal sink coupled with the source and thus feeding the latter. Hence, the particular solution to the wave equation which is given by Equation 11 represents a double-cone-shaped pulse propagating rigidly and superluminally along the axis z. In other words it represents an X-type wave as put together from (i) the cone of incoming waves collapsing into the sink, thereby generating a superluminal Huygens-type point source and from (ii) wakewave-type radiation cone of the source. Let it be recalled that the field given by Equation 11 had been found for δ-like spatial distribution of the charge. That is why the field diverges on the surface of the double cone or on any of its X-shaped generatrices given by $(z - vt) = \pm\, \rho\sqrt{v^2/c^2 - 1}$ and the field can be considered as an elementary one constituting a base for constructing various X-type waves through appropriate linear superpositions. Hence, any axisymmetric X-type wave could be correlated to its specific (continuous and time-dependent) distribution $\rho = \delta(x)\delta(y)\lambda(z, t)$ of the "charge" (or the sink-and-source) with linear density $\lambda(z, t)$ on the propagation axis, while the superluminal speed of the wave corresponds to the velocity v of propagation of that distribution along the axis, that is, $\lambda(z, t) = \lambda(z - vt)$.

Let us first choose the "charge" distribution $\lambda(z - vt) = \lambda(z_t)$ as a Lorentzian with the half width at half maximum (HWHM) denoted by Δ. In this case the field potential is given as the convolution of Equation 11 with the normalised distribution, which can be evaluated using Fourier and Laplace transform tables:

$$\Phi(\rho, z_t) \otimes \frac{1}{\pi}\frac{\Delta}{z_t^2 + \Delta^2} = -q\sqrt{\frac{2}{\pi}}\,\mathrm{Im}\left(\frac{1}{\sqrt{(\Delta - iz_t)^2 + \rho^2/\gamma^2}}\right), \tag{12}$$

(a)

(b)

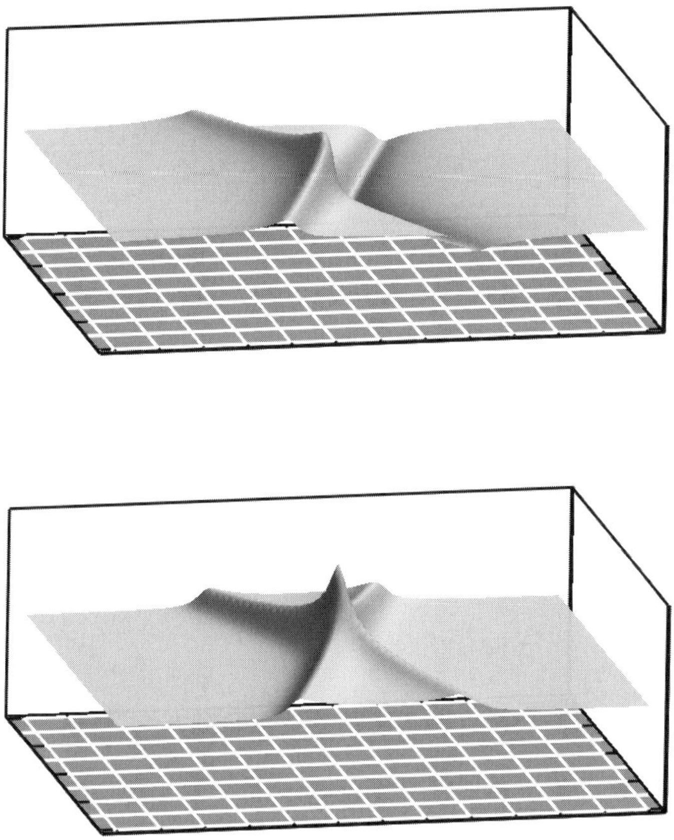

Figure 6. *Dependence on the longitudinal coordinate* $z_t = z - vt$ *(increasing from left to right) and a lateral one* $x = \pm\rho$ *of the imaginary (a) and real (b) parts of the field of the simplest X-wave. The velocity* $v = 1.005c$ *and, correspondingly, the superluminality parameter* $\gamma = 10$. *Distance between grid lines on the basal plane is* 4Δ *along the axis* z_t *and* 20Δ *along the lateral axis, the unit being the half-width* Δ.

where $\gamma = 1/\sqrt{v^2/c^2 - 1} = \cot\theta$ is the superluminal version of the Lorentz transformation coefficient and $z_t = z - vt$ as earlier. The resulting potential shown in Figure 6 (a) moves rigidly along the axis z from left to right with the same superluminal speed $v > c$. The plot (a) depicts qualitatively also the elementary potential as far as the divergences of Equation 11 are smoothed out in Equation 12.

We see that a unipolar and even "charge" distribution gives an odd and bipolar potential, as expected, while the symmetry of the plot differs from what might be expected from superimposing two plane wave pulses under the tilt angle 2θ. Indeed, in the latter case, the plane waves are depicted by each of the two diagonal branches (\ and /) of the X-shaped plot and therefore the profile of the potential on a given branch has to retain its sign and shape if one moves from one side of the central interference region to the other side along the same branch. Disappearance of the latter kind of symmetry, which can be most distinctly followed in the case of bipolar single-cycle pulses, the case shown in Figure 6a, is due to mutual interference of all the plane wave pairs forming the cone as ϕ runs from 0 to π.

Secondly, let us take the "charge" distribution as a dispersion curve with the same width parameter Δ, that is, as the Hilbert transform of the Lorentzian. Again, using Fourier and Laplace transform tables, we readily obtain:

$$\Phi(\rho, z_t) \otimes \frac{1}{\pi}\frac{z_t}{z_t^2 + \Delta^2} = q\sqrt{\frac{2}{\pi}}\mathrm{Re}\left(\frac{1}{\sqrt{(\Delta - iz_t)^2 + \rho^2/\gamma^2}}\right). \tag{13}$$

The potential given by Equation 13 and depicted in Figure 6b is (to within the accuracy of a real constant multiplier) nothing but the zeroth-order unipolar X wave, first introduced by Lu and Greenleaf in 1992 and studied in a number of papers afterwards.

Hence, we have demonstrated here how the real and imaginary parts of the simplest X-wave solution given by Equation 9 can be represented as fields generated by corresponding Huygens-type "sink-and-source charge" distributions moving superluminally along the propagation axis. Such representation may be named as Sommerfeld representation to acknowledge his unfortunate result of 1904.

2.4 Superluminality taboo

Let us make some remarks on the intriguing superluminality issues of the X-type wave pulses. Indeed, while phase velocities greater than c are well known in various fields of physics, a superluminal group velocity more often than not is considered as a taboo, because at first glance it seems to be at variance with the special theory of relativity, particularly with relativistic causality. However, since the beginning of the previous century—starting from Sommerfeld's works on plane-wave pulse propagation in dispersive media and precursors appearing in this process—it is known that group velocity need not to be a physically profound quantity and by no means should be confused with signal propagation velocity. But in the case of X-type waves not only the group velocity exceeds c but the pulse as a whole propagates rigidly faster than c.

A number of authors have recently discussed this startling but experimentally verified fact since it is not a simple task to show explicitly how the signal propagates subluminally in a superluminally moving pulse. Attention must be paid also to statements insisting that the pulse is not a real one but simply an interference pattern rebuilt at every point of its propagation axis from truly real plane wave constituents travelling at a slight tilt with respect to the axis. Such argumentation is not wrong but goes nowhere. Of course, there is a similarity between superluminality of the X wave and a faster-than-light movement of the cutting point in the scissors effect or of a bright stripe on a screen when a plane wave light pulse is falling at the angle θ onto the screen. But in the central highest-energy part of the X wave there is nothing moving at the tilt angle. The phase planes are perpendicular to the axis and the whole field moves rigidly along the axis. The Pointing vector lays also along the axis, however, the energy flux is not superluminal. Hence, to consider the X waves as something inferior compared to "real" pulses is not sound. If we thought so, by similar logic we would arrive at a conclusion that femtosecond pulses emitted by a mode-locked laser are not real but "simply an interference" between the continuous-wave laser modes. In other words, one should not ignore the superposition principle of linear fields, which implies a reversible relation between "resultant" and "constituent" fields and does not make any of the possible orthogonal bases inferior to others. Moreover, even plane waves, as far as they are truly real ones, suffer from a certain superluminality. Indeed, as is well known, the most simple physically feasible realization of a plane wave beam is the Gaussian beam with its constrained cross-section and, correspondingly, a finite energy flux. However, one can readily check on the analytical expressions for the beam (Horvath and Bor 1999) that due to the Gouy phase shift the group velocity in the waist region of the Gaussian beam is slightly superluminal. The Gouy effect in the case of ultrawideband focused fields will be considered in the next section.

In our opinion, just as it was decided by the French Academy of Sciences concerning the *perpetuum mobile* projects, there is no need to go into detailed analysis in order to prove that the X-type waves are not, and cannot be, at variance with the special theory of relativity. Indeed, the waves have been derived as solutions to the D'Alambert wave equation and corresponding electromagnetic vector fields are solutions to the Maxwell equations. Consequently, relativistic causality has been inherently built into them as it was demonstrated also in the present paper, when we developed the Sommerfeld representation based on the relativistically invariant retarded and advanced Green functions. This is enough for the proof, whereas an analysis of local evolution and propagation of a "signal mark" made, for example, by a shutter onto the X wave is not a simple task due to diffractive changes in the field behind the "mark".

In conclusion, superluminal movement of individual material particles is not allowed but excitations in an ensemble of particles or field may propagate with any speed. However, if the speed exceeds c they cannot transmit any physical signal. The last two decades have made it profoundly clear how promising and fruitful it is to study superluminal phenomena instead of considering them as some sort of triviality or taboo. We have in mind here not only localised waves, photon tunneling or propagation in inverted resonant media, etc., but also the implementation and application of entangled states of Einstein-Podolsky-Rozen pairs of particles in quantum telecommunication, cryptography, and computing.

2.5 Application prospects of X-type waves

Limiting the aperture of practically realizable X-type waves causes an abrupt decay of the interference structure of the wave after flying rigidly over a certain distance. However, the depth of invariant propagation of the central spot of the wave can be made substantial—typically of the order of 1m in the case of using common sub-10cm diameter optics—since the distance is larger by a factor of $\cot\theta = \gamma = 1/\sqrt{v^2/c^2 - 1}$ than the aperture radius. Such a type of electromagnetic pulse, enabling directed, laterally and temporally concentrated and non-spreading propagation of wavepacket energy through space-time, have a number of potential applications in various areas of science and technology. Let us briefly consider some results obtained along this line which may have prospects in optics, in which case the pulse must have a carrier, that is, at least one cycle in each conical branch and along the axis (see Figure 7). Applications of ultrasonic and terahertz versions of the localised waves are more obvious, for example, in medical imaging.

Any ultrashort laser pulse propagating in a dispersive medium (even air) suffers from

(a)

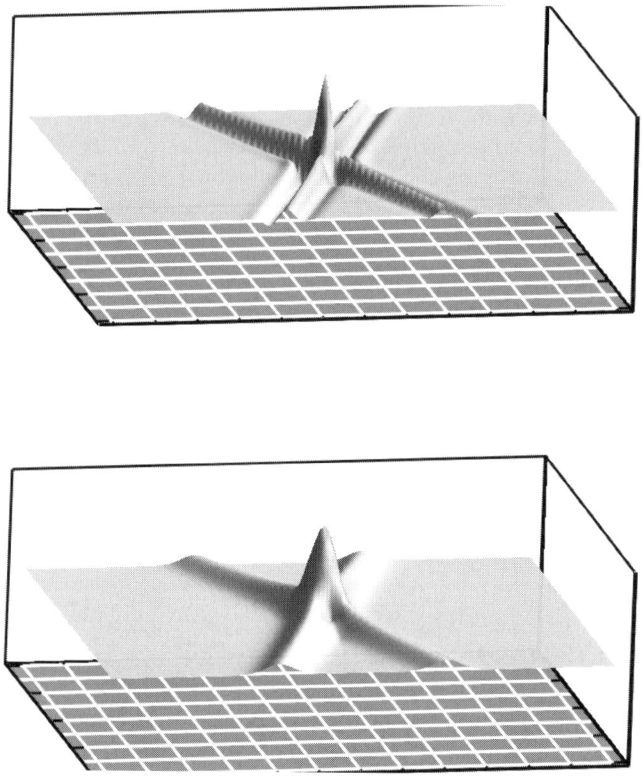

(b)

Figure 7. *The longitudinal-lateral dependences of the real part (a) and the modulus (b) of the field of the Bessel-X wave. The parameters v and γ are the same as in Figure 6. The new parameters of the wave, the wavelength $\lambda = 2\pi/k_z$ of the optical carrier being the unit, the distance between grid lines on the basal plane is λ along the axis z_t and 5λ along the lateral axis, while the half-width $\Delta = \lambda/2$. For visible light pulses λ is in sub-micrometer range, which means that the period of the cycle as well as full duration of the pulse on the propagation axis are as short as a couple of femtoseconds.*

a temporal spread; this is a well-known obstacle in femtosecond optics. For the Bessel-X wave with its composite nature, however, there exists a possibility to suppress the broadening caused by the group-velocity dispersion (Sõnajalg and Saari 1996). Namely, the dispersion of the angle θ, which is to a certain extent inherent in any Bessel-X wave generator, can be played against the dispersion of the medium with the aim of their mutual compensation. In dispersive media the frequency-dependent refractive index enters into Equation 9 since now $k_{\parallel} = (\omega/c)n(\omega)\cos\theta$. The idea is to make the Axicon angle θ frequency-dependent in such a manner that the frequency dependence of the propagation constant remains linear

$$\frac{\omega}{c} n(\omega) \cos\theta(\omega) = a + b\omega, \tag{14}$$

where a and b are arbitrary constants and the equality need not hold outside the spectral band of the optical pulse. If Equation 14 is satisfied, the group velocity dispersion for the Bessel-X pulse is zero. Moreover, Equation 14 is a sufficient condition for pulse propagation *intensity* profile invariance, as one can check by Equation 8.

This idea has been verified in an experimental setup based on a holographic optical element in the form of a circular grating by introducing an appropriate dispersion of the angle θ (Sõnajalg *et al.* 1997). Thus, an application of optical X-type waves has been worked out—a method of designing femtosecond pulsed light fields that maintain their strong (sub-millimeter range) longitudinal and lateral localization in the course of superluminal propagation into a considerable depth of a given dispersive medium.

Optical X waves (or Bessel-X waves) allow a sort of diffraction-free transmission of arbitrary 2-dimensional images (Saari 1996, Saari and Sõnajalg 1997). Despite its highly localised "diffraction-free" bright central spot, the zeroth-order monochromatic Bessel beam behaves poorly in a role of point-spread function in 2-D imaging. The reason is that its intensity decays too slowly with lateral distance, that is, as $\sim\rho^{-1}$. On the contrary, the Bessel-X wave is offering a loop hole to overcome the problem. Despite the fact that the time-averaged intensity of the Bessel-X wave possesses the same slow radial decay $\sim\rho^{-1}$ due to the asymptotic behavior along the X-branches, the instantaneous intensity has strong (e. g. Gaussian or rectangular) localization in lateral cross-section at the maximum of the pulse (see Figures 6 and 7) and therefore it might serve as a point-spread function with well-constrained support, also with the extraordinary capability of maintaining the image focused with no spread over large propagation lengths. By further developing this approach, it is possible to build a specific communication system (Lu and He 1999). It is obvious that for a majority of possible applications the spread-free central spot is the most attractive peculiarity of the axisymmetric X-type waves. In this respect a new modification of X waves, so-called focused X wave (Besieris *et al.* 1998), seems to be rather promising. Also, various higher-order non-axisymmetric waves possess additional interesting properties, for example, a sharp dark spot instead of the central maximum.

For those who are interested we give some references to other interesting applications: superluminal tunneling of X waves through planar slabs (Shaarawi and Besieris 2000), superluminal localised solutions to Maxwell equations propagating along waveguides (Zamboni-Rashed *et al.* 2002), diffraction of X waves by conducting objects (Attiya *et al.* 2002), and self-imaging pulsed fields (Reivelt 2002).

3 Temporal reshaping of focused subcycle pulses

In this section we consider ultra-wideband pulses that are not propagation-invariant but pass through a focus, where they gain their strongest lateral localization. A basic concept of wave optics is the Gaussian beam. It has many diverse applications in areas such as laser resonators, waveguides, nonlinear optics, optical information processing, etc. From the theoretical point of view the waist region of a (weakly-focused) Gaussian beam constitutes perhaps the simplest realization of the plane wave concept, since a mathematical plane wave with its infinite range and energy is a physically unrealistic object.

However, the Gaussian beam is not an exact solution to the free-space wave equation. Instead, as a result of paraxial diffraction theory, it obeys the parabolic truncated version of the equation. Therefore the paraxial approximation is insufficient in a high-aperture case—it starts breaking down for numerical apertures of about $1/\sqrt{2}$ (i.e. $f: 0.5$), which is far from being a limit for various devices, e. g., microcavities. The well-known elegant mathematical trick to get the common Gaussian beam as the field of a point source placed at an imaginary displacement along the optical axis gives nonphysical singularities if applied beyond the paraxial case. Moreover, ultrawideband pulses contain the Fourier components with frequencies far below the mean one, the lowest one being the dc-component. For the low-frequency components the product of the wavenumber and the Rayleigh range (a half of the confocal parameter) or, equivalently, the ratio of the beam waist diameter to the wavelength, may easily turn out to be of the order of unity. This means that for wideband and pulsed Gaussian beams the paraxial model is unsatisfactory even in a modest-aperture case, especially if the low-frequency tail of the spectrum has not been suppressed or cut off.

In this section, we give a simple but an exact treatment of the behaviour of subcycle pulses under arbitrarily tight focusing. The main ingredient in building the appropriate model is the source-and-sink representation possible for any free field, as shown in Subsections 2.2 and 2.3. No matter how wide the pulse bandwidth or the solid angle subtended by the focusing optics, the model remains exact in the sense of obeying the scalar HWE, yet it reduces to the common Gaussian beam in the paraxial and narrow-band limits. Naturally, the vectorial nature of electromagnetic waves becomes essential at high apertures, but for simplicity here we restrict ourselves to the scalar treatment in order to better convey the overall physical picture.

3.1 The model

Let us recall that the free-field Riemann function, Equation 10, allows us to express any solution of the HWE as the 4-dimensional convolution with an appropriately chosen "charge" distribution $\rho(t, \mathbf{r})$, whereas the quantity $\rho(t, \mathbf{r})$ expresses strength of both the source and the sink at the same point. Generally a given field does not determine uniquely the distribution of the charge generating the field. Consequently, the sinks and sources need not cover a distant surrounding surface, the picture commonly associated with the Huygens principle, but may be spatio-temporally localised in an appropriate way provided they generate the same field. So, let a point-like "charge" at the origin behave in time according to the Lorentzian curve with the width parameter Δ (HWHM), that is, let the distribution be

$$\rho(t, r) = q\delta(r)\frac{t_0}{it + \Delta}\,, \tag{15}$$

where q is the total sink-and-source strength, the parameter t_0 will be specified later on, and where the Lorentzian expressing the temporal dependence has been introduced in the form of the complex analytic signal for mathematical convenience. When expressing the source-free field through the 4D-convolution of the functions $D(t, R)$ and $\rho(t, \mathbf{r})$ as commonly done to obtain source-induced retarded potentials, the spatial integration is trivial and one obtains:

$$\begin{aligned} \phi(t, \mathbf{r}) &= \frac{c}{R}\int dt' \frac{qt_0}{it' + \Delta}\left\{\delta\left[R - c(t - t')\right] - \delta\left[R + c(t - t')\right]\right\} \\ &= \frac{q}{R}\left[\frac{ct_0}{-i(R - ct) + c\Delta} - \frac{ct_0}{i(R + ct) + c\Delta}\right] = q\frac{2ict_0}{R^2 + c^2(it + \Delta)^2}\,. \end{aligned} \tag{16}$$

Now we shift the "charge" from the origin to an imaginary location $z_0 = id$ on the z-axis. Naturally, this makes the expression of the distance R between a field point and the "charge" location a complex quantity. For convenience, we set $t_0 = d/c$ and $\Delta = t_0 + a/c$. The latter definition introduces an arbitrary contribution $a > 0$ to the pulse length but assures that the condition $\Delta > t_0$ is satisfied. This is required for a finite field at $\mathbf{r} = t = 0$. Thus, from Equation 16 the model field reads

$$\phi(t, z, \rho) = qd \frac{2i}{\rho^2 + [a - i(z - ct)]\,[2d + a + i(z + ct)]} \,. \tag{17}$$

The real and the imaginary part of this rather simple expression give us two different exact solutions of the HWE, which we will study in the next subsection.

3.2 Gouy effect in time domain

To visualise the pulsed solutions in non-paraxial conditions, we set the pulse-length parameter $a = 0.05$, where the confocal parameter $2d$ has been taken for the unit of length. The same relative unit is also used on the length scales of the forthcoming figures. In an absolute scale this unit might possess values from a few centimeters or less, to relate our results to typical terahertz pulses, down to, say, ten microns for a frontier femtosecond optical experiment. Consequently, the ratio $a/d = 0.05/0.5 = 0.1$ is rather large stressing the non-paraxiality. According to Equation 17 the field does not depend on the angle φ, that is, it is axisymmetric around the z axis and it suffices to study the behaviour of the pulse in the plane $y = 0$, which is depicted in Figure 8.

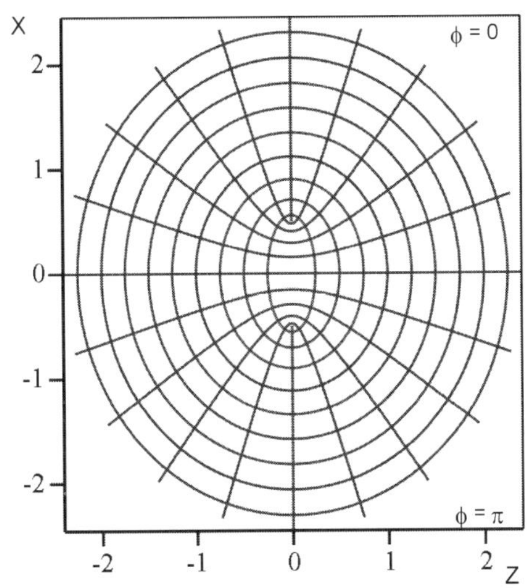

Figure 8. *The z-x plane (or two meridian planes $\phi = 0$ and $\phi = \pi$) and lines of intersection with the oblate spheroidal coordinate surfaces. Note that, for example, the pair of hyperbolas close to the axis z is reminiscent of the meridional profile of the (paraxial) Gaussian beam.*

Thus, by making use of the third axis of a 3-dimensional plot for depicting the strength of the field, we can visualise spatial dependences of Equation 17 at any given instant t as shown in Figure 9 (corresponding video clips, where the time t has been related to the frame number of animation, are available on-line through the electronic publication Saari 2001).

We can see that both pulses undergo a considerable temporal reshaping in the course of propagation, a spectacular manifestation of the Gouy phase shift, which is an almost imperceptible finesse in the case of monochromatic or narrow-band focused beams. In passing through the focus the real pulse is both time reversed and polarity inverted while the imaginary pulse is only time reversed, see Figure 10.

These results can straightforwardly be extended for the few- and many-cycle pulse cases by taking a temporal derivative of a corresponding order from Equation 17. In order to obtain linear field polarization which corresponds to the usual TEM_{00} mode in the paraxial limit, we should replace the "charge" with an electric dipole and a magnetic dipole, oriented along the x and the y axes, respectively, and combined with the sinks possessing the same dipole properties. This procedure, aimed at nonparaxial expressions for the pulses of the fields $\mathbf{E}$ and $\mathbf{H}$ (or $\mathbf{B}$), is most convenient if carried out by making use of the Hertz vector technique, taking the scalar solution, Equation 17, in the role of the magnitude of the Hertz vector.

There is an interesting direct connection between the exact solution obtained for the field undergoing focusing and for certain propagation-invariant fields. Namely, Equation 17 turns out to be a version of the so-called modified power spectrum pulse, which, in turn, is a superposition of focus wave modes (Besieris 1998).

4 Conclusion

A development can be viewed as a rising straight line. This means that, in order to succeed, one has to climb upwards as fast as one can and never look back. An opposite standpoint is that there is nothing really new under the sun, but only well forgotten things. Apparently, it is most wise to assume that things evolve along spirals. This seems to be true for physics in particular. There are many examples of spirals and fruitful revisits in the phylogenesis of physics. The wave of discoveries in nonlinear optics in the 1960s as compared to the study of relativistic nonlinear effects made possible by present-day femtosecond optics, the boom – almost three quarters of century after posing the famous Einstein-Podolski-Rozen problem – in investigation of quantum entanglement with the aim to build a quantum computer, if we list just some. Discovering and studying Bessel beams and other localised waves also belongs to this list.

Consequently, spirals should be replayed during ontogenesis as well. It means that for all of us it is wise sometimes to return to textbook truths learned in a rush while climbing the academic ladder and to rethink them carefully.

Acknowledgment

This work was supported by the Estonian Science Foundation.

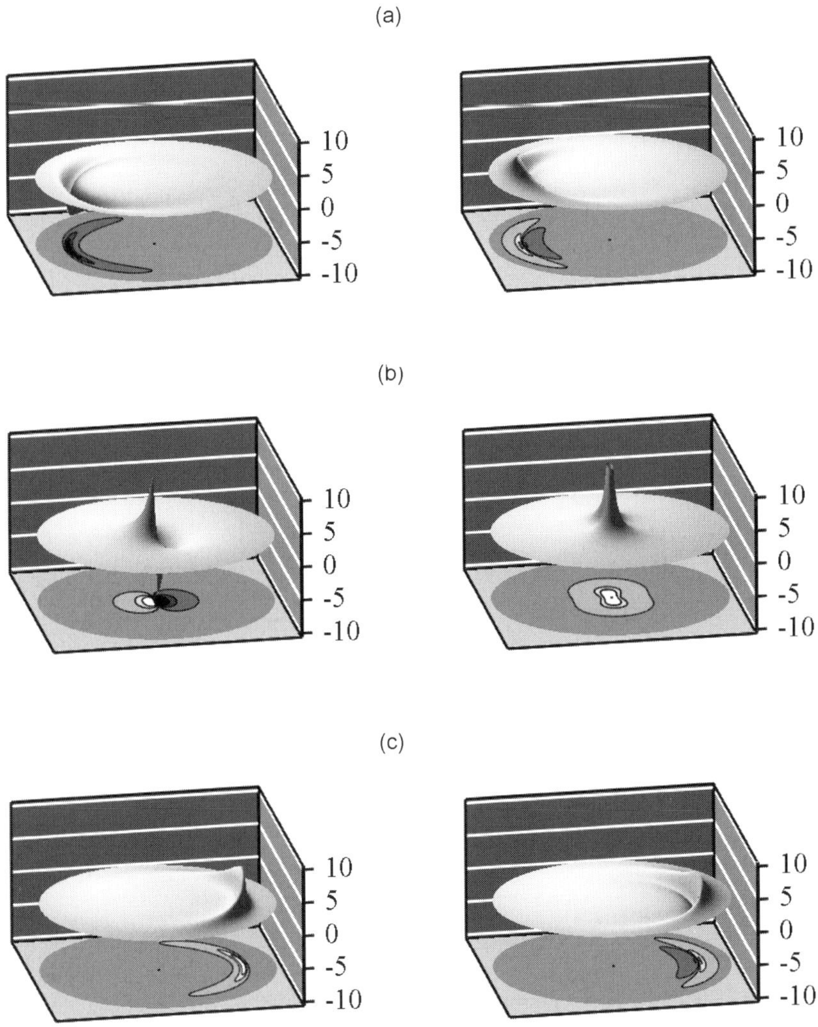

Figure 9. *The solutions given by the real part (left-hand) and the imaginary part (right-hand) of Equation 17 at different time instants* $ct = -1.75$ *(a),* $ct = 0$ *(b),* $ct = 1.75$ *(c). The propagation axis* z *has been directed from left to right. For a better visualization a "lighting" of the surface plots has been used and a grey-scale contour plot of the same data is shown at the bottom. The Rayleigh range* $d = 0.5$*, the pulse-width* $a = 0.05$ *(values given relative to the confocal parameter, see the text). The vertical scale of the amplitude has been normalised to the unit "charge"* $q = 1$*. For the imaginary part at the focus the pulse peak has been partially cut off, as its highest value (* $\simeq 19$*) falls outside the vertical scale.*

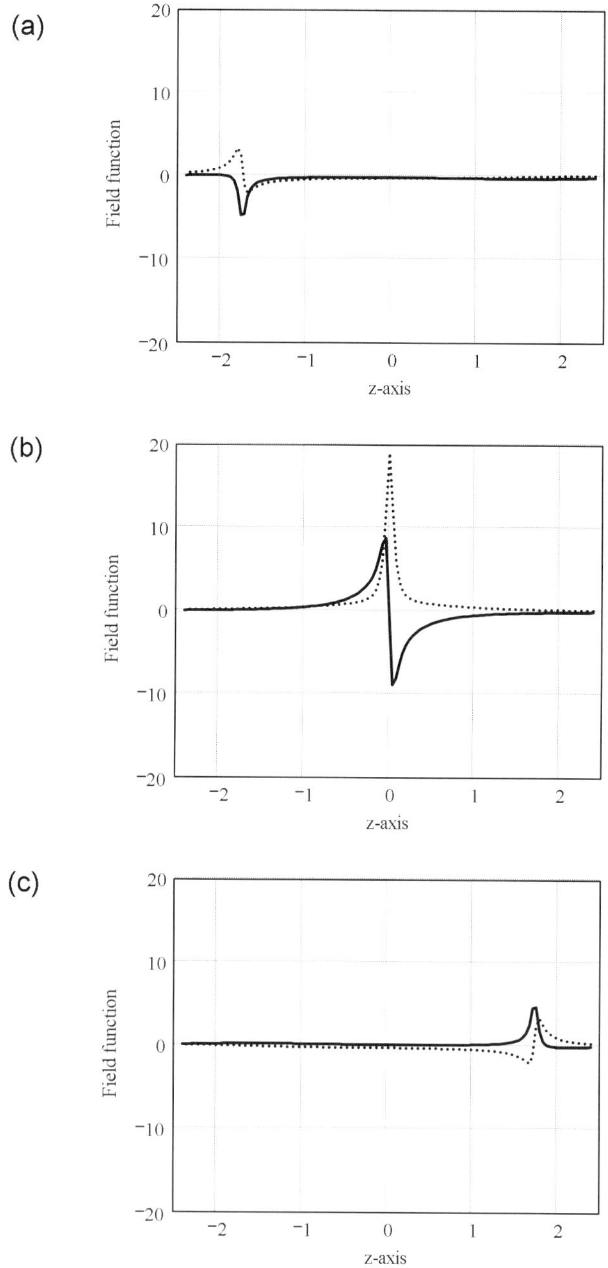

Figure 10. *On-axis behaviour of the real part (solid line) and the imaginary part (dotted) of Equation 17. Other parameters and units are the same as in Figure 9. The curves depict the distribution of the field along the z axis at instants* $ct = -1.75$ *(a),.* $ct = 0$ *(b),* $ct = 1.75$ *(c).*

References

Adler C, 1987, *American Journal of Physics* **55** 739.

Alexeev I, Kim K Y, and Milchberg H M, 2002, 'Measurement of the superluminal group velocity of an ultrashort Bessel beam pulse', to be published in *Phys. Rev. Lett* (8873901).

Bell J S, 1993, *Speakable and unspeakable in quantum mechanics* (Cambridge University Press).

Besieris I, Abdel-Rahman M, Shaarawi A, and Chatzipetros A,1998, 'Two fundamental representations of localized pulse solutions to the scalar wave equation', *Progress in Electromagnetic Research* **19** 48.

Durnin J J, Miceli Jr, and Eberly J H, 1987, 'Diffraction-free beams', *Phys. Rev. Lett.* **58** 1499.

Faingold M I, 1976, 'On superluminal source of type of moving focus', (in Russian), 276–285 in *Einsteinovski Sbornik* , (Nauka, Moscow).

Horváth Z L and Bor Zs, 1999, 'Reshaping of femtosecond pulses by the Gouy phase shift', *Phys. Rev.*, **E 60** 2337.

Jackson J D, 1999, *Classical electrodynamics* (John Wiley).

Lu J and Greenleaf J F, 1992, 'Nondiffracting X waves – exact solutions to free-space scalar wave equation and their finite aperture realizations', *IEEE Transactions on Ultrasonics, Ferroelectricity, and Frequency Control* **39** 19.

Lu J and He S, 1999, 'Optical X wave communications', *Optics Communications* **161** 187.

Mugnai D, Ranfagni A, and Ruggeri R, 2000, 'Observation of superluminal behaviors in wave propagation', *Phys. Rev. Lett.* **84** 4830.

Okun L B, 1989, *Physics Today*, June 31.

Overfelt P L, 1991, 'Bessel-Gauss pulses', *Phys. Rev.* **A 44** 3941.

Recami E, 1986, 'Classical tachyons and possible applications', *Rivista Nuovo Cimento* **9** 1.

Recami E, 1998, 'On localized X-shaped superluminal solutions to Maxwell equations', *Physica A* **252** 586.

Reivelt K and Saari P, 1997, 'Evidence of X-shaped propagation-invariant localized light waves', *Phys. Rev. Lett.* **79** 4135.

Reivelt K and Saari P, 2002, 'Experimental demonstration of realizability of optical focus wave modes', *Phys. Rev.* **E 66** 056611.

Reivelt K, 2002, 'Self-imaging of pulsed wave fields', *Optics Express* **10** 360.

Sõnajalg H and Saari P, 1996, 'Suppression of temporal spread of ultrashort pulses in dispersive media by Bessel beam generators', *Optics Letters* **21** 1162.

Sõnajalg H, Rätsep M, and Saari P, 1997, 'Demonstration of the Bessel-X pulse propagating with strong lateral and longitudinal localization in a dispersive medium', *Optics Letters* **22** 310.

Saari P, 1996, 'Spatially and temporally nondiffracting ultrashort pulses', 151–156 in*Ultrafast Processes in Spectroscopy*, editors Svelto O, De Silvestri S, Denardo G (Plenum).

Saari P, 2001, 'Superluminal localized waves of electromagnetic field in vacuo', 37–48 in *Time's Arrows, Quantum Measurement and Superluminal Behavior, Scientific Monographs: Physics Sciences Series* (Italian CNR Publ., Rome), see also Los Alamos preprint archive, http://xxx.lanl.gov/abs/physics/0103054.

Saari P, 2001a, 'Evolution of subcycle pulses in nonparaxial Gaussian beams', *Optics Express* **8** 590–598, see http://www.opticsexpress.org/abstract.cfm?URI=OPEX-8-11-590.

Saari P, Aaviksoo J, Freiberg A, Timpmann K, 1981, 'Elimination of excess pulse broadening at high spectral resolution of picosecond duration light emission', *Optics Communications* **39** 94.

Saari P and Sõnajalg H, 1997, 'Pulsed Bessel beams', *Laser Physics* **7** 32.

Salo J, Fagerholm J, Friberg A T, Salomaa M M, 2000, 'Unified description of non-diffracting X and Y waves', *Phys. Rev.* **E 62** 4261.

Shaarawi A M and Besieris I M, 2000, 'Superluminal tunneling of an electromagnetic X wave through a planar slab', *Phys. Rev.* **E 62** 7415.

Sommerfeld A, 1904, in *Göttinger Nachrichten.*

Soroko L M, 1996, *Meso-Optics* (World Scientific).

Zamboni-Rached M, Nóbrega K Z, Recami E, and Hernández-Figueroa H E, 2002, 'Superluminal X-shaped beams propagating without distortion along a coaxial guide', *Phys. Rev.* **E 71** 66.

Ziolkowski R W, 1989, 'Localized transmission of electromagnetic energy', *Phys. Rev.* **A 39** 2005.

Poster Sessions

The titles of papers presented by participants at the School, either orally or as posters, are listed below.

Ben Agate: Efficient, high repetition-rate fs blue source using a compact Cr:LiSAF laser.

Rafael Aldaz: Monolithically integrated mode-locked vertical cavity surface emitting laser.

Ramunas Aleksiejunas: All-optical non-destructive characterization of bulk semiconductors and heterostructures.

Ayi Bahtiar: Nonlinear waveguides from conjugated polymers.

Alessandro Belardini: Characterisation of the C-O properties of nonlinear polymers.

Temel Bilici: Elastic Light Scattering from sapphire microspheres for integrated optoelectronics.

Henry Bookey: Nonlinear all-optical switching in chalcogenide glass fibre.

Sebastien Bruneau: Femtosecond laser ablation: application to micromachining.

Bayram Butun: Low temperature GaAs based 1.55 micron photo-detectors.

Mariano Campoy-Quiles: Ellipsometry study of the anisotropy in PFO.

Joshua Casswell: Working towards three dimensional lightwave circuits, the STAR 3DLC project.

Maria Cataluna: Multimode dynamics in a CW-operating Ti:Sapphire laser.

Anthony Clark: GaInNAs/GaAs semiconductor optical amplifiers for high bit-rate amplification at 1.3um.

Iain Cormack: Real-time characterisation of ultrashort pulses: video-rate acquisition of a sonogram trace.

Eric Cormier: Recent progress on multi-Terawatt femtosecond high contrast laser pulses with high repetition rate.

Andrej Degtev: Software package for a semiconducter laser - LaserCADIII.

Hilmi Demir: Diffusion conduction based optically controlled optical switching..

Nuran Dogru: Mode-locked and noise characteristics of hybrid soliton pulse source utilizing fiber Bragg gratings.

Adrian Dragomir: Two-photon absorption in UV transparent materials and its application to long-period grating inscription in telecommunication fibres.

Ilya Elterman: Numerical modelling of light propagation and plane wave refraction in 2D photonic crystals.

Herve Eusebe: Ultrafast optoelectronics using Be doped LTG-GaAs.

Fedir Fedotov: Simulation of an electromagnetic signal propagation through a layer of non-linear dielectric using integral equations approach.

Andrew Ferguson: Erbium complexes as potential materials for ultra-thin film optical amplifiers at 1.5 μm.

Michael Flynn: Anomalous tuning behaviour of a synchronously pumped optical parametric oscillator.

Eric Gansen: Polarization dynamics induced by the 'virtual excitation of spin-polarized carriers, application to all-optical polarization switching.

Igor Gerasimchuk: Superlocalisation of light pulses in connected optical waveguide arrays.

Tine Greibe: Electrically pumped short-pulsed VCSEL at 980nm.

Noah Helman: Optical interconnects using short pulses.

Youfang Hu: Widely tunable gain-switched operation of an external grating-coupled surface emitting laser.

Senol Isci: Morphology dependent resonances of solid microspheres for frequency controlling in optical communications.

Yaser Kerachian: All-optical spin injections in semiconductors.

Victor Kisel: Yb-doped tungstate crystals for femtosecond pulse generation.

Tamara Knyazyan: Computer modelling of linearly chirped fiber Bragg gratings for dispersion compensation by an advanced method of single expression.

Stanislav Konorov: Coherent 4-wave mixing in a photonic band-gap planar hollow waveguide..

Andrey Konukhov: The dynamics of few femtosecond pulse Kerr-lens mode-locked laser from full space-time domain simulations.

Alexander Kovbasa: Calculation of the electromagnetic field in PC using transfer matrix method.

Nayot Kurukitkoson: Raman fibre lasers based on phosphorus doped fibre.

Justin Lawrence: Broadband polymer optical amplifiers.

Chris Leburn: GHz femtosecond lasers based on Cr:YAG.

Alexander Legatsky: Self-starting passively mode-locked femtosecond Cr^{4+}:YAG laser diode pumped by a Yb-fiber.

Paul Lloyd: Fibre-feedback optical parametric oscillators.

Valery Lobanov: Trapping of spatial QPM solitons with third harmonic multistep cascading.

Volodymyr Lysak: Group delay investigation for N-order chirping mirrors.

Liudmila Makarova: Application of quasi-periodic structures to compression of ultrashort laser pulses.

Katsiaryna Nazarava: The shift of spectral edges as the method of spectroscopy.

Nils Nielsen: Coherent pulse propagation on the A-exciton resonance in CdSe.

Marco Peccianti: Nonlocal spatial soliton interactions in nematic liquid crystal cells.

Alexander Podlipensky: Interaction of intense fs laser pulses with Ag nanoparticles embedded in silica glass.

Mike van der Poel: Coherent polariton dynamics in semiconductor microcavities.

Soma Venugopal Rao: Measurements of optical loss in $GaAs/Al_{2.}O_{3.}$ nonlinear waveguides in the infrared using femtosecond scattering technique: The role of two-photon absorption.

Luca Razzari: Numerical study of cascaded wavelength conversion in quadratic media.

Siarhei Rusau: Spatio-temporal dynamics in far-infrared laser diode.

Arvydas Ruseckas: Probing the electron delocalisation in conjugating polymers.

Vasili Savitski: PbS-quantum dots-doped glasses: picosecond transient bleaching and photonics. applications.

Steffen Sinning: Optical properties of N-implanted GaAs.

Barry Stormont: Gigahertz repetition rate from a diode pumped femtosecond laser.

Andrey Sukhorukov: Spatial optical solitons in periodic dielectric structures.

Giuseppe Talli: Effects of the amplified spontaneous emission on the gain dynamics of semiconductor optical amplifiers.

Tolga Tekin: GS-MZI as an all-optical demultiplexer for 160 Gbit/s data stream.

Karl Tillman: Ultrafast OPO incorporating a long, aperiodically poled KTP crystal..

Graham Turnbull: Emission characteristics and photonic band structure of microstructured polymer lasers.

Turgut Tut: Design, fabrication and characterization of ultrafast GaN photodetectors.

Erdem Ultanir: Spatial solitons in semiconductor optical amplifiers.

Javier Molia Vasquez: Wavelet transforms for ultrafast pulse analysis.

Huei Ching Wong: Rate equation modelling of quantum dots.

Dmitry Zelenchuk: Periodical structures in optical fibers.

Svetlana Zolotovskaya: Nonlinear spectroscopy and laser applications of copper chalcogenides in glasses.

Participants

- Ben Agate
  University of St Andrews,
  School of Physics and Astronomy,
  North Haugh,
  St Andrews, KY16 9SS,
  Scotland, UK.

- Rafael Aldaz
  Dept. of Electrical Engineering,
  CISX, Building B113-24,
  Stanford University,
  Stanford, CA 94305,
  USA.

- Ramunas Aleksiejunas
  Vilnius University,
  Inst. of Material Science and Applied Research,
  Sauletekio av 9 build.3,
  2040 Vilnius,
  Lithuania.

- Hatice Altug
  102 Hoskins Ct. Apt 6H,
  Escondido Village,
  Stanford, CA 94305,
  USA.

- Ayi Bahtiar
  University of Mainz,
  Max-Planck Institute for Polymer Research,
  Ackermannweg 10,
  D-55128, Mainz,
  Germany.

- Alessandro Belardini
  Universita di Roma,
  Dipartimento di Energetica,
  "La Sapienza", Via Scarpa 16,
  I-00161 Roma,
  Italy.

- Matthew Bell
  University of St Andrews,
  School of Physics and Astronomy,
  North Haugh,
  St Andrews, KY16 9SS,
  Scotland, UK.

- Aparna Bhatnagar
  N-138,
  450 Via Palou,
  Ginzton Labs,
  Stanford CA 94305-4088,
  USA.

- Temel Bilici
  Boston University,
  Dept. of Electrical and Computer Engineering,
  8 St Mary's Street,
  Boston, MA 02215,
  U S A

- Henry Bookey
  Physics Dept,
  Heriot-Watt University,
  Riccarton,
  Edinburgh EH14 4AS,
  UK.

- Alessio Bosco
  University La Sapienza,
  Energetics Department,
  Via Scarpa 16,
  I-00161 Roma,
  Italy.

- Donal Bradley
  Imperial College,
  Department of Physics,
  Blackett Laboratory,
  Prince Consort Road,
  London,SW7 2AY
  UK.

- Tom Brown
  University of St Andrews,
  School of Physics and Astronomy,
  North Haugh,
  St Andrews, KY16 9SS,
  Scotland, UK.

- Sebastien Bruneau
  Laboratoire LP3 UMR 6182,
  Parc Scientifique et Technologique de Luminy,
  163 Avenue de Luminy,
  case 917,
  13288 Marseille cedex 9
  France.

- Todd Bryden
  Dept. of Chemistry,
  Imperial College,
  S Kensington,
  London SW7 2AY,
  UK.

- Bayram Butun
  Bilkent University,
  E E Department,
  TR-06533 Bilkent,
  Ankara,
  Turkey.

- Mariano Campoy-Quiles
  Blackett Laboratory,
  Imperial College,
  Prince Consort Road,
  London,SW7 2AY,
  UK.

- Alison Carleton
  University of St Andrews,
  School of Physics and Astronomy,
  North Haugh,
  St Andrews, KY16 9SS,
  Scotland, UK.

- Joshua Caswell
  Heriot-Watt University,
  Dept of Physics,
  Riccarton,
  Edinburgh, EH14 4AS,
  UK.

- Maria Ana Cataluna
  University of St Andrews,
  School of Physics and Astronomy,
  North Haugh,
  St Andrews, KY16 9SS,
  Scotland, UK.

- Eda Cetinorgu
  University of Cukurova,
  Physics Department,
  01330, Adana,
  Turkey.

- Anthony Clark
  University of Strathclyde,
  Institute of Photonics,
  Wolfson Centre,
  106 Rottenrow,
  Glasgow,G4 0NW
  Scotland, UK.

- Iain Cormack
  ICFO Institut de Ciensies Fotonique,
  Jordi Girona 29,
  Nexus 2
  08034 Barcelona,
  Spain

- Eric Cormier
  University of Bordeaux I,
  CELIA,
  351 cours de la Liberation,
  33405, Talence,
  France.

- Andrej Degtev
  Kharkov National University of Radio Electronics,
  14 Lenin Av.
  Kharkov, 61166,
  Ukraine.

- Andrej Dementjev
  Institute of Physics,
  Savanoriu pr. 231,
  Vilnius
  Lithuania.

- Hilmi Demir
  Stanford University,
  Ginzton Laboratory,
  450 Via Palou,
  Stanford, CA, 94305-4085,
  USA.

- Nuran Dogru
  University of Gaziantep,
  Electrical and Electronics
  Department,
  27310 Gaziantep,
  Turkey.

- Adrian Dragomir
  Tasmanstraat 65,
  2815 VK,
  Den Haag
  Holland

- Ilya Elterman
  Department of Laser
  and Computer Physics,
  Saratov State University,
  Astrahanskaya 83,
  Saratov,
  Russia.

- Herve Eusebe
  Universite de Savoie,
  LAHC, Campus Scientifique,
  bat Le Chablais,
  73376 Le Bourget-du-Lac,
  France.

- Fedir Fedotov
  pr Gagarina 171A kV88,
  Kharkov-124 61124,
  Ukraine.

- Julia Fenn
  Avenida Medina Azahara 20, 3°, 1°,
  Cordoba 14005,
  Spain

- Andrew Ferguson
  Imperial College,
  Dept of Chemistry,
  Exhibition Road,
  London, SW7 2AY,
  UK.

- David Finlayson
  University of St Andrews,
  School of Physics and Astronomy,
  North Haugh,
  St Andrews, KY16 9SS,
  Scotland, UK.

- Michael Flynn
  University of St Andrews,
  School of Physics and Astronomy,
  North Haugh,
  St Andrews, KY16 9SS,
  Scotland, UK.

- Eric Gansen
  University of Iowa,
  Laboratory for Photonics and Quantum Electronics,
  Iowa City,
  Iowa, 52242-1000,
  USA.

- Igor Gerasimchuk
  Departamento de Propiedades Opticas, Magneticas y de Transporte,
  Instituto de Ciencia de Materiales de Madrid (ICMM) CSIC, Cantoblanco
  28049 Madrid,
  Spain

- Alvero GOMEZ Inlesis
  University of St Andrews,
  School of Physics and Astronomy,
  North Haugh,
  St Andrews, KY16 9SS,
  Scotland, UK.

- Tine Greibe
  Technical University of Denmark,
  Research Centre COM,
  DTU - bldg. 345v,
  Denmark.

- Takis Hadjifotiou
  Nortel Networks,
  London Road,
  Harlow,
  Essex, CM17 9NA,
  UK.

- George Heliotis
  Blackett Laboratory,
  Imperial College,
  Prince Consort Road,
  London SW7 2BZ,
  UK.

- Noah Helman
  Stanford University,
  Department of Applied Physics,
  Stanford,
  CA, 94305-4085,
  USA.

- Youfang Hu
  Photonics Research Group,
  Electronic Engineering,
  Aston University,
  Birmingham B4 7ET,
  UK.

- Jorn Hvam
  Technical University of Denmark,
  Mikroelektronik Centret Bygning,
  345 ST DK- 2800, Lyngby,
  Denmark.

- Senol Isci
  TUBITAK-UEKAE,
  National Institute of Electronics and Cryptology.
  P.K: 74 41470,
  Gebze, KOCAELI,
  Turkey

- Tim Karle
  University of St Andrews,
  School of Physics and Astronomy,
  North Haugh,
  St Andrews, KY16 9SS,
  Scotland, UK.

- Yaser Kerachian
  University of Toronto,
  Department of Physics,
  60 St. George St.,
  Toronto,
  Ontario, M5S 1A7,
  Canada.

- Victor Kisel
  Belarus State Polytechnical Academy,
  International Laser Centre,
  65, F. Scoryna prospect,
  Minsk 220000,
  Belarus.

- Tamara Knyazyan
  State Engineering University of Armenia,
  Fiber Optics Communication Lab.,
  105 Terian Str.,
  Yerevan, 375009,
  Armenia.

- Stanislav Konorov
  Moscow State University,
  119992 International Laser Centre,
  Moscow,
  Russia.

- Andrey Konukhov
  Institute of Radio-Engineering & Electronics of RAS,
  Zelyonaya 38,
  Saratov, 410019,
  Russia.

- Maria Kotlyar
  University of St Andrews,
  School of Physics and Astronomy,
  North Haugh,
  St Andrews, KY16 9SS,
  Scotland, UK.

- Jaroslav Kovac
  Slovak University of Technology,
  International Laser Centre,
  Ilkovicov3, 812 19,
  Bratislava,
  Slovak Republic.

- Alexander Kovbasa
  Kharkov National University of Radio Electronics,
  14 Lenin Av.
  Kharkov, 61166,
  Ukraine.

- Thomas Krauss
  University of St Andrews,
  School of Physics and Astronomy,
  North Haugh,
  St Andrews, KY16 9SS,
  Scotland, UK.

- Nayot Kurukitkoson
  Aston University,
  Photonics Research Group,
  School of Engineering & Applied Science,
  Aston Triangle,
  Birmingham B4 7ET,
  UK.

- Alexander Lagatsky
  University of St Andrews,
  School of Physics and Astronomy,
  North Haugh,
  St Andrews, KY16 9SS,
  Scotland, UK.

- Justin Lawrence
  University of St Andrews,
  School of Physics and Astronomy,
  North Haugh,
  St Andrews, KY16 9SS,
  Scotland, UK.

- Chris Leburn
  University of St Andrews,
  School of Physics and Astronomy,
  North Haugh,
  St Andrews, KY16 9SS,
  Scotland, UK.

- Paul Lloyd
  20 Northfield Road,
  London E6 2AJ,
  UK

- Valery Lobanov
  Radiophysics Department,
  Moscow State University,
  Leninskie Gory,
  Moscow, 119992,
  Russia.

- Panagiotis Loukakos
  Hellas Institute of Electronic Structure and Laser,
  P O Box 1527,
  Heraklion 711 10,
  Crete,
  Greece.

- Volodymyr Lysak
  14 Lenin av.,
  Lab. Photonics,
  Kharkov National University of Radio Electronics,
  Kharkov 61166
  Ukraine.

- Liudmila Makarova
National Academy of Sciences,
Institute of Molecular & Atomic Physics,
Belarus.

- Antonio Mandatori
University of Roma,
Dipartimento di Energetica,
"La Sapienza", Via Scarpa 16,
I-00161 Roma,
Italy.

- Maxim Markosov
pr.KlMA 5/34 k6
St Petersburg 199155,
Russia

- Michael Mazilu
University of St Andrews,
School of Physics and Astronomy,
North Haugh,
St Andrews, KY16 9SS,
Scotland, UK.

- Luciano Mescia
Politecnico di Bari,
Department of Electrical Engineering and Electronics,
Via And Orabona,
4 70125 Bari,
Italy.

- Alan Miller
University of St Andrews,
School of Physics and Astronomy,
North Haugh,
St Andrews, KY16 9SS,
Scotland, UK.

- David A. B. Miller
Stanford University,
Ginzton Laboratory,
450 Via Palou,
Stanford,
CA, 94305-4085,
USA.

- Javier Molina Vazquez
Physics Dept.
Heriot-Watt University,
Riccarton,
EH14 4AS, Edinburgh
Scotland, UK.

- Dietmar Nau
University of Bonn,
Institute of Applied Physics,
Wegelerstrasse 8,
53115 Bonn,
Germany.

- Katsiaryna Nazarava
International Sakharov Environmental University,
23 Dolgobrodskaya St.
220009, Minsk,
Belarus.

- Nils Nielsen
Max-Planck-Institut fur Festkorperforschung,
Heisenbergstrasse 1,
D-70569 Stuttgart,
Germany.

- Henrik Nilson
Physics Dept,
Heriot-Watt University,
Riccarton,
Edinburgh EH14 4AS,
Scotland, UK.

- Jean-Louis Oudar
Laboratoire Photonique et Nanostructure,
CNRS,
route de Nozay,
91460, Marcousis,
France.

- Marco Peccianti
University of Rome "Roma Tre",
Via della Vasca Navale 84 00146,
Roma,
Italy.

- Richard Penty
Cambridge University Engineering Department,
Trumpington Street,
Cambridge CB2 1PZ,
UK.

- Alexander Podlipensky
Martin-Luther-University
Halle-Wittenberg,
Physics Department,
Hoher Weg 8,
06099 Halle (Saale),
Germany.

- Miguel Carlos Ramon
  Blackett Laboratory,
  Imperial College,
  Prince Consort Road,
  London SW7 2AY,
  UK.

- Edik Rafailov
  A.F. Ioffe Physico-Technical Institute,
  26 Polytekhnicheskaya,
  St. Petersburg 194021,
  Russia.

- Luca Razzari
  Universita'Di Pavia,
  Dipartimento Di Electronica,
  Via Ferrata 1,
  27100 Pavia,
  Italy.

- Derryck Reid
  School of Engineering and Physical Sciences,
  Heriot-Watt University,
  Edinburgh EH14 4AS,
  Scotland, UK.

- Charis Reith
  University of St Andrews,
  School of Physics and Astronomy,
  North Haugh,
  St Andrews, KY16 9SS,
  Scotland, UK.

- Siarhei Rusau
  Institute of Physics,
  National Academy of Sciences of Belarus,
  F.Skaryna Ave. 68,
  220072, Minsk,
  Belarus.

- Arvydas Ruseckas
  University of St Andrews,
  School of Physics and Astronomy,
  North Haugh,
  St Andrews, KY16 9SS,
  Scotland, UK.

- Peeter Saari
  University of Tartu,
  Institute of Physics,
  Riia 142
  Tartu 51014,
  Estonia.

- Solomon Mois Saltiel
  Sofia University,
  Faculty of Physics,
  5 James Bourchier Blvd.,
  1164 Sofia,
  Bulgaria.

- Ifor Samuel
  University of St Andrews,
  School of Physics and Astronomy,
  North Haugh,
  St Andrews, KY16 9SS,
  Scotland, UK.

- Vasili Savitski
  Belarus State Polytechnical Academy,
  International Laser Centre,
  65, F. Scoryna prospect,
  Minsk, 220000,
  Belarus.

- Wilson Sibbett
  University of St Andrews,
  School of Physics and Astronomy,
  North Haugh,
  St Andrews, KY16 9SS,
  UK.

- Sinning, Steffen
  Research Centre Rossendorf,
  Dept: FWIH, PF: 51 01 19,
  01314 Dresden,
  Germany.

- George Stegeman
  University of Central Florida,
  School of Optics / CREOL,
  4000 Central Florida Parkway,
  Orlando, FL, 32816-2700,
  USA.

- Robert Stegeman
  University of Central Florida,
  School of Optics / CREOL,
  4000 Central Florida Parkway,
  Orlando, FL, 32816-2700,
  USA.

- Barry Stormont
  University of St Andrews,
  School of Physics and Astronomy,
  North Haugh,
  St Andrews, KY16 9SS,
  Scotland, UK.

- Andrey Sukhorukov
  Nonlinear Physics Group,
  RSPhysSE,
  Australian National University,
  Canberra
  ACT 0200 Australia.

- Clementine Symonds
  Institut d' Electronique Fondamental,
  Bat 220,
  Universite Paris-Sud
  91405 Orsay,
  France

- Giuseppe Talli
  University of Essex,
  Department of Electronic Systems Engineering,
  Wivenhoe Park,
  Colchester CO4 3SQ,
  UK.

- Tolga Tekin
  Heinrich-Hertz-Institut,
  Fraunhofer Institute for telcommunications,
  Einsteinufer 37,
  Berlin 10587,
  Germany

- Karl Tillman
  Dept. of Physics,
  Heriot-Watt University,
  Riccarton,
  Edinburgh EH14 4AS,
  Scotland, UK.

- Graham Turnbull
  University of St Andrews,
  School of Physics and Astronomy,
  North Haugh,
  St Andrews, KY16 9SS,
  Scotland, UK.

- Turgut Tut
  Bilkent University,
  Physics Department,
  TR-06533 Bilkent,
  Ankara,
  Turkey.

- Erdem Ultanir
  Creol/School of Optics,
  University of Central Florida,
  Parkway,
  Orlando, FL 32816-2700
  USA.

- Mike Van Der Poel
  Technical University of Denmark,
  Research Centre COM,
  DTU - bldg. 345v,
  Denmark.

- Soma Venugopal Rao
  Centre for Ion Beam Applications,
  Dept. of Physics,
  Blk S12, 2 Science Drive 3,
  National University of Singapore,
  Singapore 117542

- Osamu Wada
  Kobe University,
  Faculty of Engineering,
  1-1 Rokkodai,
  Nada,
  Kobe, 657-8501,
  Japan.

- William Whelan-Curtin
  University of St Andrews,
  School of Physics and Astronomy,
  North Haugh,
  St Andrews, KY16 9SS,
  Scotland, UK.

- Steven White
  University of St Andrews,
  School of Physics and Astronomy,
  North Haugh,
  St Andrews, KY16 9SS,
  Scotland, UK.

- Michael Wiemer
  P O Box 17416,
  Stanford, CA, 94309,
  USA.

- Huei Ching Wong
  University of Bristol,
  Dept. Electrical and Electronic Eng.,
  Merchant Venturers Building,
  Woodland Road,
  Bristol, BS8 1UB,
  UK.

- Philip Wood
  University of St Andrews,
  School of Physics and Astronomy,
  North Haugh,
  St Andrews, KY16 9SS,
  Scotland, UK.

- Lijun Wu
  Construction Building 801,
  Xia Gang avenue 983
  Guang Zhou Development Zone,
  Guang Zhou 510730,
  P.R. China

- Dmitry Zelenchuk
  Physics Department,
  Rostov State University,
  5 Zorge Street,
  Rostov-on-Don, 344090,
  Russia.

- Aleksei M. Zheltikov
  M V Lomonosov Moscow State University,
  Physics Faculty,
  International Laser Centre, Vorobevy,
  Moscow, 119899,
  Russia.

- Svetlana Zolotovskaya
  Belarus State Polytechnical Academy,
  International Laser Centre,
  65, F. Scoryna prospect,
  Minsk, 220000,
  Belarus

Index